MICROWAVE NOISE IN SEMICONDUCTOR DEVICES

MICROWAVE NOISE IN SEMICONDUCTOR DEVICES

HANS LUDWIG HARTNAGEL
University of Technology, Darmstadt, Germany

RAMŪNAS KATILIUS
ARVYDAS MATULIONIS
Semiconductor Physics Institute, Vilnius, Lithuania

A Wiley-Interscience Publication
JOHN WILEY & SONS, INC.
New York / Chichester / Weinheim / Brisbane / Singapore / Toronto

This book is printed on acid-free paper. ∞

Copyright © 2001 by John Wiley & Sons, Inc. All rights reserved.

Published simultaneously in Canada.

No part of this publication may be reproduced, stored in a retrieval system or transmitted in any form or by any means, electronic, mechanical, photocopying, recording, scanning or otherwise, except as permitted under Section 107 or 108 of the 1976 United States Copyright Act, without either the prior written permission of the Publisher, or authorization through payment of the appropriate per-copy fee to the Copyright Clearance Center, 222 Rosewood Drive, Danvers, MA 01923, (978) 750-8400, fax (978) 750-4744. Requests to the Publisher for permission should be addressed to the Permissions Department, John Wiley & Sons, Inc., 605 Third Avenue, New York, NY 10158-0012, (212) 850-6011, fax (212) 850-6008, E-mail: PERMREQ@WILEY.COM.

For ordering and customer service, call 1-800-CALL-WILEY.

Library of Congress Cataloging-in-Publication Data:

Hartnagel, Hans, 1934-
    Microwave noise in semiconductor devices / Hans Ludwig Hartnagel, Ramūnas Katilius, Arvydas Matulionis.
       p. cm.
    Includes bibliographical references and index.
    ISBN 0-471-38432-1 (cloth : alk. paper)
      1. Electronic circuits–Noise. 2. Solid state electronics. 3. Semiconductors. 4. Hot carriers. 5. Microwave circuits. I. Katilius, Ramūnas, 1935-II. Matulionis, Arvydas, 1940-III. Title.

TK7867.5 .H36 2001
621.381'33—dc21   00-043688

Printed in the United States of America.
10 9 8 7 6 5 4 3 2 1

CONTENTS

PREFACE		xi
LIST OF SYMBOLS		xv
1	**INTRODUCTION**	**1**
	1.1 Electronic Noise / 1	
	1.2 Trends in High-Speed Electronics / 2	
	1.3 Main Sources of Noise in Semiconductor Structures / 4	
	1.4 Microwave Noise in a Field-Effect Transistor / 5	
	1.5 Noise Characterization / 7	
	1.6 Approaches to Response Modeling / 11	
	1.7 Microwave-Noise Modeling / 13	
2	**KINETIC THEORY OF NONEQUILIBRIUM PROCESSES**	**16**
	2.1 Boltzmann Equation; Nonequilibrium States / 17	
	2.2 Relaxation and Response Problems / 21	
3	**FLUCTUATIONS: KINETIC THEORY**	**25**
	3.1 Fluctuations in Electron Distribution / 26	
	3.2 Equal-Time Correlation / 28	
	3.3 Spectral Intensities of Fluctuations / 30	

3.4 Interelectron Collisions Neglected / 31
3.5 "Kinetic" Derivation of Nyquist Relation / 32

4 EFFECT OF INTERELECTRON COLLISIONS ON FLUCTUATION PHENOMENA — 37

4.1 Additional Correlation Due to Interelectron Collisions / 37
4.2 Expression for Spectral Intensity of Current Fluctuations / 41
4.3 Criteria for Applicability of Kinetic Theory of Fluctuations / 42

5 BOLTZMANN–LANGEVIN EQUATION — 44

5.1 Langevin Approach / 44
5.2 Derivation of the Boltzmann–Langevin Equation / 47

6 FLUCTUATIONS AND DIFFUSION — 50

6.1 Fluctuation–Diffusion Relationship / 50
6.2 Violation of Fluctuation–Diffusion Relation; Additional Correlation Tensor / 52
6.3 Drift–Diffusion Equation; Generalized Chapman–Enskog Procedure / 55

7 FEATURES OF HOT-ELECTRON FLUCTUATION SPECTRA — 62

7.1 Current-Fluctuation Spectrum at Equilibrium / 63
7.2 Convective Noise / 64
7.3 Fluctuation Peculiarities Under Optical-Phonon Emission / 70
7.4 Fluctuations in Streaming-Motion Regime / 72
7.5 Intervalley Noise / 75
7.6 Historical Comments on the Theory of Fluctuations / 76

8 EXPERIMENTAL TECHNIQUES — 81

8.1 Early Experiments on Hot-Electron Noise / 82
8.2 E-Spectroscopy and ω-Spectroscopy of Noise / 83
8.3 Measurable Quantities and Definitions / 84
8.4 Waveguide-Type Short-Time-Domain Gated Radiometer / 86

8.5 Small-Signal Response / 91
8.6 Coaxial-Type Gated Setup for Noise Measurements / 92
8.7 Nanosecond-Time-Domain Gated Radiometric Setup / 95

9 HOT-ELECTRON MICROWAVE NOISE IN ELEMENTARY SEMICONDUCTORS   96

9.1 Transverse Noise Temperature / 96
9.2 Energy Relaxation Time Estimated by Noise Technique / 98
9.3 Convective Noise in p-Type Germanium / 99
9.4 Resonance Due to Streaming Motion / 103
9.5 Intervalley Fluctuations in Si and Ge / 105
9.6 Standard ω-Spectroscopy of Hot-Electron Noise in Silicon / 108
9.7 E-Spectroscopy of Hot-Electron Noise / 110
9.8 Longitudinal and Transverse Diffusion Coefficients / 113

10 HOT-ELECTRON MICROWAVE NOISE IN GaAs AND InP   116

10.1 Intervalley Noise in GaAs and InP / 116
10.2 Thermally Quenched Hot-Electron Intervalley Noise / 120
10.3 Terminated Runaway of Lucky Electrons / 122
10.4 Doping-Dependent Hot-Electron Noise / 123
10.5 Resonant Impurity Scattering in GaAs / 127

11 LENGTH-DEPENDENT HOT-ELECTRON NOISE   128

11.1 Short-Channel Effect on Hot-Electron Noise / 129
11.2 Suppression of Intervalley Noise / 133
11.3 Doping-Dependent Short-Channel Effect / 136
11.4 Doping-Stimulated Intervalley Noise / 139
11.5 Γ–X and Γ–L Intervalley Transfer Times for GaAs / 141

12 HOT-ELECTRON NOISE IN DOPED SEMICONDUCTORS: THEORY   147

12.1 Effective Electron Temperature and Its Fluctuations / 147
12.2 Fluctuations in Drifted Maxwellian Approximation / 154
12.3 Fluctuations in Weakly Heated Electron Gas of Intermediate Density / 156

13 ELECTRONIC NOISE IN STANDARD-DOPED n-TYPE GaAs   161

13.1 Monte Carlo Simulation / 161
13.2 Experimental Data / 165
13.3 Analytic Approach / 167

14 ELECTRON DIFFUSION IN STANDARD-DOPED n-TYPE GaAs   175

14.1 Electron Diffusion Coefficients in Doped GaAs / 176
14.2 Degree of Violation of Price's Relation / 177

15 ELECTRONIC SUBBANDS IN QUANTUM WELLS   181

15.1 Band Structure of Compound Semiconductors / 182
15.2 Heterojunctions / 183
15.3 Empty Quantum Wells / 185
15.4 Occupied Quantum Wells / 188
15.5 Effective Barrier Height / 194

16 HOT-ELECTRON NOISE IN AlGaAs/GaAs 2DEG CHANNELS   197

16.1 2DEG Channels / 198
16.2 Hot-Electron Real-Space Transfer Noise / 199
16.3 Longitudinal Fluctuations Due to Transverse Tunneling / 204
16.4 Intersubband Transfer Noise / 207

17 HOT-ELECTRON NOISE IN InP-BASED 2DEG CHANNELS   209

17.1 Suppression of Real-Space-Transfer Noise / 209
17.2 Energy Relaxation Time in InGaAs Quantum Well Channels / 212
17.3 Real-Space Transfer Noise at Low Electric Fields / 216

18 CUTOFF FREQUENCIES OF FAST AND ULTRAFAST PROCESSES   220

18.1 Hot-Electron Energy Relaxation / 220
18.2 Interwell and Intervalley Transfer / 224
18.3 Noise Sources for Equivalent Circuits / 227

19 SPATIALLY INHOMOGENEOUS FLUCTUATIONS   230

19.1 Spatially Inhomogeneous Fluctuations of Carrier Distribution / 230
19.2 Drift–Diffusion Equation with Fluctuations / 237
19.3 Experimental Detection of Spatially Inhomogeneous Fluctuations / 241
19.4 Space-Dependent Fluctuations of Effective Electron Temperature / 243

20 MONTE CARLO APPROACH TO MICROWAVE NOISE IN DEVICES   245

20.1 Microscopic Approaches / 245
20.2 Planar-Doped-Barrier Diode / 247
20.3 Ultrafast Field-Effect Transistors / 253
20.4 Comparison to Experiment: InGaAs-Channel T-Gate HEMT / 255
20.5 Concluding Remarks / 256

BIBLIOGRAPHY   258

INDEX   281

PREFACE

This book is about high-frequency fluctuations and electronic noise in semiconductors, semiconductor structures, and devices for microwave electronics. It is intended to bridge the apparently existing gap between the contemporary microscopic approach (the fundamental kinetic theory of hot-electron fluctuations) and traditional approaches to noise problems in engineering of microwave devices. The information presented is based on substantial progress in theoretical and experimental investigation of fluctuation processes in nonequilibrium. Microwave fluctuation and noise phenomena are considered in terms of ultrafast kinetic processes specific to modern semiconductor structures for microwave devices.

Current advances in information technology stimulate development of ultrafast devices which are able to process large amounts of data during short periods of time. Ecological limitations together with requirements specific for mobile long-distance communication impose a strong stress on the signal-to-noise ratio. As a result, low-noise high-speed performance of semiconductor devices appears to be of primary importance. This book considers available data on the speed–noise tradeoff in semiconductors and semiconductor structures of direct interest for high-speed electronics.

Modern electronic components are based on heterostructures where reduced-dimensional electron-wave processes play the dominant role. The relevant fluctuation effects need to be known and used in view of minimum noise behavior in signal-processing circuits. This book deals with new experimental results and contemporary microscopic understanding of electronic processes and the associated fluctuations in dimensionally quantizing structures such as lattice-matched and pseudomorphic channels containing the two-dimensional electron gas. This more sophisticated approach tends to substitute presently used semi-empirical hit-and-miss techniques.

Until now, noise effects in low-noise devices and circuits were commonly treated neglecting hot-electron effects in biased channels indispensable for ultrafast operation. This book presents the first systematic treatment of fluctuations and microwave response of semiconductors and semiconductor structures under nonequilibrium conditions.

Three aspects of noise problems are dealt with, and the book is organized as follows. Chapters 2–7, 12, and 19 (constituting a deliberately dispersed Part One: "Noise Theory") present systematically the kinetic theory of fluctuations and its corollaries. Applications are considered when microwave noise properties of semiconductors are explained and predicted in Chapters 8–11, 13, and 14 (Part Two: "Noise in Semiconductors"). Low-dimensional structures and devices are treated in Chapters 15–18 and 20 (Part Three: "Noise in Semiconductor Structures and Devices").

Part One is intended for both physicists, interested in stochastic and kinetic (transport) phenomena, and material scientists and electronics engineers, requiring broad awareness of the physical background of noise phenomena in up-to-date semiconductor structures and devices. Part Two presents methods of measurement of microwave noise followed by a panorama of experimental results on hot-carrier noise in semiconductors, including peculiarities of noise phenomena in doped channels and in very short channels at very high electric fields. The possibility of fluctuation spectroscopy resolving microwave noise sources of different origin is evidenced. This part is intended for semiconductor physicists and materials scientists, as well as for electronics engineers, providing an introduction to Part Three. There, practicing electronics engineers and materials scientists will find a description of noise phenomena in two-dimensional structures and devices, with emphasis on the physical understanding of noise sources and ways to reduce them. In particular, a discussion of high-field, low-noise operation possibilities in up-to-date structures and devices is presented.

Numerous illustrations (over 100) present recent experimental data for up-to-date semiconductor structures designed for ultrafast electronics, together with the results of microscopic simulation where available. Examples include transition from shot-noise to hot-electron noise in forward-biased diodes, fluctuations due to real-space transfer of charges between parallel two-dimensional electron gas channels, resonant-tunneling transfer of hot electrons, and the results of performance simulation of a high-speed field-effect transistor, among others. Time constants of ultrafast kinetic processes resolved using microwave noise techniques are listed in the tables. References include a historical survey and recent publications.

Two of us (R. K. and A. M.) are indebted to Professors Sh. Kogan (Moscow–Los Alamos), A. Shulman (Moscow), C. Jacoboni (Modena), J. P. Nougier (Montpellier), and L. Reggiani (Lecce), as well as to our colleagues in the Semiconductor Physics Institute (Vilnius), especially to Dr. J. Liberis and Dr. V. Aninkevičius, with all of whom we have had the pleasure of discussing one or another of the related topics. One of us (R. K.) would like to express his gratitude to Professors V. L. Gurevich and S. V. Gantsevich of the A. F. Ioffe

Physical–Technical Institute (St. Petersburg) for long and fruitful collaboration in the field.

The idea of this book came to the authors while participating in a large cooperative effort on fluctuation investigation with a number of research institutions under the Copernicus funding of the European Commission. In the process of writing, two of us (R. K. and A. M.) enjoyed the support of the Lithuanian National Foundation for Science and Education.

The authors would like to thank Dr. Ilona Matulionienė for her devoted assistance in the process of preparing the manuscript.

HANS LUDWIG HARTNAGEL
RAMŪNAS KATILIUS
ARVYDAS MATULIONIS

LIST OF SYMBOLS

a	quantum well width
$\mathcal{B}_\mathbf{p}(\mathbf{q}, \omega)$	operator of spatially inhomogeneous response
D^{eq}, D_0	zero-field electron diffusion coefficient
$D_{\alpha\beta}$	tensor of electron diffusion coefficients
$D_\parallel(\mathbf{E})$	hot-electron diffusion coefficient in direction of steady current
$D_\perp(\mathbf{E})$	hot-electron diffusion coefficient in transverse direction
dS/do	cross section of electromagnetic-wave scattering from electron system
e	elementary charge
$\mathcal{E}_g$	energy gap
$\mathcal{E}_{L\Gamma}$, $\mathcal{E}_{X\Gamma}$	intervalley separation energies
$\mathcal{E}_j$	jth subband energy
$\mathcal{E}_F$	Fermi energy
$\mathcal{E}^{thr}$	threshold energy
$\mathcal{E}^{ii}$	impact ionization energy
$\mathbf{E}$	electric field
f	frequency
f^{cutoff}	cutoff frequency
$\overline{F}_\mathbf{p}(t)$	time-dependent one-particle distribution function
$\overline{F}_\mathbf{p}$	steady-state distribution function
$F_\mathbf{p}^M$	displaced Maxwellian distribution
$\hbar$	Planck constant
$\hbar\omega^{acoust}$	acoustic phonon energy
$\hbar\omega^{opt}$	optical phonon energy

xvi LIST OF SYMBOLS

I	steady current
$I_{\mathbf{p}}$	linearized operator of Boltzmann equation (relaxation operator)
$(I_{\mathbf{p}} - i\omega)$	response operator
$\tilde{I}_{\mathbf{p}}$	relaxation operator in absence of electron–electron collisions
$I_{\mathbf{p}}^{ee}(\overline{F})$	linearized electron–electron collision operator
$I_{\mathbf{p}}^{th}$	operator of interaction with thermal bath
$\mathbf{j}$	steady current density
$\mathbf{j}^{D}$	diffusion flux of electrons
$\mathbf{k}$	wave vector
k_B	Boltzmann constant
$l \equiv v\tau$	electron mean free path
L	distance between electrodes, channel length
L_1, L_2, L_{tot}	critical lengths
m	electron effective mass
n	electron density
$n_0 = N/\mathcal{V}_0$	average electron density
n_c	critical sheet density of electrons
N	total number of electrons
N^{ion}	density of ionized impurities
N_A	acceptor density
N_D	donor density
NF	noise figure
$\mathbf{p}$	quasi-momentum
P	consumed power per electron; rate of energy loss
$\mathbf{q}$	wave vector of space-dependent fluctuation
Q	channel cross-section area; interface area
r_D	Debye screening length
r_0	classical (Thomson) radius of electron
R	resistance
$S_X(\omega), (X^2)_\omega$	spectral intensity of random variable $X(t)$
$S_\mathcal{V}(\omega)$	spectral intensity of velocity fluctuations
$S_\mathcal{V}^{inter}$	spectral intensity of velocity fluctuations due to intervalley (interwell) transfer
$S_I(\omega)$ and $S_V(\omega)$	spectral intensity of current and voltage fluctuations
$S^{\text{"therm"}}$	spectral intensity of "thermal" fluctuations of hot electrons
t	time
t^{tr}	transit time
T_0	absolute temperature
T_e	electron temperature
T_n	equivalent noise temperature
$T_{n\parallel}$ and $T_{n\perp}$	longitudinal and transverse noise temperatures

LIST OF SYMBOLS   xvii

U	applied voltage
$\mathbf{V}, \mathbf{v}^{dr}$	translational velocity of electron system (drift velocity)
$\mathbf{V}'$	differential drift velocity
$\mathbf{v_p}$	electron velocity in state $\mathbf{p}$
$V(z)$	quantum-well potential
V_{DS}	drain-source voltage
$\mathcal{V}_0$	volume
$W_{\mathbf{p}'}^{\mathbf{p}}$	transition probability for electron collision with thermal bath
$W_{\mathbf{p}'\mathbf{p}'_1}^{\mathbf{p}\mathbf{p}_1}$	transition probability for an electron–electron collision
x, y, z	space coordinates
$Z(f)$	AC impedance
$\delta F_{\mathbf{p}}(t)$	fluctuation of distribution function
$\overline{\delta F_{\mathbf{p}} \delta F_{\mathbf{p}_1}}$	equal-time correlation function of distribution function fluctuations
$\overline{\delta F_{\mathbf{p}}(t) \delta F_{\mathbf{p}_1}}$	time-displaced correlation function of distribution function fluctuations
$\overline{\delta F_{\mathbf{p}}(\mathbf{r}) \delta F_{\mathbf{p}_1}(\mathbf{r}_1)}$	equal-time two-point correlation function
$\overline{\delta F_{\mathbf{p}}(\mathbf{r},t) \delta F(\mathbf{r}_1, \mathbf{p}_1)}$	time-displaced two-point correlation function
$(\delta F_{\mathbf{p}} \delta F_{\mathbf{p}_1})_\omega$	spectral intensity of distribution function fluctuations
$\delta \mathbf{j}(t)$	fluctuation of current density
$\overline{\delta j_\alpha(t) \delta j_\beta}$	time-displaced correlation function of current density fluctuations
$(\delta j_\alpha \delta j_\beta)_\omega$	spectral intensity of current density fluctuations
$\delta \mathbf{J}_{\mathbf{p}}(t)$	random flux in Boltzmann–Langevin equation
$\Delta \mathcal{E}_c$	conduction band discontinuity
$\Delta P_n(f)$	noise power emitted into frequency band Δf; available noise power
Δt^{dur}	duration of collision
$\Delta_{\alpha\beta}$	tensor of additional correlation due to interelectron collisions
$\overline{\varepsilon}$	electron average energy
$\varepsilon_{\mathbf{p}}$	electron energy in state $\mathbf{p}$
μ^{eq}, μ_0	low-field low-frequency mobility
ξ_j	jth envelope function
$\sigma_{\alpha\beta}(\omega)$	AC small-signal conductivity
$\sigma_\parallel(\omega)$ and $\sigma_\perp(\omega)$	longitudinal and transverse AC small-signal conductivities
$\tilde{\sigma}$	"chord" conductivity
τ	quasi-momentum relaxation time
$\overline{\tau}$	effective quasi-momentum relaxation time
τ^{en}	energy relaxation time
$\overline{\tau}^{en}$	effective energy relaxation time
τ^{inter}	intervalley (interwell) relaxation time

$\bar{\tau}^{inter}$	effective intervalley (interwell) relaxation time
τ^{ee}	interelectron relaxation time
τ^{gr}	generation-recombination relaxation time
τ_T	electron temperature relaxation time
τ_m	relaxation time for mth electronic processes
ϕ	effective barrier height
$\varphi_{\mathbf{pp}_1}$	mean correlated occupancy of two states
ω	circular frequency

CHAPTER 1

INTRODUCTION

1.1 ELECTRONIC NOISE

Semiconductors have opened new possibilities to satisfy modern needs of society in information, especially in the field of data processing and storage. A specific branch, called high-speed electronics, is developed to solve the problems arising when large amounts of data are processed during short periods of time. High speed of operation is also attractive for telecommunication: Many noninteracting broad-band channels can work simultaneously in the range of microwave and millimeter-wave frequencies.

The progress in high-speed electronics strongly depends on the solutions combining a high speed of operation with a reduced power consumption. However, the operation of extremely low-power electric signals faces another problem; that is, it suffers from a possible loss of information or its distortion when the signal power is comparable to that of the spontaneous fluctuations present in the circuit. As a result, an optimal design should consider all limitations, including those associated with fluctuations.

Fluctuations in electric circuits result from the discrete nature of charge carriers and their chaotic motion. Electric fluctuations are irregular temporary deviations of variables (current, voltage, resistance, frequency, etc.) either from their long-time averages or from some regular time-dependent values of information-bearing signals. The fluctuations manifest themselves in everyday life as an acoustic noise in a telephone or a radio receiver, or as an irregular flickering on a television screen known as a "snowfall" flicker, and on numerous other occasions. Electric noise in microelectronics is caused by fluctuations of the variables associated with mobile charge carriers: electrons and holes. For simplicity, we shall often write electrons instead of charge carriers, unless the mentioning of holes is necessary for specific reasons.

Noise is commonly viewed by researchers as the limiting factor: "When one tries to measure or amplify small signals, one usually arrives at a lower limit set by the spontaneous fluctuations in current, voltage, and temperature of the system under test. These spontaneous fluctuations are referred to as noise. Noise is an important problem in science and engineering, since it sets lower limits to the accuracy of any measurement and to the strength of signals that can be processed electronically" (van der Ziel [1]). Electronic noise is understood best in an electron gas kept at thermodynamic equilibrium with the phonons, the latter being in equilibrium to the ambient bodies. This knowledge is sufficient to estimate the ultimate accuracy of electrical measurements and signal processing under near-equilibrium conditions. However, advances in communication and information technologies strongly rely on solutions based on essential deviations from the equilibrium. The universal thermodynamic relations fail at nonequilibrium conditions typical for ultrafast devices, and a more sophisticated approach is necessary for considering the limitations due to fluctuations under nonequilibrium conditions in semiconductor structures and devices designed for high-speed electronics.

However, there is also another aspect of the problem, now fully realized and widely exploited. Fluctuations in macroscopic observables result from microscopic random processes: Every source of fluctuations is associated with some microscopic mechanism accompanied by dissipation, and measuring fluctuations *out of equilibrium* provides one with new information about the system — new as compared to that obtainable while measuring the average values of the observables. The obtained knowledge of microscopic origin of the excess electronic noise helps to control it, and it suggests how to eliminate the dominant sources of noise through improvement of material technology and circuit design, thus contributing to development of highly sensitive devices for high-speed electronics.

1.2 TRENDS IN HIGH-SPEED ELECTRONICS

Figure 1.1 illustrates the main trends in high-speed semiconductor electronics. Compound semiconductor technologies are in strong competition with silicon technology for market in wireless modems, ultra-high-capacity networks, long-range radar systems, satellite television, satellite-to-satellite telecommunication, short-range radar systems for traffic safety, and other applications exploiting ultrafast transistors and circuits operating in analog and digital modes at microwave and millimeter-wave frequencies.

Silicon metal–oxide–semiconductor (Si MOS) technology has, in general, hardly any competitor at frequencies below 1 GHz in a low-noise circuit used to amplify weak electric and electromagnetic signals often referred to as preamplifier. However, the progress toward higher frequencies depends strongly on GaAs field-effect transistor (MESFET) and heterostructure field-effect transistor

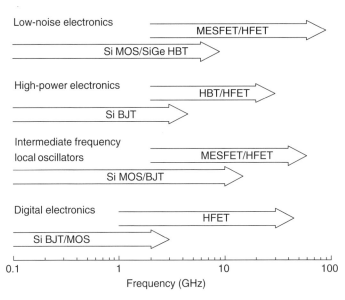

Figure 1.1. Current status in high-speed electronics. Most impressive progress toward higher frequencies depends on HFET and HBT technologies. HFETs are of exceptional importance for low-signal low noise microwave and millimeter-wave circuits.

(HFET) technologies (Fig. 1.1). The main advantages of heterostructure FETs are low noise and high speed of operation resulting from high mobility of the electrons separated from the donors and confined in the quantum well.

High-power circuits are used to generate electromagnetic signals for long-distance telecommunication and other applications. Circuits based on bipolar junction transistors (BJTs) are among the most efficient ones (Fig. 1.1). This feature is advantageous in order to minimize the excess heat and the associated thermal walkout. However, at microwave and millimeter-wave frequencies, the circuit efficiency drops, and the Si BJT power circuits become inferior as compared with those exploiting GaAs-based and InP-based heterobipolar transistors (HBTs).

Local oscillators generate standard electromagnetic signals at intermediate frequencies required for signal-mixing circuits. A high stability of the frequency is most important for the local oscillators. The low-frequency noise is up-converted to sidebands of the carrier frequency and causes phase noise in the signal-mixing circuit. As a result, sources of low-frequency noise restrict the overall sensitivity of the system and set a limit in small-signal processing. Therefore, the low-frequency noise has to be as low as possible. The silicon BJT features very low noise at a low frequency and supports excellent stability of the oscillation at the intermediate frequency (Fig. 1.1), eliminating often the need for a dielectric resonator. As a result, the size of microwave telecommunication equipment can be reduced essentially. The MESFETs and HFETs show a higher noise at low frequencies as compared to silicon transistors. Nevertheless, HFET oscillators for signal-mixing circuits have no competitor at millimeter-wave frequencies.

4 INTRODUCTION

High-speed analog–digital and digital–analog converters and other digital circuits are quite important in many applications. The heterostructure technologies support the best solutions at the highest frequencies (Fig. 1.1).

The discussion of the main trends in high-speed electronics helps to indicate applications where the low-noise performance is of primary importance. The fluctuations at microwave and millimeter-wave frequencies decide the minimum noise figure in receiver circuits. These fluctuations set the upper limit for the amplifier sensitivity and the lowest limit for the magnitude of information bearing signals to be amplified without a loss of information. The low-frequency noise causes phase noise in the oscillator circuit and restricts the sensitivity of the system in small-signal mixer applications. Investigating the origin of diverse sources of fluctuations is important for the progress in the mentioned directions of high-speed electronics.

1.3 MAIN SOURCES OF NOISE IN SEMICONDUCTOR STRUCTURES

The most important sources of electronic noise are thermal, shot, generation–recombination, and hot-electron noise [1].

Thermal Noise

Any resistance shows spontaneous fluctuations of current caused by the universal thermal motion of the mobile electrons. The resultant noise is called thermal noise or *Johnson noise* [2]. No bias is needed for the fluctuating electromotive force to appear. The Nyquist theorem [3] for the thermal noise states that the available noise power of a thermal-noise source is the *universal* function of the absolute temperature. In particular, while every source of fluctuation is associated with the mechanism of dissipation, the Nyquist theorem relates the *spectral intensity of current fluctuations* in a conductor and the *dissipative part of its conductance*. The whole argument in terms of thermodynamic equilibrium collapses in a biased semiconductor: New sources of noise appear.

Shot Noise

Shot noise is important in structures where the current is controlled by a barrier: a p–n junction, a Schottky barrier, a heterojunction, a tunneling structure, a nonuniformity of doping, a nonohmic contact, and so on. The universal Schottky formula [4] for the shot noise tells us that the spectral intensity of current fluctuations is proportional to the current: Details of the electron transport have no influence on the noise, provided that no interaction between the electrons takes place. The shot noise is white over wide range of frequencies.

1/F Noise

This noise is produced by conductance fluctuations leading to fluctuations of current or voltage in a current-carrying state. The $1/f$ noise dominates at low frequencies and decays as $1/f$ with frequency; the associated fluctuations are called *1/f fluctuations*. The origin of $1/f$ noise is still often open to debate [5, 6].

Generation–Recombination Noise

In trap-containing semiconductors, random transitions between the band and the traps cause fluctuations of the number of mobile electrons [1, 6–8]. The resultant fluctuations of conductance manifest themselves as excess noise, provided that the steady current is not zero. The generation–recombination noise can be decomposed into the noise sources associated with different trap levels. Random band-to-band transitions leading to formation and annihilation of electron–hole pairs also cause generation–recombination noise.

Hot-Electron Noise

The electron gas in a current-carrying state can be pretty far from the thermal equilibrium, provided that the biasing electric field is of sufficiently high strength. The customary name for this situation is *hot electrons*. The nonequilibrium distribution of electrons maintained by the electric field is realized in a great number of semiconductor devices. Taking into account the present tendencies in electronics, one can predict a further growth of the role of hot electrons in small-size and low-dimensional structures.

The hot-electron noise is associated with ultrafast kinetic processes of dissipation taking place inside the conduction band (energy relaxation, intervalley transfer, impact ionization, etc.). In contrast to thermal and shot noise, no universal system-independent expression is available for hot-electron noise: The characteristics of hot-electron noise, like fingerprints, depend on the given nonequilibrium conditions [9–12]. It is often practiced to decompose hot-electron noise into the sources of noise featuring essentially different spectral properties. The cutoff frequencies of the associated kinetic processes lie in the frequency range from 10^9 to 10^{13} Hz. Thus, the hot-electron noise is most important at frequencies corresponding to the microwave and millimeter-wave range.

Fast and ultrafast kinetic processes in modern quantum-electronic structures produce a variety of hot-electron noise phenomena specific to electron confinement in quantum wells. For example, if two narrow quasi-two-dimensional electron gas layers are positioned at two different energy levels but spatially very close to each other, the electron transport along the conduction layers can lead to electron heating with the resultant partial transfer of the hot electrons into the neighboring layer. This hot-electron real-space transfer produces its own source of noise. There are other examples of how modern semiconductor structures produce their own noise behavior.

1.4 MICROWAVE NOISE IN A FIELD-EFFECT TRANSISTOR

As mentioned, the field-effect transistors (MESFETs and HFETs) are active devices demonstrating excellent noise properties at microwave and

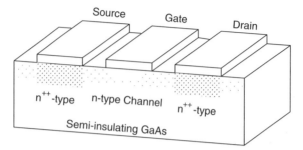

Figure 1.2. A schematic view of a metal-semiconductor field-effect transistor (MESFET) on a semi-insulating substrate. The source and drain electrodes form ohmic contacts to the n-type channel. The gate electrode forms a Schottky barrier.

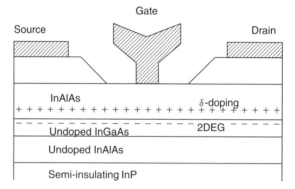

Figure 1.3. A schematic cross-sectional view of a high-electron-mobility transistor (HEMT). The high-mobility 2DEG channel is located in the undoped InGaAs layer. The mobile electrons for the 2DEG channel are supplied by the donor-doped InAlAs layer.

millimeter-wave frequencies. A field-effect transistor consists of a conductive channel supplied with two ohmic electrodes (the source and the drain) and the gate electrode controlling the channel current (Fig. 1.2). A typical GaAs MESFET has the channel of n-type GaAs forming a Schottky barrier with the gate electrode. Heterostructure FETs exploit two-dimensional electron gas (2DEG) channels. The HFETs are also known as HEMTs (high-electron-mobility transistors) to emphasize the feature of special importance. Usually, the 2DEG channel is located in the narrow-gap semiconductor separated from the doped wide-gap semiconductor (the supply layer) by the heterojunction barrier and the undoped spacer (Fig. 1.3). As a result, high mobility of electrons is combined with a high density of 2DEG. This supports a high drift velocity as required for ultrafast operation. Data on the electron drift velocity and the gate length allow one to estimate the cutoff frequency of the transistor. Frequencies over 300 GHz were predicted [13] for a 0.1-μm gate length InAlAs/InGaAs/InAlAs HEMT in contrast to a 150-GHz cutoff frequency predicted for a GaAs MESFET of the same dimensions. Recent experimental data are in agreement with these predictions, with the difference in performance between the two types of FETs tending even to disappear [14].

There are several main sources of microwave noise in a MESFET. The Johnson noise dominates in the electrode resistance and the contact resistance of the source and the drain where the electric field is extremely low. The shot noise originates in the Schottky barrier under the gate; it is caused by the leaky barrier. Finally, the hot-electron fluctuations in the channel cause fluctuations of the channel current.

Measurements of noise characteristics as a function of current, frequency, and lattice temperature help to distinguish different sources of excess noise and suggest how to eliminate some of them. In particular, technological means can be used to eliminate the shot noise in the gate circuit by eliminating the gate leakage current. High conduction of the source and drain electrodes and perfect ohmic contacts are desirable in order to minimize the contribution of the Johnson noise. The generation–recombination noise can be reduced by getting rid of the trapping centers. The hot-electron noise is the most inevitable feature of a FET: High electric field assures the required speed of operation of the transistor, and neither the channel can be eliminated nor the electric field can be switched off without a loss of the transistor function.

Within the structure location the fluctuations can be described in terms of equivalent noise sources: Linear resistors subjected to a low electric field produce frequency-independent Johnson noise, ideal capacitors and inductors produce no noise, diodes and barriers produce shot noise, and nonlinear resistors produce hot-electron noise. The noise signals act together with the frequency-dependent impedance of the capacitors and the inductors and cause frequency-dependent noise of the transistor or the transistor-containing circuit.

1.5 NOISE CHARACTERIZATION

The fluctuating — that is, time-dependent current — causes emission of electromagnetic waves into an open space or into the load (a coaxial cable, a waveguide, a coplanar line, etc.). In other words, the device feeds the noise power into the load. Measurements of the emitted noise power is of special importance at microwave and millimeter-wave frequencies because the current fluctuation spectra (available from the theory or circuit analysis) are not measured in a straightforward way in this frequency range. It is a convention to perform noise power experiments for matched impedance of the device output and the load, unless stated otherwise. Under this condition, the noise power is called the power available at the noise source or the available noise power [1, 6].

Equivalent Noise Temperature

The available noise power $\Delta P_n(f)$ emitted by a source of noise into the fixed frequency band Δf around the frequency f can be estimated by comparing it with the power radiated into the same frequency band by an absolutely black body kept at the known temperature. In case of equal powers in the given frequency band, one can say that the equivalent noise temperature of the noise source at this frequency equals the absolute temperature of the reference black body. Thus the equivalent noise temperature, or the noise temperature, $T_n(f)$, multiplied by the Boltzmann constant k_B is, by definition, the noise power, per unit band around the frequency f, dissipated by the source of noise into the matched load:

$$T_n(f) = \frac{\Delta P_n(f)}{k_B \Delta f}. \tag{1.1}$$

Any deviation from equilibrium introduces excess noise with the spectral features specific to the given nonequilibrium conditions. Consequently, the available noise power depends on frequency, bias, and so on. This forces one to introduce frequency- and bias-dependent equivalent noise temperature. In particular, the hot-electron noise temperature, $T_n(f)$, represents a property of an electron gas differing from its energy temperature T_e that enters the expression for the electron average energy:

$$\bar{\varepsilon} = \tfrac{3}{2} k_B T_e \tag{1.2}$$

in a nondegenerate three-dimensional electron gas, and

$$\bar{\varepsilon} = k_B T_e \tag{1.3}$$

in a nondegenerate two-dimensional electron gas.

Spectral Intensity of Current Fluctuations

The available noise power can be expressed in terms of the spectral intensity of the current (voltage) fluctuations and the impedance. Under thermal equilibrium, the spectral intensity of voltage fluctuations S_V and the AC impedance of the noise source Z are interrelated through the *Nyquist theorem* [3]. In the classical limit we have

$$S_V^{eq} = 4 k_B T_0 \operatorname{Re}\{Z\} \tag{1.4}$$

(the quantum correction factor is neglected: the inequality $\hbar \omega \ll k_B T_0$ is assumed).

Under equilibrium the noise temperature equals the absolute temperature of the source of noise:

$$T_n^{eq} = T_0. \tag{1.5}$$

Equation (1.5) means that the equilibrium noise spectrum is white in the range of frequencies $\omega = 2\pi f \ll k_B T_0/\hbar$, and the available power per unit bandwidth is related to the energy through the equipartition law. In principle, the noise power available at the equilibrium source can serve for the establishment of the absolute scale of temperature (see reference 15, p. 21). According to the Nyquist theorem (1.4), an independent determination of the spectral intensity of voltage fluctuations does not add the information that is not available from the response (impedance) data.

Unlike this, the measuring of excess noise provides one with the complementary information on the associated kinetic processes not available from the response characteristics. In other words, under nonequilibrium conditions, a two-terminal noisy element is characterized by two physical quantities: for example, the noise temperature (power) and the spectral intensity of current or voltage fluctuations, S_I or S_V. At nonequilibrium conditions the noise temperature $T_n(f)$ takes the place of the absolute temperature T_0 in Eq. (1.4). Two equivalent expressions are available:

$$S_V(f) = 4k_B T_n(f) \operatorname{Re}\{Z(f)\} \tag{1.6}$$

and

$$S_I(f) = 4k_B T_n(f) \operatorname{Re}\{Z^{-1}(f)\}, \tag{1.7}$$

where $Z(f)$ is the AC impedance of the noise source around the DC bias point. The real (dissipative) parts of the sample impedance and admittance enter Eqs. (1.6) and (1.7).

Noise Figure

A noisy two-port circuit — for example, a transistor or a circuit of a low-noise amplifier — is often characterized by its noise figure NF. It is reasonable to express the noise contribution inherent to the circuit in terms of the noise-to-signal ratio at the output port and the noise-to-signal ratio at the input port. Under certain convention, the ratio of these two quantities, or its logarithmic function in decibels, is the noise figure. It can be written as a function of two noise temperatures, $T'_n(f)$ and T_0:

$$NF(f) = 10 \log_{10} \frac{T'_n(f)}{T_0}, \tag{1.8}$$

where $T'_n(f)$ represents the output noise power reduced back to the circuit input (that is, divided by the circuit gain, $T'_n = T_n/G$, where G is the gain) and T_0 represents the input noise power produced by the input-matched resistor kept at the conventional $T_0 = 290$ K temperature. A noise-free circuit is characterized with $NF = 0$ dB.

A so-called noise mismatch of the circuit output and the load cause an increase in the noise figure. The minimum noise figure NF_{min} is reached under the optimal load and the optimal bias. The obtained value NF_{min} is the noise characteristic independent of external conditions and port terminations. However, knowledge of the minimum noise figure is not sufficient for the complete description of noise behavior of a two-port circuit: Four independent parameters are required [16–18].

Noise–Speed Tradeoff

Because the acceptable noise figure of a transistor is below 2 dB, a standard Si BJT is not a promising device at frequencies exceeding several GHz (dash-dot curve, Fig. 1.4). An essentially lower noise in a wider range of frequencies (1.5 dB at 5 GHz) is demonstrated by a Si HBT containing a SiGe base (squares [19]). The minimum noise figure of an AlGaAs/InGaAs HBT does not exceed 2 dB at frequencies up to 22 GHz (closed circles [20]). The increased interest in the HBTs is supported by their ability to combine low microwave noise with exceptionally low $1/f$ noise, leading to excellent low-phase-noise high-power performance.

Gallium arsenide MESFETs can be considered as low-noise devices in a wide range of frequencies (Fig. 1.4). The main improvement of the noise–speed tradeoff is achieved through reduction of the gate length: a 0.12-μm gate length GaAs MESFET suitable for millimeter-wave applications shows 0.9-dB noise at 18 GHz (down triangles [21]). Another way to improve low-noise performance is to use high-mobility 2DEG channels. The commercially available low-noise GaAs-based HEMTs show 1.5-dB noise at 26 GHz (dotted curve [22]). The InP HEMTs with lattice-matched InGaAs 2DEG channels and (0.15- to 0.2-μm gates show an excellent (0.9- to 0.8-dB minimum noise figure near 60-GHz frequency (open diamonds [23], centered up-triangle [24]). The minimum noise figure $NF_{min} = 2.6$ dB at 94 GHz has been reported for a 0.1-μm gate device

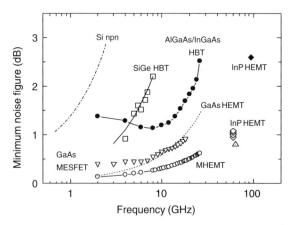

Figure 1.4. Noise–speed tradeoff of low-noise microwave and millimeter-wave transistors. Dash–dot curve represents silicon $n-p-n$ BJT; □ represents SiGe HBT [19]; ● represents AlGaAs/InGaAs HBT [20]; ▽ represents GaAs ion-implanted MESFET [21]; dotted curve represents GaAs HEMT [22]; ○ represents GaAs-based metamorphic HEMT (MHEMT) [22]; ◊ represents InP HEMT [23]; centered △, InP HEMT [24]; ♦, InP HEMT [25].

(closed diamond [25]). A metamorphic HEMT using GaAs-based InGaAs 2DEG channel supports 12-dB gain associated with 0.6-dB noise figure at 25 GHz (open circles [22]). A low-noise amplifier built using 0.1-μm gate metamorphic GaAs-based InGaAs HEMT technology demonstrates 18-dB gain and 4.2-dB noise figure at 82 GHz and 10.4-dB gain at 110 GHz [26].

There is no simple and direct relation of the noise properties with the standard material properties such as mobility, doping level, electron density, and so on. For example, it is evident that a weaker scattering implies stronger hot-electron effects: In particular, a higher hot-electron temperature is reached at a given electric field, provided that the electron mobility is higher. On the other hand, an HEMT features a lower noise temperature at microwave and millimeter-wave frequencies as compared to that of a MESFET exploiting a low-mobility channel. A satisfactory understanding of this behavior and other hot-electron noise effects requires a complete physical treatment of the associated dissipative processes in biased semiconductors and semiconductor structures. In particular, the optimization of the low-noise high-speed performance calls for an investigation of the ultrafast kinetic processes responsible for the fluctuations in the microwave and millimeter-wave frequency range.

1.6 APPROACHES TO RESPONSE MODELING

An efficient design of a device and a circuit for low-noise high-speed applications is based on accurate modeling of the transistor, or other active element, and its noise behavior is determined by various noise contributions at low and high frequencies. An adequate response model is a prerequisite for the noise modeling.

Equivalent Circuit Approach

Traditionally, static and high-frequency response characteristics of semiconductor devices are modeled using equivalent circuit approaches, where the electrical behavior of the device at the connecting terminals is represented by a circuit consisting of lumped two-terminal elements standing for ideal resistance, capacitance, inductance, and so on.

In the *physics-based equivalent circuit approach*, the modeling of an active device starts with a choice of a lumped-element equivalent-circuit model consisting of linear and nonlinear elements. The lumped elements are related to the physical mechanisms in the device [16]. The linear and bias-independent extrinsic elements are extracted from small-signal response experiments—for example, S-parameter measurements. To determine the nonlinear elements of the inner transistor, deembedding procedures are applied to the measured small-signal S-parameters [27]. The nonlinear elements can be modeled by analytical functions of voltages of the inner transistor. Examples for this type of model are given by Ebers–Moll, Gummel–Poon, Pucel–Haus–Statz, VBIC, and so on [28].

In the *empirical equivalent circuit models*, curve-fitting functions such as polynomials, cubic splines, and so on, are used. They are strictly analytical and useful in commercial circuit simulators. The representatives are Curtice, Tajima, Materka, Statz, MOS3, BSIM3, and Root (tabular model) models [27]. These models suffer from the absence of relations to physical processes, material parameters, and geometric data of the active devices. Sometimes, unphysical values for equivalent circuit elements should be introduced. As a consequence, these models lack predictive features.

Physical Approach

Physical modeling is based on a physical description of the active device in terms of electron transport taking into account material properties and geometric structure [14, 29, 30]. Different approaches for the calculation of potentials, carrier densities, band profiles, currents, and so on, are used to obtain static and high-frequency response characteristics.

Microscopic physical models are more complex than equivalent circuit models and normally require the application of numerical methods. Recent advances in computer technology and powerful numerical methods have allowed microscopic physical models to address device structures in great detail. Simulation software is now available commercially and is utilized increasingly in the semiconductor industry.

The majority of semiconductor devices can be modeled using semiclassical kinetic methods, where the direct influence of quantum phenomena are assumed to be of minor importance. This approach is applicable to all devices where the critical dimensions are significantly greater than a de Broglie wavelength. Examples of contemporary semiconductor structures that can be accurately represented using semiclassical methods are: $p-n$ junction diodes, bipolar junction transistors, heterojunction bipolar transistors with layer thickness of greater than 0.1 µm, and MESFETs with gate lengths of greater than 50 nm. In the case of strongly quantized structures, it is necessary to solve coupled Schrödinger and Poisson equations self-consistently. At the present time the best example of a device requiring this type of treatment is the HEMT, where the charge-control model requires at least a one-dimensional solution of the Schrödinger–Poisson equations to consider the subband structure taking into account subband occupation and charge distribution self-consistently.

Recent advances are based on elaborated models with particular reference to structures containing two- and three-dimensional electron gases subjected to a high electric field. Semiconductor device models are given in terms of a set of equations describing electron transport on the level of the Boltzmann transport equation. Numerical simulation of kinetic processes inside the conduction band is often obtained using the Monte Carlo technique (For detailed consideration of the procedures and results, see monographs of Jacoboni and Lugli [31], Laux and Fischetti [32], and Moglestue [33]). The relatively advanced state of development of physical models means that a high degree of quantitative agreement between simulated and measured data is now possible, and physical models are increasingly finding a role in computer-aided design applications.

Drift–diffusion models are based on the current continuity equation and the drift–diffusion equations for electrons and holes. The equation were derived by Wannier [34] from the time-dependent Boltzmann transport equation. The traditional drift–diffusion models neglect hot-electron effects and assume that the average electron temperature throughout the device is fixed to the lattice temperature. In high electric fields, electron heating and the associated hot-electron effects become important. The modified drift–diffusion models take into account dependence on electric field of the electron drift velocity and the electron diffusion coefficient, with their steady-state values being available from experiment or Monte Carlo simulation.

1.7 MICROWAVE-NOISE MODELING

When the fluctuation–dissipation relation (Nyquist theorem) fails in a nonequilibrium state, the noise spectra depend on external forces and details of the interaction of the electrons with the thermal bath. Calculation, or measurement, of the spectral intensity of noise proves to be an independent problem that cannot be reduced to the problem of calculation, or measurement, of the response of the system to the external perturbation. Fluctuation characteristics of a nonequilibrium system in fact are new independent kinetic (transport) characteristics of the system. Once the response characteristics are available, the fluctuations can be taken into account, and noise characteristics can be modeled. Different fast and ultrafast dissipation mechanisms, including relaxation of quasi-momentum and energy, intervalley relaxation, as well as free-carrier number relaxation, reflect themselves in the noise spectrum pattern of a semiconductor having a biasing current. In the case of semiconductor structures (structures with quantum wells, etc.), the intersubband, real-space transfer and other kinetic processes impresses their fingerprints on the nonequilibrium noise spectra of the structures. Since there is no necessary connection between the electric noise and the impedance of the nonequilibrium system, one must examine the noise in terms of its internal mechanisms. There are several approaches to account for the strength and location of diverse noise sources important at microwave and millimeter-wave frequencies.

Equivalent Noise Sources

In the lumped-element equivalent circuit approach, the independent sources of noise are represented either by the fluctuating electromotive force associated with a noise-free impedance in series or by the noise current source coupled with a noise-free admittance in parallel. The thermal noise and the shot noise are the traditional microwave relevant sources; within their location in the structure they are considered as frequency-independent noise sources in the frequency range of interest for the high-speed electronics. Sources of hot-electron noise associated with fast and ultrafast relaxation processes feature their specific spectral properties determined by the corresponding relaxation time constants.

Extensive GaAs field-effect transistor and HEMT noise measurements are used to compare different noise models based on lumped-element equivalent circuits [35]. The models under investigation differ in the number of independent parameters needed to calculate the four noise parameters for the two-port device. The two-parameter Pospieszalski model [36] is found to be the most suitable one. The lumped-element circuit models are attractive because of their simplicity.

Nevertheless, the approach has serious drawbacks. For example, an unsolved problem is correlation of the noise sources. For example, most models represent a channel of a transistor by several lumped resistors connected in series: In a real transistor a strong correlation of the sources of hot-electron noise in the channel takes place. The correlation of the fluctuations in the barrier and in the channel is essential as well. Solutions of the correlation problems are beyond the lumped-element circuit approach.

Drift–Diffusion Equations with Fluctuations

Correlation in time and space is dealt with using so-called drift–diffusion equation with fluctuations [37, 38]. The approach is similar to the well-known method of investigation of hydrodynamic fluctuations in gases and fluids. In the case of carrier drift–diffusion equation, the corresponding Langevin forces often are expressible in terms of diffusion coefficients of hot carriers. This property is referred to as the second fluctuation–dissipation theorem. The approach is valid for long-wavelength fluctuations. The modified quasi-hydrodynamic description of fluctuations in a wider region of wavelengths and frequencies is also possible in the so-called effective electron temperature approximation applicable at high densities of electrons.

Microscopic Physical Models of Noise

The semiclassical quantitative description of a nonequilibrium state is accessible provided the standard kinetic (Boltzmann) theory is applicable. The macroscopic quantities then are expressible in terms of the distribution functions of free electrons, holes, phonons, magnons, and other excitations. The apparatus of the kinetic theory was generalized to describe fluctuations by Kadomtsev [39], Lax [40], Price [41] and Gantsevich, Gurevich, and Katilius [37]. The Langevin approach was made suitable for nonequilibrium conditions by Lax [40], and Kogan and Shulman [42]. These achievements have shaped the modern microscopic theory of fluctuations in a nonequilibrium state. The criteria of applicability of the theory coincide with those of the Boltzmann equation: If the kinetic equation for the one-particle distribution function can be worked out and solved, then the kinetic equations for fluctuations can also be written and, though more complicated, successfully solved. These results, in principle, have completed the theory of classical fluctuations in a weakly interacting many-particle system — in the same sense in which the Boltzmann equation exhausts the theory of transport in such a system. Many pages in this book will be devoted to applications of the general theory of fluctuations to a gas of electrons in semiconductors and semiconductor structures, especially of "hot" electrons.

The frequencies associated with the characteristic relaxation times of hot electrons in semiconductors and semiconductor structures lie in the frequency range from 10^9 to 10^{13} Hz. Therefore, it is quite natural that the investigation of the noise spectrum in the microwave and millimeter-wave frequency range has proved to be a powerful diagnostic tool of the hot-electron state realized by creating in a structure the electric field of sufficiently high strength. The remarkable progress has been made, and continues to be being made, in the theoretical and experimental investigations of noise and fluctuation phenomena in semiconductors and semiconductor structures under hot-electron conditions at high frequencies [9–12].

The hot-electron fluctuation characteristics are quite sensitive to the details of the scattering mechanisms and the band structure. The satisfactory agreement of the available experimental results with the microscopic interpretation supports an improved understanding of the different scattering mechanisms that mobile electrons undergo in their motion in semiconductors or semiconductor structures. On the other hand, the sensitivity of the microwave fluctuations to subtle details of the band structure and the scattering mechanisms make it impossible to predict noise characteristics of update low-noise ultrafast devices without addressing to microscopic level of understanding of nonequilibrium kinetic processes of hot electrons in semiconductors and semiconductor structures subjected to a high electric field.

CHAPTER 2

KINETIC THEORY OF NONEQUILIBRIUM PROCESSES

The purpose of this chapter is to present a summary of the well-known kinetic description of nonequilibrium states of a gas of charge carriers (mobile electrons) in a semiconductor. The technique to describe a response of the nonequilibrium gas to a weak additional external perturbation will also be presented, being indispensable while formulating, in the following chapters, the kinetic theory of fluctuations.

Let us begin with few general remarks concerning an open system subjected to a continuous energy flow, some energy being added from the external world and then dissipated out to the external world. The situation can be induced by the "driving force" transferring energy from outside to the system. In the typical nonequilibrium conditions, the system is displaced from thermal equilibrium with the "environment" (the thermal bath), so that the system does not have the energy distribution corresponding to the temperature of the thermal bath. The system may be driven into a state stationary in time for an interval long enough for some measurements to be made, provided that the energy is steadily passed on to the thermal bath. Such a virtually stationary state of the system substantially displaced from the equilibrium with the thermal bath will be a subject of our interest.

The statistical description of a many-particle system in the thermal equilibrium state is based on a few very general properties or principles, such as the Gibbs distribution, the possibility to introduce thermodynamic potentials, and the validity of fluctuation–dissipation theorems. For nonequilibrium states, these principles do not work. In particular, thermodynamic potentials cannot be introduced. Contrary to the universal functional form of the Gibbs distribution, the nonequilibrium distribution functions depend on both the external forces creating the bias from thermal equilibrium and on details of the interaction of the system with the thermal bath. Accordingly, the distribution functions are much more dependent on the experimental situation and should be found for each case. In other words, instead of thermodynamics, other concepts and theoretical apparatus had to be and were invented and used.

As mentioned in Chapter 1, a quantitative description of nonequilibrium states is accessible, provided that the standard kinetic theory is applicable to the system (see, e.g., reference 43). The macroscopic quantities then are expressible in terms of the one-particle distribution function obeying the kinetic, or transport, equation introduced by Ludwig Boltzmann to describe a neutral gas with pair, or two-particle, collisions and then adapted to a gaseous plasma as well as a gas of excitations in solids.

In the case of a semiconductor or a semiconductor structure, the nonequilibrium distribution of the carriers can be realized with comparative ease by applying the electric field of sufficiently high strength. The gas of carriers, which determines electrical properties of the structure including its electric noise characteristics, in a current-carrying state can be far from equilibrium with the lattice (the *hot carriers*). The losses of the carrier energy and momentum due to interaction with the lattice (the thermal bath) are compensated by the work done by the external electric field. A gas of electrons in semiconductors and semiconductor structures, including the gas of "hot" electrons, obeys the Boltzmann equation. The nonequilibrium distributions of carriers maintained by the electric field have been widely investigated both theoretically and experimentally during a number of decades. Taking into account the present tendencies, one can predict a further growth of the role of nonequilibrium states, especially of those very far from equilibrium.

2.1 BOLTZMANN EQUATION; NONEQUILIBRIUM STATES

Boltzmann Equation

The gas of free electrons can be easily displaced from the equilibrium with the thermal bath by application of a DC electric field creating a steady current in the system. Mobile electrons are accelerated by the electric field; but due to collisions with phonons and impurities, they lose the gained energy and quasi-momentum, transferring them to the crystal lattice that acts as the thermal bath.

Here we shall restrict ourselves with spatially homogeneous (uniform, space independent) systems and electron distributions. This enables us to deal with quasi-momentum space (**p**-space) instead of quasi-momentum–coordinate space ((**p**, **r**)-space, often referred to also as μ-space). Coordinate-dependent states and processes will be investigated in Chapter 6.

At thermal equilibrium (i.e., in the absence of an external field), the equilibrium distribution of free electrons ensures the equality between the mean rates of the flow *into* a unit cell of the quasi-momentum space and *out* of it, with the flow being caused by collisions. The applied electric field upsets the balance and gives rise to a redistribution of electrons in the momentum space: A flow in the quasi-momentum space now is caused not only by the electron interaction with the thermal bath and other electrons, but also by the electric field. A redistribution is governed by the time-dependent *Boltzmann equation* (see, e.g., reference 43):

$$\frac{\partial}{\partial t}\overline{F}_{\mathbf{p}}(t) = -e\mathbf{E}\frac{\partial}{\partial \mathbf{p}}\overline{F}_{\mathbf{p}}(t) - I_{\mathbf{p}}^{\text{th}}\overline{F}_{\mathbf{p}}(t) - I_{\mathbf{p}}^{\text{ee}}\{\overline{F}(t), \overline{F}(t)\}. \quad (2.1)$$

Here $\overline{F}_{\mathbf{p}}(t)$ is the time-dependent one-particle distribution function (the overbar here and in the following designates the average over the "ensemble of systems"). The Boltzmann equation (2.1) simply says that the mean rate of change of the number of electrons in the unit cell of the quasi-momentum space around the state $\mathbf{p}$ is equal to the net flow into this cell, with the net flow being the sum of contributions caused by (i) the electric field, (ii) the interaction with the thermal bath ("one-particle collisions"), and (iii) the interaction among the electrons (two-particle collisions). Each of the contributions (ii) and (iii), in turn, is the difference between the corresponding mean rates of the flow *into* and *out* of the unit cell $\mathbf{p}$. (In the case of coordinate-dependent distributions, additional terms appear in the Boltzmann equation describing convection of electrons and action of the local electric fields created by redistribution of electrons in the space, see Chapter 6.)

Collision Integrals

Let us denote as $W_{\mathbf{p}'}^{\mathbf{p}}$ the transition probability for a collision with the thermal bath which carries the electron from the state $\mathbf{p}$ to the state $\mathbf{p}'$. With the ensemble-average number of electrons in the state $\mathbf{p}$ being $\overline{F}_{\mathbf{p}}(t)$, the $\mathbf{p} \to \mathbf{p}'$ *transition mean rate* is $W_{\mathbf{p}'}^{\mathbf{p}}\overline{F}_{\mathbf{p}}(t)$ (for the sake of simplicity, we deal here with the nondegenerate statistics). The mean rate of decrease of the particle number in the state $\mathbf{p}$ due to "one-particle" collisions (i.e., the "one-particle collision integral"), is given by the difference between the mean rates of flow *out* of and *into* the state $\mathbf{p}$ due to these collisions:

$$\sum_{\mathbf{p}'}[W_{\mathbf{p}'}^{\mathbf{p}}\overline{F}_{\mathbf{p}}(t) - W_{\mathbf{p}}^{\mathbf{p}'}\overline{F}_{\mathbf{p}'}(t)].$$

In the case of nondegenerate statistics, the one-particle collision integral is a *linear* functional of $\overline{F}$, so that the *linear operator* $I_{\mathbf{p}}^{\text{th}}$ can be introduced to describe statistical interaction of electrons with the thermal bath:

$$I_{\mathbf{p}}^{\text{th}}\Phi_{\mathbf{p}} \equiv \sum_{\mathbf{p}'}[W_{\mathbf{p}'}^{\mathbf{p}}\Phi_{\mathbf{p}} - W_{\mathbf{p}}^{\mathbf{p}'}\Phi_{\mathbf{p}'}]. \quad (2.2)$$

The statistical interaction *among* the electrons is described by the two-particle collision integral

$$I_{\mathbf{p}}^{\text{ee}}\{\overline{F}, \overline{F}\} = \sum_{\mathbf{p}_1 \mathbf{p}' \mathbf{p}'_1}[W_{\mathbf{p}'\mathbf{p}'_1}^{\mathbf{p}\mathbf{p}_1}\overline{F}_{\mathbf{p}}(t)\overline{F}_{\mathbf{p}_1}(t) - W_{\mathbf{p}\mathbf{p}_1}^{\mathbf{p}'\mathbf{p}'_1}\overline{F}_{\mathbf{p}'}(t)\overline{F}_{\mathbf{p}'_1}(t)], \quad (2.3)$$

where $W_{\mathbf{p'p'_1}}^{\mathbf{pp_1}}$ is the transition probability for the electron–electron collision that carries two electrons from the states $\mathbf{p}$ and $\mathbf{p_1}$ to the states $\mathbf{p'}$, $\mathbf{p'_1}$.

Identities for Transition Probabilities

The correctly calculated transition probabilities entering the collision integrals are related by some identities. First of all, note that

$$W_{\mathbf{p'p'_1}}^{\mathbf{pp_1}} = W_{\mathbf{p'p'_1}}^{\mathbf{p_1p}} = W_{\mathbf{p'_1p'}}^{\mathbf{pp_1}} = W_{\mathbf{p'_1p'}}^{\mathbf{p_1p}}. \tag{2.4}$$

The one-particle collision integral (2.2) should vanish after the insertion of the thermal-equilibrium distribution

$$\overline{F}_{\mathbf{p}}^{\mathrm{eq}} \propto \exp(-\varepsilon_{\mathbf{p}}/T_0), \tag{2.5}$$

where $\varepsilon_{\mathbf{p}}$ is the electron energy, and T_0 is the temperature of the thermal bath in energy units; thus,

$$I_{\mathbf{p}}^{\mathrm{th}}\overline{F}_{\mathbf{p}}^{\mathrm{eq}} = 0, \quad \text{or} \quad \sum_{\mathbf{p'}} W_{\mathbf{p'}}^{\mathbf{p}} \exp[-\varepsilon_{\mathbf{p}}/T_0] = \sum_{\mathbf{p'}} W_{\mathbf{p}}^{\mathbf{p'}} \exp[-\varepsilon_{\mathbf{p'}}/T_0]. \tag{2.6}$$

The two-particle collision integral (2.3) vanishes after substitution of the "displaced" Maxwellian distribution

$$F_{\mathbf{p}}^{\mathrm{M}} \propto \exp[-(\varepsilon_{\mathbf{p}} - \mathbf{p} \cdot \mathbf{V})/T] \tag{2.7}$$

with *arbitrary* values of the temperature T and the translational or drift velocity of the electron system, $\mathbf{V}$. Indeed, interelectron collisions conserve the kinetic energy and quasi-momentum of the electron system. The transition probabilities in Eq. (2.3) contain δ-functions ensuring conservation of energy and quasi-momentum in a single collision:

$$\mathbf{p} + \mathbf{p_1} = \mathbf{p'} + \mathbf{p'_1}, \qquad \varepsilon_{\mathbf{p}} + \varepsilon_{\mathbf{p_1}} = \varepsilon_{\mathbf{p'}} + \varepsilon_{\mathbf{p'_1}}. \tag{2.8}$$

This leads to the expression

$$I_{\mathbf{p}}^{\mathrm{ee}}\{F^{\mathrm{M}}, F^{\mathrm{M}}\} = \sum_{\mathbf{p_1}} \exp\left[-\frac{\varepsilon_{\mathbf{p}} + \varepsilon_{\mathbf{p_1}} - (\mathbf{p}+\mathbf{p_1})\cdot\mathbf{V}}{T}\right] \sum_{\mathbf{p'p'_1}} \left(W_{\mathbf{p'p'_1}}^{\mathbf{pp_1}} - W_{\mathbf{pp_1}}^{\mathbf{p'p'_1}}\right), \tag{2.9}$$

which equals zero thanks to *Stueckelberg's normalization property* [44]:

$$\sum_{\mathbf{p'p'_1}} W_{\mathbf{p'p'_1}}^{\mathbf{pp_1}} = \sum_{\mathbf{p'p'_1}} W_{\mathbf{pp_1}}^{\mathbf{p'p'_1}} = 1. \tag{2.10}$$

In Eq. (2.10), the sum encompasses the probability $W_{pp_1}^{pp_1}$ for the particles to remain unscattered.

Of course, the displaced Maxwellian distribution (2.7) by no means turns into zero the sum of the other terms on the right-hand side of Boltzmann equation (2.1) (the field term plus the one-particle collision integral).

Nonequilibrium Steady-State

An applied external force — a DC field **E** — displaces electron gas from the equilibrium with the lattice. The action of the field is described by the "field term"

$$e\mathbf{E} \cdot \partial/\partial \mathbf{p} \quad (2.11)$$

in the Boltzmann equation (2.1). Interaction with scatterers aims at restoring the equilibrium, being represented in Eq. (2.1) by the collision integral I^{th}. The evolution of the electron gas prescribed by the Boltzmann equation (2.1) may lead to a nonequilibrium *steady* state for which the time derivative in Eq. (2.1) vanishes. Accordingly, the *steady-state distribution function* $\overline{F}_\mathbf{p}$ should satisfy the *time-independent (stationary) Boltzmann equation*

$$e\mathbf{E}\frac{\partial}{\partial \mathbf{p}}\overline{F}_\mathbf{p} + I_\mathbf{p}^{th}\overline{F}_\mathbf{p} + I_\mathbf{p}^{ee}\{\overline{F}, \overline{F}\} = 0 \quad (2.12)$$

[we confine ourselves to cases in which the solution of Eq. (2.12) is unique and stable]. By making equal zero the *sum* of three items on the left-hand side of Eq. (2.12) (but by no means making equal zero each collision term taken separately), the steady-state distribution function $\overline{F}_\mathbf{p}$ can bear no resemblance to an equilibrium distribution. The mean energy per electron can exceed by two orders of magnitude, or even more, the electron mean kinetic energy in equilibrium ($3T_0/2$ for the three-dimensional motion, T_0 for the two-dimensional one). The shape of the distribution in the quasi-momentum space can vary from fairly isotropic to highly peaked along the field direction ("needle-shaped"). The steady non-equilibrium distributions maintained by a DC electric field in semiconductors and semiconductor structures ("hot electrons") have been widely investigated both theoretically and experimentally (see references 45–50).

Observables

The observables characterizing the system of mobile electrons as a whole are expressible in terms of the distribution function and can be calculated, provided that the latter is known. For example, the steady current density is

$$\bar{\mathbf{j}} = \frac{e}{\mathcal{V}_0}\sum_\mathbf{p} \mathbf{v}_\mathbf{p}\overline{F}_\mathbf{p} = en_0\mathbf{V}, \quad (2.13)$$

where the velocity of the electron in the state $\mathbf{p}$ is denoted as $\mathbf{v_p}$; that is $\mathbf{v_p} \equiv \partial \varepsilon_\mathbf{p}/\partial \mathbf{p}$, where $\varepsilon_\mathbf{p}$ is the electron energy. Here $n_0 = N/\mathcal{V}_0$ is the electron density, the distribution function $\overline{F}_\mathbf{p}$ is normalized to the total number of free electrons N,

$$\sum_\mathbf{p} \overline{F}_\mathbf{p} = N, \qquad (2.14)$$

and $\mathcal{V}_0$ is the volume of the system. The *steady value of the drift velocity of the electron system as a whole* is given by

$$\mathbf{V} = \frac{1}{N} \sum_\mathbf{p} \mathbf{v_p} \overline{F}_\mathbf{p}. \qquad (2.15)$$

Because electrons are substantially redistributed in the quasi-momentum space due to the action of the applied electric field, the current density $\bar{\mathbf{j}}$ and the drift velocity $\mathbf{V}$ cease to be linear functions of the field (deviations from the Ohm law occur). To predict the current–voltage characteristic, one should find the steady-state distribution function $\overline{F}_\mathbf{p}$ by solving the time-independent Boltzmann equation (2.12) and then calculating the current density (2.13).

One should emphasize that nowadays the distribution function $\overline{F}_\mathbf{p}$ itself is open to a direct experimental observation. In particular, the experiments on hot-carrier luminescence [51] and on light scattering from free carriers [52–54] are very informative (see also references 55–62).

2.2 RELAXATION AND RESPONSE PROBLEMS

Evolution of Small Deviations from Steady-State

Because of the interelectron collision integral (2.3), Eqs. (2.1) and (2.12) are nonlinear with respect to the distribution function $\overline{F}_\mathbf{p}$. The evolution of *small deviations* from a steady-state is governed by the Boltzmann equation *linearized* with respect to the small deviation from the steady distribution, $\Delta \overline{F}_\mathbf{p}(t)$:

$$\overline{F}_\mathbf{p}(t) = \overline{F}_\mathbf{p} + \Delta \overline{F}_\mathbf{p}(t). \qquad (2.16)$$

Inserting Eq. (2.16) into Eq. (2.1), linearizing the latter with respect to $\Delta \overline{F}_\mathbf{p}(t)$, and using the stationary Boltzmann equation (2.12), we obtain the equation that governs the evolution of the deviation $\Delta \overline{F}_\mathbf{p}(t)$:

$$\left(\frac{\partial}{\partial t} + I_\mathbf{p} \right) \Delta \overline{F}_\mathbf{p}(t) = 0, \qquad (2.17)$$

where $I_\mathbf{p}$ is the *linearized* operator of the Boltzmann equation, given by the expression

$$I_\mathbf{p} \equiv e\mathbf{E} \frac{\partial}{\partial \mathbf{p}} + I_\mathbf{p}^{\mathrm{th}} + I_\mathbf{p}^{\mathrm{ee}}(\overline{F}). \qquad (2.18)$$

Here $I_{\mathbf{p}}^{ee}(\overline{F})$ is the *linearized electron–electron collision operator* obtainable by linearization of the two-particle collision integral (2.3):

$$I_{\mathbf{p}}^{ee}(\overline{F})\psi_{\mathbf{p}} \equiv I_{\mathbf{p}}^{ee}\{\overline{F}, \psi\} + I_{\mathbf{p}}^{ee}\{\psi, \overline{F}\}. \tag{2.19}$$

In the explicit form, Eq. (2.19) reads as

$$I_{\mathbf{p}}^{ee}(\overline{F})\psi_{\mathbf{p}} \equiv \sum_{\mathbf{p}_1 \mathbf{p}' \mathbf{p}'_1} [W_{\mathbf{p}'\mathbf{p}'_1}^{\mathbf{p}\mathbf{p}_1}(\overline{F}_{\mathbf{p}}\psi_{\mathbf{p}_1} + \overline{F}_{\mathbf{p}_1}\psi_{\mathbf{p}}) - 2W_{\mathbf{p}\mathbf{p}_1}^{\mathbf{p}'\mathbf{p}'_1}\overline{F}_{\mathbf{p}'}\psi_{\mathbf{p}'_1}], \tag{2.20}$$

where the operator $I_{\mathbf{p}}^{ee}(\overline{F})$ (and hence the operator $I_{\mathbf{p}}$) is the functional of the steady-state distribution $\overline{F}$.

As demonstrated by Eq. (2.17), the operator $I_{\mathbf{p}}$ governs the relaxation of small deviations from a steady distribution and thus can be called the *relaxation operator*. When electric field is applied, the relaxation operator consists of two distinct parts: the "field operator" (2.11), $e\mathbf{E}\partial/\partial\mathbf{p}$, which describes the effect of streaming due to the external force, and the operator

$$I_{\mathbf{p}}^{coll} = I_{\mathbf{p}}^{th} + I_{\mathbf{p}}^{ee}(\overline{F}), \tag{2.21}$$

which describes the effect of collisions. In cases where electron–electron collisions are negligible (e.g., for a semiconductor with low carrier density), the collision operator $I_{\mathbf{p}}^{coll}$ is none other than the linear electron–lattice collision operator, with the operator $I_{\mathbf{p}}^{th}$ given by Eq. (2.2). Hence, in these cases the relaxation operator is given by the expression

$$\tilde{I}_{\mathbf{p}} = e\mathbf{E}\frac{\partial}{\partial\mathbf{p}} + I_{\mathbf{p}}^{th}, \tag{2.22}$$

where we introduce a special notation, $\tilde{I}_{\mathbf{p}}$, for the relaxation operator with electron–electron collisions omitted. In such a case of electrons interacting only with the thermal bath but not with each other, the Boltzmann equation itself is linear:

$$\frac{\partial}{\partial t}\overline{F}_{\mathbf{p}}(t) = -\left(e\mathbf{E}\frac{\partial}{\partial\mathbf{p}} + I_{\mathbf{p}}^{th}\right)\overline{F}_{\mathbf{p}}(t), \tag{2.23}$$

so that the relaxation operator *coincides* with the Boltzmann equation operator. The equation for the steady-state distribution function $\overline{F}_{\mathbf{p}}$ in such a case is

$$\tilde{I}_{\mathbf{p}}\overline{F}_{\mathbf{p}} = 0, \tag{2.24}$$

so that the steady-state distribution function $\overline{F}_{\mathbf{p}}$ at the same time is an eigenfunction of the relaxation operator $\tilde{I}_{\mathbf{p}}$, corresponding to its zero eigenvalue. In general, due to nonlinearity of the Boltzmann equation (2.12), this is not the case: The solution of the nonlinear Boltzmann equation (2.12) is not the eigenvalue of the relaxation operator (2.18). On the other hand, differentiating Eq. (2.12) with respect to the total number of free electrons N, we have

$$I_\mathbf{p} \partial \overline{F}_\mathbf{p}/\partial N = 0, \qquad (2.25)$$

where the function $\partial \overline{F}_\mathbf{p}/\partial N$ is the eigenfunction of the relaxation operator (2.18) corresponding to its zero eigenvalue.

We use the distribution function normalized to N [see Eq. (2.14)]; thus we have

$$\sum_\mathbf{p} \partial \overline{F}_\mathbf{p}/\partial N = 1. \qquad (2.26)$$

The function $\partial \overline{F}_\mathbf{p}/\partial N$ coincides with $\overline{F}_\mathbf{p}/N$ only in the equilibrium state or, as we have seen, if interelectron collisions can be neglected. The function $\partial \overline{F}_\mathbf{p}/\partial N$ plays a special role in the theory of spatially inhomogeneous response and fluctuations (e.g., in the theory of collision-controlled scattering of electromagnetic waves; see references 37 and 63). We shall deal with the function $\partial \overline{F}_\mathbf{p}/\partial N$ in Chapters 6 and 19.

Response Problem

The relaxation operator $I_\mathbf{p}$ determines also a *response* of the electron system in the nonequilibrium steady-state *to a weak additional external perturbation*. With the steady distribution of the electrons being settled by combined action of collisions and the DC field, the distribution function variation $\Delta \overline{F}_\mathbf{p}(t)$ set by a *small additional* AC field applied to the system,

$$\Delta \mathbf{E}(t) = \Delta \mathbf{E}_0^\omega \exp(-i\omega t), \qquad (2.27)$$

is governed by the equation

$$(-i\omega + I_\mathbf{p})\Delta \overline{F}_\mathbf{p} = -e\Delta \mathbf{E} \cdot \frac{\partial \overline{F}_\mathbf{p}}{\partial \mathbf{p}}. \qquad (2.28)$$

Equation (2.28) is none other than the Boltzmann equation (2.1) linearized with respect to a weak additional harmonic perturbation. The operator $(-i\omega + I_\mathbf{p})$ can be called *the response operator*. The formal solution of Eq. (2.28) can be written as

$$\Delta \overline{F}_\mathbf{p} = -e\Delta \mathbf{E} \cdot (-i\omega + I_\mathbf{p})^{-1} \frac{\partial \overline{F}_\mathbf{p}}{\partial \mathbf{p}}, \qquad (2.29)$$

introducing in this way the *inverse response operator*.

Response Current. Small-Signal Conductivity

A variation in the electron distribution, $\Delta \overline{F}_{\mathbf{p}}(t)$, produces the current density variation

$$\Delta \bar{\mathbf{j}} = \frac{e}{V_0} \sum_{\mathbf{p}} \mathbf{v}_{\mathbf{p}} \Delta \overline{F}_{\mathbf{p}}. \tag{2.30}$$

The current density variation caused by a small additional AC field applied to the system may be written as

$$\Delta \bar{j}_\alpha = \sigma_{\alpha\beta}(\omega) \Delta E_\beta, \tag{2.31}$$

where $\sigma_{\alpha\beta}(\omega)$ is the AC *small-signal conductivity*. (Note that here and henceforth we use the so-called Einstein summation convention while dealing with vector and tensor components.) By inserting expression (2.29) into Eq. (2.30) we obtain the formal expression for the AC small-signal conductivity:

$$\sigma_{\alpha\beta}(\omega) = -\frac{e^2}{V_0} \sum_{\mathbf{p}} v_{\mathbf{p}\alpha} (-i\omega + I_{\mathbf{p}})^{-1} \frac{\partial \overline{F}_{\mathbf{p}}}{\partial p_\beta}. \tag{2.32}$$

The *differential* conductivity $\partial \bar{j}_\alpha / \partial E_\beta$ is nothing but the zero-frequency limit of the AC small-signal conductivity (2.32):

$$\sigma_{\alpha\beta} \equiv \sigma_{\alpha\beta}(\omega)|_{\omega=0} = -\frac{e^2}{V_0} \sum_{\mathbf{p}} v_{\mathbf{p}\alpha} I_{\mathbf{p}}^{-1} \frac{\partial \overline{F}_{\mathbf{p}}}{\partial p_\beta}. \tag{2.33}$$

The small-signal conductivity offers an example of a *kinetic coefficient*. This term is used for coefficient of proportionality between a macroscopic flux (of particles, energy, etc.) and an applied external "force" of small enough strength for the flux to remain proportional to it. We shall use expression (2.32) for the AC small-signal conductivity while comparing response and fluctuation phenomena in the following chapter.

CHAPTER 3

FLUCTUATIONS: KINETIC THEORY

Fluctuation processes have been intensively investigated during the last three decades. Owing to the various developments in theory, instrumentation, computers, and other advanced techniques, a profound understanding of fluctuation phenomena in nonequilibrium systems has been achieved. It has become evident that phenomena such as the noise of hot carriers in semiconductors, light scattering from nonequilibrium carriers (or from a gas or fluid with gradients), and nonequilibrium diffusion in these systems, as well as a number of other nonequilibrium phenomena, have major similarities from a fundamental kinetic point of view. A unified description of all these physical processes on the basis of the currently available kinetic theory of nonequilibrium fluctuations could be presented but would be beyond the scope of this book. Our goal in this and following chapters is to overview contemporary theoretical and experimental research of the spectra of electric noise and fluctuations, first of all of hot carriers, in semiconductors and semiconductor structures.

Fluctuation phenomena in a dissipative system displaced into a nonequilibrium steady-state are known to be quite sensitive to details of kinetic processes in the system, often being much more sensitive than average values of physical quantities. Measuring fluctuations often supplies us with a knowledge that is much harder to obtain in other ways. In particular, data on electric noise in semiconductors and semiconductor structures subjected to a high electric field contain valuable information on charge carrier scattering processes and band structure of the crystal not easily obtainable from charge transport characteristics. The area of modern physics which can be called fluctuation spectroscopy, or noise spectroscopy, of solid state, especially of effects nonlinear in applied electric field strength and carrier density, is based on this distinctness of fluctuation phenomena.

Since the beginning of the twentieth century, when A. Einstein (see reference 64) and M. von Smoluchowski [65] developed the theory of Brownian motion, fluctuation science has been one of the most important integral parts of statistical physics and physical kinetics. Fluctuations in a physical system take place due to its discrete nature and the thermal motion in it. Investigation of fluctuations is indispensable if one seeks a perfect understanding of the connection between the microscopic and macroscopic properties of the physical system.

In equilibrium, thermal noise is related to energy through the equipartition theorem, which states that every physical system at temperature T_0 contains an average amount of kinetic energy of $k_B T_0/2$ per degree of freedom, where k_B is the Boltzmann constant. The theorem is true for macroscopic degrees of freedom as well, with the macroscopic thermal motion at thermal equilibrium on average containing a "microscopic" amount of energy $k_B T_0/2$. The transformation from the total average energy (including the fluctuation components at all frequencies) governed by equipartition and considered in thermodynamics, to the noise power in a given frequency band, was achieved by Nyquist [3]. In Nyquist's derivation, some trace may be seen of Rayleigh's [66] application of the equipartition theorem to the standing-wave modes of black-body radiation. In some sense, the available noise power is a special low-frequency case of the black-body radiation (cf. reference 15). At thermal equilibrium the power spectrum of noise in the range of frequencies $\hbar\omega \ll k_B T_0$ (classical noise) has no features. The Nyquist theorem, or the more general fluctuation–dissipation, or Callen and Welton, theorem [67] (see also [68 and 69]) relates noise in some variables to the admittance of the system for the same variables, so that the measurement of fluctuations in equilibrium gives the same information as the measurement of the related transport coefficient.

However, this is true only for systems that are in thermal equilibrium. Fluctuations in the system substantially displaced from the equilibrium do not obey the fluctuation–dissipation theorem. The noise produced in a nonequilibrium state of the system will be different from that in the equilibrium state.

On the other hand, it is the failure of fluctuation–dissipation theorem that makes the fluctuation spectroscopy a valuable tool for the diagnostics of a nonequilibrium system. This tool has a fundamental as well as a practical aspect. The experimental investigation of hot-electron noise — the fluctuation spectroscopy of hot electrons — is a very active field at the moment.

The theory of classical fluctuations about the nonequilibrium state described by the Boltzmann equation is expounded in review articles [70–73] and even textbooks and monographs [43, 74, 75]). The theory was developed by efforts of many workers, an extensive bibliography can be found in the above mentioned references.

3.1 FLUCTUATIONS IN ELECTRON DISTRIBUTION

Fluctuations in Occupation Numbers

The actual values of observables at a given instant, of course, are not equal to their ensemble-average values (averaged over the process realizations), but, due to stochastic character of collisions in the system, somehow or other *fluctuate* around the average values. The classical fluctuations of observables are expressible in terms of *fluctuations in occupation numbers of one-electron states* — just as the ensemble-average values of observables are expressible in terms of the mean occupation numbers, that is, of the one-particle distribution function $\overline{F}_\mathbf{p}$. Because distribution function $\overline{F}_\mathbf{p}$ is the ensemble average or, for a stationary process, the time average of an actual distribution of electrons among the states specified by the quasi-momentum $\mathbf{p}$, the actual distribution at the time t, denoted as $F_\mathbf{p}(t)$, is not identical with its average $\overline{F}_\mathbf{p}$ — that is, there are *fluctuations* around the average:

$$\delta F_\mathbf{p}(t) = F_\mathbf{p}(t) - \overline{F}_\mathbf{p}, \qquad \overline{\delta F_\mathbf{p}(t)} = 0. \tag{3.1}$$

The fluctuations can be characterized by the *ensemble-average of the product* of fluctuations δF at two different moments, t_1 and $t_1 + t$:

$$\overline{\delta F_\mathbf{p}(t_1 + t)\delta F_{\mathbf{p}_1}(t_1)} = \overline{F_\mathbf{p}(t_1 + t) F_{\mathbf{p}_1}(t_1)} - \overline{F}_\mathbf{p}\overline{F}_{\mathbf{p}_1}. \tag{3.2}$$

For a steady-state, the averaged distribution function $\overline{F}_\mathbf{p}$ is independent of time, and the *time-displaced correlation function* (3.2) depends only on the difference of the time arguments $t_1 + t$ and t_1. In that case, an average in Eq. (3.2), instead of being taken over an ensemble, can be taken over time t_1 for the fixed value of the displacement in time, t. We shall deal only with fluctuations around steady-states, and, correspondingly, denote the time-displaced correlation function (3.2) as $\overline{\delta F_\mathbf{p}(t)\delta F_{\mathbf{p}_1}}$.

The concept of fluctuations of occupancies of one-particle states — "fluctuations of distribution function" — was introduced by Hashitsume [76] and Kadomtsev [39].

Fluctuations of Observables

The characteristics of space-independent fluctuations of observables are expressible through the correlation function (3.2). For the tensor of time-displaced correlation functions of the spatially homogeneous fluctuations of current density, we have

$$\overline{\delta j_\alpha(t)\delta j_\beta} = (e/\mathcal{V}_0)^2 \sum_{\mathbf{p}\mathbf{p}_1} v_{\mathbf{p}\alpha} v_{\mathbf{p}_1\beta} \overline{\delta F_\mathbf{p}(t)\delta F_{\mathbf{p}_1}}, \tag{3.3}$$

where $\alpha, \beta, \ldots$ enumerate vector components, with $\delta\mathbf{j}(t)$ being a current density fluctuation averaged over the volume $\mathcal{V}_0$:

$$\delta\mathbf{j}(t) = (e/\mathcal{V}_0)\sum_\mathbf{p} \mathbf{v}_\mathbf{p}\delta F_\mathbf{p}(t) \tag{3.4}$$

[cf. expression (2.13) for the steady value of current density]. Let us remind the reader that $\mathbf{v}_\mathbf{p}$ stands for the velocity of an electron in the state $\mathbf{p}$; that is $\mathbf{v}_\mathbf{p} \equiv \partial\varepsilon_\mathbf{p}/\partial\mathbf{p}$, where $\varepsilon_\mathbf{p}$ is the electron energy.

Equation for Time-Displaced Correlation Function

Now the question is: What equation governs the time-displaced two-particle correlation function $\overline{\delta F_{\mathbf{p}}(t)\delta F_{\mathbf{p}_1}}$, provided that the one-particle distribution function $\overline{F}_{\mathbf{p}}$ satisfies the Boltzmann equation?

The answer is pretty simple: The equation in question is

$$\left(\frac{\partial}{\partial t}+I_{\mathbf{p}}\right)\overline{\delta F_{\mathbf{p}}(t)\delta F_{\mathbf{p}_1}} = 0, \qquad t > 0, \tag{3.5}$$

where $I_{\mathbf{p}}$ is the linearized operator of the Boltzmann equation, given by Eq. (2.18) and referred to in Chapter 2 as the relaxation operator. Equation (3.5) coincides in form with the Eq. (2.17), which governs the time evolution of a small but *macroscopic* deviation, $\Delta\overline{F}_{\mathbf{p}}(t)$, of the distribution function from the steady-state one.

Let us discuss the physical meaning of Eq. (3.5) and the reason for its formal coincidence with Eq. (2.17). The fluctuation $\delta F(t_1)$ — that is, the deviation in the distribution at some moment t_1 — is created by the statistical nature of the system (by the statistical character of collision events). Hence, the deviation $\delta F(t_1 + t)$, in general, will not be equal to zero even if, by accident, $\delta F(t_1) = 0$. For $\delta F(t_1) \neq 0$, the deviation $\delta F(t_1 + t)$ would consist of two parts: (1) the remainder of the deviation $\delta F(t_1)$ left in the course of its evolution and (2) "new" fluctuations that have arisen *after* the moment t_1. Insofar as the "new" fluctuations are independent of the existing ones (in the stable *many-particle* system such an influence is negligible), the average taken over many realizations of some given "initial" (at the moment t_1) deviation δF removes the effect of "new" fluctuations. The product $\delta F(t)\delta F$ averaged in such a manner becomes a smooth function of t and describes the "average" time evolution of the initial deviation δF. There is no difference from the physical point of view between a fluctuational deviation and a deviation caused by some "macroscopic" impetus (small enough for linearization to be possible). Hence, it is not surprising that, after the described elimination of "new" stochastic deviations, some given fluctuational deviation in the electron distribution evolves just in the same manner as a macroscopic one.

The ensemble (or time) averaging denoted by the bar in the definition of the correlation function covers not only the averaging over many realizations of a given "initial" deviation δF but also the averaging over these "initial" deviations. This averaging, obviously, does not touch the law of evolution. Thus, Eq. (3.5) for the time-displaced two-particle correlation function simply follows from the Onsager regression hypothesis [77, 78] applied to fluctuations of the distribution of particles. [For a rigorous derivation of Eq. (3.5) based only upon the criteria of applicability of the Boltzmann equation, see reference 79.]

3.2 EQUAL-TIME CORRELATION

Once Eq. (3.5) for the time-displaced correlation function is established, the problem is reduced to the derivation of an initial condition for this equation — the equal-time correlation function (the "second moment")

$$\overline{\delta F_{\mathbf{p}} \delta F_{\mathbf{p}_1}} \equiv \overline{\delta F_{\mathbf{p}}(t) \delta F_{\mathbf{p}_1}}|_{t=0}. \qquad (3.6)$$

At first we calculate the diagonal part of this function,

$$\overline{\delta F_{\mathbf{p}} \delta F_{\mathbf{p}}} \equiv \overline{\delta F_{\mathbf{p}}^2} = \overline{F_{\mathbf{p}}^2} - \overline{F}_{\mathbf{p}}^2, \qquad (3.7)$$

for a system of Fermi particles. Since the occupation number of a quantum state can take only two values, 0 and 1, the averaged squared occupancy equals the average occupancy: $\overline{F_{\mathbf{p}}^2} = \overline{F}_{\mathbf{p}}$, so that $\overline{\delta F_{\mathbf{p}}^2} = \overline{F}_{\mathbf{p}}(1 - \overline{F}_{\mathbf{p}})$. For a nondegenerate statistics $\overline{F}_{\mathbf{p}} \ll 1$, and

$$\overline{\delta F_{\mathbf{p}}^2} = \overline{F}_{\mathbf{p}}. \qquad (3.8)$$

In the classical approach the same result follows from the Poisson probability distribution (see reference 68).

The problem of the off-diagonal terms of the equal-time correlation function is less trivial. After detaching the diagonal part, we introduce a function (the "factorial cumulant"):

$$\varphi_{\mathbf{p}\mathbf{p}_1} \equiv \overline{\delta F_{\mathbf{p}} \delta F_{\mathbf{p}_1}} - \overline{F}_{\mathbf{p}} \delta_{\mathbf{p}\mathbf{p}_1}. \qquad (3.9)$$

Just as the one-particle distribution function $\overline{F}_{\mathbf{p}}$ describes the mean occupancy of a particle state, the function $\varphi_{\mathbf{p}\mathbf{p}_1}$ describes the mean correlated occupancy of two different states. If the occupation numbers of the different states are completely uncorrelated, the function $\varphi_{\mathbf{p}\mathbf{p}_1}$ vanishes. This is how matters stand for particles moving independently of one another and interchanging with an inexhaustible reservoir (the "grand canonical ensemble"). Each state then is occupied with a Poisson probability, and there is no correlation between the occupancies of different states. This immediately leads to the following expression for the second moment:

$$\overline{\delta F_{\mathbf{p}} \delta F_{\mathbf{p}_1}} = \overline{F}_{\mathbf{p}} \delta_{\mathbf{p}\mathbf{p}_1}, \quad \text{i.e.,} \quad \varphi_{\mathbf{p}\mathbf{p}_1} = 0. \qquad (3.10)$$

On the contrary, fixation of the total number of particles N in the system leads to the probability game of distributing N balls among a set of boxes, with the box $\mathbf{p}$ having the probability $\overline{F}_{\mathbf{p}}/N$, which leads to the standard multinomial distribution whose second moments are

$$\overline{\delta F_{\mathbf{p}} \delta F_{\mathbf{p}_1}} = \overline{F}_{\mathbf{p}} \delta_{\mathbf{p}\mathbf{p}_1} - \overline{F}_{\mathbf{p}} \overline{F}_{\mathbf{p}_1}/N, \quad \text{i.e.,} \quad \varphi_{\mathbf{p}\mathbf{p}_1} = -\overline{F}_{\mathbf{p}} \overline{F}_{\mathbf{p}_1}/N. \qquad (3.11)$$

So, the constraint $\sum_{\mathbf{p}} F_{\mathbf{p}} = N = \text{const}$ leads to the correlation between the occupancies of different states—to the existence of the cross-correlation. This simplest example demonstrates that, in principle, the equal-time cross-correlation can exist, and shows the possibility of a nonvanishing φ. In more complicated cases—for instance, when electron–electron collisions are involved—the function φ cannot be found by using simple probability problem considerations, and one needs a more powerful method of finding this function. We shall return to this problem in Chapter 4.

In situations where electron–electron collisions can be neglected (e.g., for lightly doped semiconductors), Eqs. (3.5) and (3.11) exhaust the theory of spatially homogeneous fluctuations of the distribution function around a nonequilibrium steady-state.

3.3 SPECTRAL INTENSITIES OF FLUCTUATIONS

Fourier Analysis. Wiener–Khintchine Theorem

In Chapter 1 we used the term "spectral intensity (density) of fluctuations" on an intuitive basis, avoiding an explicit definition. To be able to give such a definition, we should perform a Fourier analysis of a random process. Let the stationary random process be described by a random variable $X(t)$, the average of it being zero: $\overline{X(t)} = 0$. According to the *Wiener–Khintchine theorem, the spectral density*, or *the spectral intensity*, of a stationary random variable $X(t)$, understood as the squared sum of its Fourier components lying in an infinitesimal frequency interval, averaged and divided by the frequency interval, is directly expressible through the Fourier transform of the time-displaced correlation function of the variable $X(t)$:

$$S_X(\omega) \equiv 2\langle X^2\rangle_\omega = 2\int_{-\infty}^{+\infty} \overline{X(t)X}\exp(i\omega t)\,dt. \tag{3.12}$$

By inversion, one obtains from Eq. (3.12)

$$\overline{X(t)X} = \frac{1}{2\pi}\int_{-\infty}^{+\infty} \langle X^2\rangle_\omega \exp(-i\omega t)\,d\omega. \tag{3.13}$$

This means that the spectral intensity $S_X(\omega)$ and the time-displaced correlation function $\overline{X(t)X}$ are interconnected by the Fourier transformation.

The spectral intensity is the *directly measurable* characteristic of the random process, provided that the registration system (comprising a quadratic detector) is tuned at the chosen frequency (for details see reference 1, p. 12). Because of the inversion relation (3.13), one can recover the time-displaced correlation function from the spectrum. In calculations, one usually evaluates the correlation function and then finds $S_X(\omega)$ with the help of Eq. (3.12). In measurements one determines $S_X(\omega)$ and evaluates the correlation function with the help of Eq. (3.13).

Spectral Intensity of Distribution Function Fluctuations

Let us define the spectral intensity of distribution function fluctuations around the steady-state as a Fourier transform of the time-displaced correlation function of fluctuations in occupation numbers of one-electron states:

$$(\delta F_\mathbf{p}\delta F_{\mathbf{p}_1})_\omega = \int_{-\infty}^{\infty} \overline{\delta F_\mathbf{p}(t)\delta F_{\mathbf{p}_1}}\exp(i\omega t)\,dt. \tag{3.14}$$

Since the solution of Eq. (3.5) for $t > 0$ has the form

$$\overline{\delta F_{\mathbf{p}}(t)\delta F_{\mathbf{p}_1}} = \exp(-I_{\mathbf{p}}t)\overline{\delta F_{\mathbf{p}}\delta F_{\mathbf{p}_1}}, \tag{3.15}$$

and the identity

$$\overline{\delta F_{\mathbf{p}}(-t)\delta F_{\mathbf{p}_1}} \equiv \overline{\delta F_{\mathbf{p}}(t_1-t)\delta F_{\mathbf{p}_1}(t_1)} \equiv \overline{\delta F_{\mathbf{p}}(t_2)\delta F_{\mathbf{p}_1}(t_2+t)} = \overline{\delta F_{\mathbf{p}_1}(t)\delta F_{\mathbf{p}}} \tag{3.16}$$

is obvious, the Fourier transform (3.14) breaks down into two parts:

$$(\delta F_{\mathbf{p}}\delta F_{\mathbf{p}_1})_\omega = \frac{1}{-i\omega + I_{\mathbf{p}}}\overline{\delta F_{\mathbf{p}}\delta F_{\mathbf{p}_1}} + \frac{1}{i\omega + I_{\mathbf{p}_1}}\overline{\delta F_{\mathbf{p}}\delta F_{\mathbf{p}_1}}. \tag{3.17}$$

The operator identity

$$\frac{1}{-i\omega + I_{\mathbf{p}}} + \frac{1}{i\omega + I_{\mathbf{p}_1}} = \frac{1}{(-i\omega + I_{\mathbf{p}})(i\omega + I_{\mathbf{p}_1})}(I_{\mathbf{p}} + I_{\mathbf{p}_1}) \tag{3.18}$$

enables one to represent Eq. (3.17) also as follows

$$(\delta F_{\mathbf{p}}\delta F_{\mathbf{p}_1})_\omega = \frac{1}{(-i\omega + I_{\mathbf{p}})(i\omega + I_{\mathbf{p}_1})}(I_{\mathbf{p}} + I_{\mathbf{p}_1})\overline{\delta F_{\mathbf{p}}\delta F_{\mathbf{p}_1}}. \tag{3.19}$$

This is the most general expression, provided that the Boltzmann equation is valid, with the operator $I_{\mathbf{p}}$ being the linearized Boltzmann equation operator, introduced in Section 2.2 [see the relaxation operator, Eq. (2.18)].

Spectral Intensity of Current Fluctuations

The spectral intensities of spatially homogeneous fluctuations of current density are given by the expression [cf. Eq. (3.3)]

$$(\delta j_\alpha \delta j_\beta)_\omega = (e/\mathcal{V}_0)^2 \sum_{\mathbf{p}\mathbf{p}_1} v_{\mathbf{p}\alpha} v_{\mathbf{p}_1\beta}(\delta F_{\mathbf{p}}\delta F_{\mathbf{p}_1})_\omega. \tag{3.20}$$

In the most general case, expression (3.19) (or (3.17)) for the spectral intensity of distribution function fluctuations should be inserted into Eq. (3.20).

3.4 INTERELECTRON COLLISIONS NEGLECTED

Spectral Intensities of Distribution Function Fluctuations

In the important case of a lightly doped semiconductor containing low density of the electron gas, the *electron–electron collisions can be neglected*. Then, as will be confirmed in the following chapter, the equal-time correlation function

$\overline{\delta F_{\mathbf{p}} \delta F_{\mathbf{p}_1}}$ is given by Eq. (3.11), so that we have from Eq. (3.17)

$$(\delta F_{\mathbf{p}} \delta F_{\mathbf{p}_1})_\omega = \left[\frac{1}{-i\omega + \tilde{I}_{\mathbf{p}}} + \frac{1}{i\omega + \tilde{I}_{\mathbf{p}_1}} \right] (\overline{F}_{\mathbf{p}} \delta_{\mathbf{p}\mathbf{p}_1} - \overline{F}_{\mathbf{p}} \overline{F}_{\mathbf{p}_1}/N), \qquad (3.21)$$

or, from Eq. (3.19),

$$(\delta F_{\mathbf{p}} \delta F_{\mathbf{p}_1})_\omega = \frac{1}{(-i\omega + \tilde{I}_{\mathbf{p}})(i\omega + \tilde{I}_{\mathbf{p}_1})} (\tilde{I}_{\mathbf{p}} + \tilde{I}_{\mathbf{p}_1})(\overline{F}_{\mathbf{p}} \delta_{\mathbf{p}\mathbf{p}_1} - \overline{F}_{\mathbf{p}} \overline{F}_{\mathbf{p}_1}/N). \qquad (3.22)$$

Here $\tilde{I}_{\mathbf{p}} \equiv e\mathbf{E}\partial/\partial\mathbf{p} + I_{\mathbf{p}}^{\text{th}}$ is the relaxation operator in the absence of electron–electron collisions [see Eq. (2.22)], where $I_{\mathbf{p}}^{\text{th}}$ is the linear electron–lattice collision operator (2.2). The steady-state distribution function $\overline{F}_{\mathbf{p}}$ now is the solution of the *linear* equation (2.24):

$$\tilde{I}_{\mathbf{p}} \overline{F}_{\mathbf{p}} \equiv \left(e\mathbf{E} \frac{\partial}{\partial \mathbf{p}} + I_{\mathbf{p}}^{\text{th}} \right) \overline{F}_{\mathbf{p}} = 0; \qquad (3.23)$$

so that the relaxation operator coincides with the Boltzmann equation operator.

Equation (3.21), or (3.22), in conjunction with Eq. (3.23), exhaust the theory of fluctuations around a nonequilibrium steady-state in the case of one-particle collisions. These equations enable one to calculate spectra of fluctuations of macroscopic observables in a semiconductor with low carrier density.

Spectral Intensity of Current Fluctuations

The spectral intensities of spatially homogeneous current density fluctuations are given by the expression (3.20). In cases where electron–electron collisions can be neglected, by inserting Eq. (3.21) into Eq. (3.20) we obtain the expression

$$(\delta j_\alpha \delta j_\beta)_\omega = (e/\mathcal{V}_0)^2 \left[\sum_{\mathbf{p}} v_{\mathbf{p}\alpha} (-i\omega + \tilde{I}_{\mathbf{p}})^{-1} (v_{\mathbf{p}\beta} - V_\beta) \overline{F}_{\mathbf{p}} + \text{H.C.} \right], \qquad (3.24)$$

where the second term in the brackets, denoted as H.C., is the Hermitian conjugate (i.e., transposed and complex conjugate) to the first one. Let us recall that $\mathbf{V}$, given by Eq. (2.15),

$$\mathbf{V} = \sum_{\mathbf{p}} \mathbf{v}_{\mathbf{p}} \overline{F}_{\mathbf{p}}/N,$$

is the steady-state value of the drift velocity of the electron system as a whole.

3.5 "KINETIC" DERIVATION OF NYQUIST RELATION

In the previous sections we have developed a powerful analytic apparatus enabling one to analyze and calculate noise characteristics of a nonequilibrium system, provided that the average behavior of the system is governed by the Boltzmann equation, with its applicability supposing a weak interaction within the system. Of course, the method works in the case of an equilibrium state as well. Moreover, near-equilibrium kinetic processes, independently of the interaction strength, possess few specific properties. These properties are: (i) validity of *Onsager's principle of symmetry of kinetic coefficients*, (ii) conclusion that spectral intensities of fluctuations are purely real, and (iii) validity of the *fluctuation–dissipation relations* connecting the frequency dependence of the spectral intensity of fluctuations with that of the corresponding kinetic coefficient. These theorems ensure that the conductivity is a symmetric tensor, while the spectral intensity of current fluctuations is purely real and proportional to the real part of the conductivity (the *Nyquist theorem*). As we already mentioned, these properties are inherent in near-equilibrium transport and fluctuations at equilibrium, independently of a strength of interaction of carriers with the thermostat and among themselves. In other words, the properties survive even when the Boltzmann equation is not applicable, with the kinetic coefficients and spectral intensities of fluctuations not being easily calculable. It is instructive to elucidate how our results obtained in the case of weak interaction, for which we have the analytic apparatus to calculate all the necessary quantities, accede to the above-mentioned general (in the sense of interaction) theorems.

Corollaries of Time-Reversal Symmetry of Fluctuations in Equilibrium

Let us note that, generally speaking, the spectral intensities of distribution function fluctuations as defined by Eq. (3.14) are complex quantities. Indeed, in a general case of nonequilibrium, $\overline{\delta F_{\mathbf{p}}(t)\delta F_{\mathbf{p}_1}} \neq \overline{\delta F_{\mathbf{p}}(-t)\delta F_{\mathbf{p}_1}}$, and hence $(\delta F_{\mathbf{p}}\delta F_{\mathbf{p}_1})_\omega \neq (\delta F_{\mathbf{p}}\delta F_{\mathbf{p}_1})_{-\omega} = (\delta F_{\mathbf{p}}\delta F_{\mathbf{p}_1})^*_\omega$. However, at *thermal equilibrium*, due to symmetry of the equations of mechanics with respect to the time reversal (see, e.g., reference 68), one has (in the absence of an external magnetic field)

$$\overline{\delta F_{\mathbf{p}}(t)\delta F_{\mathbf{p}_1}}^{\text{eq}} = \overline{\delta F_{\mathbf{p}}(-t)\delta F_{\mathbf{p}_1}}^{\text{eq}} \qquad (3.25)$$

and then, due to Eq. (3.16),

$$\overline{\delta F_{\mathbf{p}}(t)\delta F_{\mathbf{p}_1}}^{\text{eq}} = \overline{\delta F_{\mathbf{p}_1}(t)\delta F_{\mathbf{p}}}^{\text{eq}}. \qquad (3.26)$$

Definition (3.14) of the Fourier transforms, with Eqs. (3.25) and (3.26) in mind, leads to

$$(\delta F_{\mathbf{p}}\delta F_{\mathbf{p}_1})^{\text{eq}}_\omega = (\delta F_{\mathbf{p}}\delta F_{\mathbf{p}_1})^{\text{eq}}_{-\omega} = (\delta F_{\mathbf{p}_1}\delta F_{\mathbf{p}})^{\text{eq}}_\omega. \qquad (3.27)$$

The time reversal appears in the Fourier components as a change of ω into $-\omega$ (of course, if the quantity itself remains unchanged by the time reversal). Since the spectral intensities (3.14) are complex quantities only due to presence of the imaginary quantity $i\omega$ in their definition, it follows from Eq. (3.27) that *in* the equilibrium the spectral intensities of occupancy fluctuations are purely real quantities:

$$(\delta F_\mathbf{p} \delta F_{\mathbf{p}_1})^{eq}_\omega = [(\delta F_\mathbf{p} \delta F_{\mathbf{p}_1})^{eq}_\omega]^*. \tag{3.28}$$

Therefore, the time symmetry, at equilibrium, of occupancy fluctuations means that corresponding spectral intensities are purely real and symmetric with respect to suffices $\mathbf{p}$ and $\mathbf{p}_1$. Let us show that, conjointly with these immanent for thermal equilibrium properties, the kinetic theory of fluctuations, as formulated in the previous sections, at equilibrium leads to the results compatible with the theorems mentioned at the beginning of this section and valid independently of the strength of interaction — that is of the applicability of the kinetic theory. In other words, we shall demonstrate that Onsager's principle of symmetry of kinetic coefficients and the Nyquist theorem connecting the spectral intensity of current fluctuations with the AC conductivity emerge quite naturally in the context of the kinetic approach applied to a near-equilibrium situation.

We shall make use of the structure of the expression for the spectral intensity of the occupancy fluctuations obtained in Section 3.3. It is obvious that Eq. (3.17) can be represented as

$$(\delta F_\mathbf{p} \delta F_{\mathbf{p}_1})^{eq}_\omega = a_{\mathbf{p}\mathbf{p}_1}(\omega) + a^*_{\mathbf{p}_1\mathbf{p}}(\omega) = a_{\mathbf{p}\mathbf{p}_1}(\omega) + a_{\mathbf{p}_1\mathbf{p}}(-\omega) \tag{3.29}$$

where [cf. with Eqs. (3.14) and (3.15)]

$$a_{\mathbf{p}\mathbf{p}_1}(\omega) \equiv \int_0^\infty a_{\mathbf{p}\mathbf{p}_1}(t) e^{i\omega t}\, dt, \tag{3.30}$$

with $a(t)$ being a purely real function. It turns out to be possible (see reference 68) to get some very general relations for the function $a(\omega)$ given by Eq. (3.30) by using the mathematical techniques of the theory of functions of a complex variable. For this purpose, one should regard ω as a complex variable ($\omega = \omega' + i\omega''$) and investigate the properties of the function $a(\omega)$ in the upper half-plane of this variable. It could be shown (see reference 68) that the following relation holds:

$$a(\omega) = \frac{1}{i\pi} \fint_{-\infty}^{+\infty} \frac{a(\omega')}{\omega' - \omega}\, d\omega', \tag{3.31}$$

where by a stroke through the integral sign the principal part of the integral taken over the real axis is indicated. Separating the real and imaginary parts in Eq. (3.31), we obtain the following two relations:

$$\mathrm{Re}\{a(\omega)\} = \frac{1}{\pi} \fint_{-\infty}^{+\infty} \frac{\mathrm{Im}\{a(\omega')\}}{\omega' - \omega}\, d\omega', \tag{3.32}$$

$$\mathrm{Im}\{a(\omega)\} = -\frac{1}{\pi} \fint_{-\infty}^{+\infty} \frac{\mathrm{Re}\{a(\omega')\}}{\omega' - \omega}\, d\omega'. \tag{3.33}$$

These relations between the real and imaginary parts of the function defined by Eq. (3.30) are referred to as the *Kramers–Kronig relations*.

We shall use the Kramers–Kronig relations to draw definite conclusions about the symmetry properties of the quantities $a_{\mathbf{p}\mathbf{p}_1}(\omega)$ entering Eq. (3.29). Because the spectral intensities of distribution function fluctuations are purely real [see Eq. (3.28)] and symmetric in the suffices $\mathbf{p}$ and $\mathbf{p}_1$ [i.e., in one-particle states discriminated by the values of quasi-momentum, see Eq. (3.27)], it follows from Eq. (3.29) that;

$$a_{\mathbf{p}\mathbf{p}_1}(\omega) - a^*_{\mathbf{p}\mathbf{p}_1}(\omega) = a_{\mathbf{p}_1\mathbf{p}}(\omega) - a^*_{\mathbf{p}_1\mathbf{p}}(\omega)$$

or

$$\mathrm{Im}\{a_{\mathbf{p}\mathbf{p}_1}(\omega)\} = \mathrm{Im}\{a_{\mathbf{p}_1\mathbf{p}}(\omega)\}, \tag{3.34}$$

and we conclude that the imaginary part of $a_{\mathbf{p}\mathbf{p}_1}(\omega)$ is symmetric. But the real and imaginary parts of each of the quantities $a_{\mathbf{p}\mathbf{p}_1}(\omega)$ are related to each other by linear integral equations, namely, the relations of Kramers and Kronig, Eqs. (3.32) and (3.33). Therefore the symmetry of $\mathrm{Im}\{a_{\mathbf{p}\mathbf{p}_1}\}$ requires also the symmetry of $\mathrm{Re}\{a_{\mathbf{p}\mathbf{p}_1}\}$ and thus of the whole $a_{\mathbf{p}\mathbf{p}_1}$. We therefore find the final answer

$$a_{\mathbf{p}\mathbf{p}_1}(\omega) = a_{\mathbf{p}_1\mathbf{p}}(\omega) \tag{3.35}$$

or, if we recall Eqs. (3.17) and (3.29),

$$(-i\omega + I^{\mathrm{eq}}_{\mathbf{p}})^{-1}\overline{\delta F_{\mathbf{p}}\delta F_{\mathbf{p}_1}}^{\mathrm{eq}} = (i\omega + I^{\mathrm{eq}}_{\mathbf{p}_1})^{-1}\overline{\delta F_{\mathbf{p}}\delta F_{\mathbf{p}_1}}^{\mathrm{eq}}, \tag{3.36}$$

where we introduced a special notation for the relaxation operator near the equilibrium state:

$$I^{\mathrm{eq}}_{\mathbf{p}} \equiv I^{\mathrm{th}}_{\mathbf{p}} + I^{\mathrm{ee}}_{\mathbf{p}}(\overline{F}^{\mathrm{eq}}) \tag{3.37}$$

[cf. Eq. (2.21)].

Fluctuation–Dissipation Relations. Nyquist Theorem

Later we shall see that in equilibrium the equal-time correlation function of occupancy fluctuations is given by Eq. (3.11),

$$\overline{\delta F_{\mathbf{p}}\delta F_{\mathbf{p}_1}}^{\mathrm{eq}} = \overline{F}^{\mathrm{eq}}_{\mathbf{p}}\delta_{\mathbf{p}\mathbf{p}_1} - \overline{F}^{\mathrm{eq}}_{\mathbf{p}}\overline{F}^{\mathrm{eq}}_{\mathbf{p}_1}/N, \tag{3.38}$$

even when interelectron collisions play a role. By inserting Eq. (3.38) into Eq. (3.17), then Eq. (3.17) into Eq. (3.20), and taking into account Eq. (3.36) we obtain the expression

$$(\delta j_\alpha \delta j_\beta)^{\mathrm{eq}}_\omega = (\delta j_\beta \delta j_\alpha)^{\mathrm{eq}}_\omega = (e/V_0)^2 2\,\mathrm{Re}\left\{\sum_{\mathbf{p}} v_{\mathbf{p}\alpha}(-i\omega + I^{\mathrm{eq}}_{\mathbf{p}})^{-1} v_{\mathbf{p}\beta}\overline{F}^{\mathrm{eq}}_{\mathbf{p}}\right\}. \tag{3.39}$$

In the equilibrium the spectral intensities of current fluctuations are the components of a *purely real symmetric tensor*. Let us compare the obtained expression for spectral intensity of current density fluctuations at equilibrium, Eq. (3.39), with the earlier derived expression for the AC conductivity, Eq. (2.32). The AC conductivity describes the effect of an external perturbing AC field, linear in the strength of the field, on the macroscopic system. In the absence of other perturbations (i.e., for the Ohmic conductivity) Eq. (2.32) takes the form

$$\sigma_{\alpha\beta}^{eq}(\omega) = (e^2/V_0 T_0) \sum_{\mathbf{p}} v_{\mathbf{p}\alpha}(-i\omega + I_{\mathbf{p}}^{eq})^{-1} v_{\mathbf{p}\beta} \overline{F}_{\mathbf{p}}^{eq}; \qquad (3.40)$$

here we have used the fact that, thanks to the special form of the equilibrium distribution, $\overline{F}_{\mathbf{p}}^{eq} \propto \exp(-\varepsilon_{\mathbf{p}}/T_0)$, we obtain

$$\frac{\partial \overline{F}_{\mathbf{p}}^{eq}}{\partial p_\alpha} = -\frac{v_{\mathbf{p}\alpha}}{T_0} \overline{F}_{\mathbf{p}}^{eq}. \qquad (3.41)$$

By comparing Eqs. (3.39) and (3.40) we find that

$$(\delta j_\alpha \delta j_\beta)_\omega^{eq} = 2(T_0/V_0) \operatorname{Re}\{\sigma_{\alpha\beta}^{eq}(\omega)\}. \qquad (3.42)$$

This important relation, connecting the spectral intensities of current density fluctuations at equilibrium with conductivities, is the classical limit of the *Nyquist theorem* [3], in turn being a special case of the more general *Callen–Welton relations* [67], called also the *fluctuation–dissipation theorem*. Callen–Welton relations connect the fluctuation characteristics of physical quantities at equilibrium with the dissipative properties of the system when external disturbances act on it (see also reference 68). We noticed at the beginning of this section that the relation (3.42) is valid independently of the strength of interaction within the system. The relation (3.42) is much more general than the expressions (3.39) and (3.40) which led us to it. The aim of our derivation was to show that our method aimed to calculation of fluctuation characteristics in far-from-equilibrium states gives results compatible with what is true at equilibrium as well.

CHAPTER 4

EFFECT OF INTERELECTRON COLLISIONS ON FLUCTUATION PHENOMENA

Two-particle (electron–electron) collisions, if essential, strongly influence the fluctuation processes near the nonequilibrium state (Leontovich [80], Green [81], Ludwig [82], Gantsevich et al. [37, 79, 83], Kogan and Shulman [42, 84]; see also reference 70). Two-particle collisions can create a correlation between the occupancies of different one-particle states so that the equal-time mean correlated occupancy of two different states $\varphi_{\mathbf{pp}_1}$ [Eq. (3.9)] ceases to be expressible in a simple way in terms of the particle distribution function as was the case when two-particle collisions were neglected [Eq. (3.11)]. In general, just as the particle distribution function $\overline{F}_{\mathbf{p}}$ — the mean occupancy of the state $\mathbf{p}$ — obeys the Boltzmann equation, the function $\varphi_{\mathbf{pp}_1}$ describing the mean correlated part of the occupancies of the states $\mathbf{p}$ and $\mathbf{p}_1$ satisfies the equation we are now going to formulate.

4.1 ADDITIONAL CORRELATION DUE TO INTERELECTRON COLLISIONS

Evolution of Equal-Time Correlation

Let us imagine that at some moment a correlation between the occupancies of states is created that differs from the stationary one. The question is how the correlation will develop in time — that is, how some given *initial* function $\varphi_{\mathbf{pp}_1}$ evolves under the action of the field and scattering.

We know that the average change of occupancy of the state due to the field and scattering is described by the action of the linearized operator of the Boltzmann equation, $I_{\mathbf{p}}$ [see Eq. (2.18)]. In the case of the correlated occupancy of two states, $\mathbf{p}$ and $\mathbf{p}_1$, the change in occupancy of any of them yields a variation of the function $\varphi_{\mathbf{p}\mathbf{p}_1}$. As far as the occupancies of two states change independently, the effect should be additive. So we can conclude that the rate of change, due to the average action of field and scattering, of the nonstationary equal-time correlation function $\varphi_{\mathbf{p}\mathbf{p}_1}(t)$ is given by the action of the sum of the relaxation operators,

$$(I_{\mathbf{p}} + I_{\mathbf{p}_1})\varphi_{\mathbf{p}\mathbf{p}_1}.$$

The linearized electron–electron collision operator, $I_{\mathbf{p}}^{ee}(\overline{F})$, stands alongside the field term $e\mathbf{E}\partial/\partial\mathbf{p}$ and the operator $I_{\mathbf{p}}^{th}$, representing scattering on the thermal bath, in the expression (2.18) for the relaxation operator $I_{\mathbf{p}}$:

$$I_{\mathbf{p}} \equiv e\mathbf{E}\frac{\partial}{\partial\mathbf{p}} + I_{\mathbf{p}}^{th} + I_{\mathbf{p}}^{ee}(\overline{F}).$$

In this way electron–electron collisions appear in Eq. (3.5) for the time-displaced correlation function and in the same way will enter, through the relaxation term $(I_{\mathbf{p}} + I_{\mathbf{p}_1})\varphi_{\mathbf{p}\mathbf{p}_1}$, the equation for $\varphi_{\mathbf{p}\mathbf{p}_1}$. The electron–electron collisions take part in the relaxation of the already existing correlation of the occupancies — an obvious and rather trivial fact. However, we claim that participation in relaxation processes is not the only role played by electron–electron collisions in the fluctuation phenomena.

Source of Correlation

Indeed, two-particle collisions *create* a correlation between the occupancies of different states — a circumstance that is of prime importance for theory of fluctuations in a nonequilibrium state. A specific feature of two-particle collisions (as compared to "one-particle" scattering by the thermal bath) is that *two* states become occupied (and two others become unoccupied) *simultaneously*, as a result of the same collision. Each such collision results in the transfer of a pair of electrons from the vicinities of the points $\mathbf{p}$ and $\mathbf{p}_1$ to those of the points $\mathbf{p}'$ and $\mathbf{p}'_1$ in the quasi-momentum space. This is the way *a correlation between the simultaneous occupancies of different states* is generated by two-particle collisions. Consequently, a term describing such a generation should appear in the equation for the equal-time correlated-occupancy function φ.

Such a "generation-of-correlation" term, describing the mean rate of *simultaneous* change of occupation numbers of the states $\mathbf{p}$ and $\mathbf{p}_1$, can be easily constructed. The mean number of collisions (per unit time interval) resulting in the simultaneous appearance of a pair of electrons in the states $\mathbf{p}$, $\mathbf{p}_1$ is given by the expression

$$\sum_{\mathbf{p}'\mathbf{p}'_1} W_{\mathbf{p}\mathbf{p}_1}^{\mathbf{p}'\mathbf{p}'_1} \overline{F}_{\mathbf{p}'} \overline{F}_{\mathbf{p}'_1}, \tag{4.1}$$

where W is the electron–electron collision probability. On the other hand, any collision between the electrons occupying the states $\mathbf{p}$ and $\mathbf{p}_1$ leads to a simultaneous diminution of the corresponding occupation numbers. The mean number (per unit time interval) of such events,

$$\sum_{\mathbf{p'p'_1}} W^{\mathbf{pp_1}}_{\mathbf{p'p'_1}} \overline{F}_\mathbf{p} \overline{F}_{\mathbf{p_1}}, \tag{4.2}$$

should be subtracted from expression (4.1) to obtain the mean total simultaneous flow into the states $\mathbf{p}$ and $\mathbf{p_1}$. That is, the "correlating" term we are seeking is

$$-I^{ee}_{\mathbf{pp_1}}\{\overline{F},\overline{F}\} \equiv \sum_{\mathbf{p'p'_1}} W^{\mathbf{p'p'_1}}_{\mathbf{pp_1}} \overline{F}_{\mathbf{p'}} \overline{F}_{\mathbf{p'_1}} - \sum_{\mathbf{p'p'_1}} W^{\mathbf{pp_1}}_{\mathbf{p'p'_1}} \overline{F}_\mathbf{p} \overline{F}_{\mathbf{p_1}}$$

$$\equiv \sum_{\mathbf{p'p'_1}} W^{\mathbf{p'p'_1}}_{\mathbf{pp_1}} (\overline{F}_{\mathbf{p'}} \overline{F}_{\mathbf{p'_1}} - \overline{F}_\mathbf{p} \overline{F}_{\mathbf{p_1}}), \tag{4.3}$$

where we have made use of Stueckelberg's property (2.10). The reason for using the symbol $I^{ee}_{\mathbf{pp_1}}$ for the correlating term (4.3) is obvious. By summing Eq. (4.3) over $\mathbf{p_1}$, one obtains the mean flow into the state $\mathbf{p}$ caused by electron–electron collisions, so we have

$$\sum_{\mathbf{p_1}} I^{ee}_{\mathbf{pp_1}}\{\overline{F},\overline{F}\} = I^{ee}_{\mathbf{p}}\{\overline{F},\overline{F}\}, \tag{4.4}$$

where $I^{ee}_{\mathbf{p}}$ is the usual two-particle collision integral (2.3) entering the Boltzmann equation [see Eqs. (2.1) and (2.12)].

Describing the mean rate of the simultaneous change in the occupancies of different states, expression (4.3) presents the rate of change of the mean correlated occupancy function $\varphi_{\mathbf{pp_1}}$ due to generation of correlation by two-particle collisions. So, it should play the role of a "source" term in the equation that the function φ satisfies.

Equation for Equal-Time Correlation Function

Thus we come to the conclusion that, the electron–electron collisions being taken into account, the function φ is governed by the following equation:

$$\frac{\partial}{\partial t}\varphi_{\mathbf{pp_1}} + (I_\mathbf{p} + I_{\mathbf{p_1}})\varphi_{\mathbf{pp_1}} = -I^{ee}_{\mathbf{pp_1}}\{\overline{F},\overline{F}\}. \tag{4.5}$$

We arrive at the following picture. A correlation between the occupancies of states $\mathbf{p}$ and $\mathbf{p_1}$ comes into being as a result of two-particle collisions. Since the occupancy of state $\mathbf{p}$ as well as that of $\mathbf{p_1}$ changes in time under the action of the external field and scattering, the correlation of occupancies of the states $\mathbf{p}$ and $\mathbf{p_1}$ will turn eventually into the correlation of occupancies of some other states. Thus, once created, the correlated occupancy spreads out in the quasi-momentum space. These processes — a creation of correlation and its transference in the quasi-momentum space — are taken into account in Eq. (4.5) by the "source" term $I^{ee}_{\mathbf{pp_1}}$

and the "relaxation" term $(I_\mathbf{p} + I_{\mathbf{p}_1})\varphi_{\mathbf{pp}_1}$, correspondingly. Equation (4.5) clearly demonstrates the two ways in which the electron–electron collisions enter the kinetics of fluctuations:

(i) The linearized electron–electron collision operator is included in the operators $I_\mathbf{p}$ and $I_{\mathbf{p}_1}$, and
(ii) The very existence of the "extra source" term $I^{ee}_{\mathbf{pp}_1}$ creating the equal-time correlation is based on the existence of the electron–electron collisions.

The *steady* value of the function φ should satisfy the equation

$$(I_\mathbf{p} + I_{\mathbf{p}_1})\varphi_{\mathbf{pp}_1} = -I^{ee}_{\mathbf{pp}_1}\{\overline{F}, \overline{F}\}. \tag{4.6}$$

This equation is not easy to solve. By creating a true physical correlation among the occupancies of one-particle states, two-particle collisions exclude the possibility of a simple form of the function φ. At present the methods for solving Eq. (4.6) are developed only in special cases. The solutions for some cases of interest will be discussed in detail in Chapter 12.

Kinetic Correlation

The extra, or additional, correlation created by two-particle collisions can also be called a "kinetic" correlation. The quite important property of the extra source term $I^{ee}_{\mathbf{pp}_1}$ is that it *vanishes in the thermal equilibrium state*:

$$-I^{ee}_{\mathbf{pp}_1}\{\overline{F}^{eq}, \overline{F}^{eq}\} \equiv \sum_{\mathbf{p}'\mathbf{p}'_1} W^{\mathbf{p}'\mathbf{p}'_1}_{\mathbf{pp}_1}(\overline{F}^{eq}_{\mathbf{p}'}\overline{F}^{eq}_{\mathbf{p}'_1} - \overline{F}^{eq}_\mathbf{p}\overline{F}^{eq}_{\mathbf{p}_1}) = 0. \tag{4.7}$$

Indeed, for the equilibrium distribution $\overline{F}^{eq}_\mathbf{p} \propto \exp(-\varepsilon_\mathbf{p}/T_0)$ the "in" and "out" terms cancel each other in Eq. (4.7) due to energy conservation in the electron–electron collision. Thus, disengaging ourselves from the correlations created by the constraints of the type of $N = $ const, we see that within the limits of applicability of the Boltzmann equation the true physical correlations do not exist in the thermal equilibrium state (appearing only in higher-order approximations in the small interaction constant). On the contrary, a kinetic correlation appearing in the nonequilibrium state does not directly depend on the order of magnitude of the interaction constant (the linearized operators $I^{ee}_\mathbf{p}(\overline{F})$ and $I^{ee}_{\mathbf{p}_1}(\overline{F})$ on the left-hand side of Eq. (4.6) as well as the "source" term $I^{ee}_{\mathbf{pp}_1}$ on the right-hand side of Eq. (4.6) are proportional to this constant). In some sense, a gas in which binary collisions are essential is *less ideal* being in a nonequilibrium state than in the thermal equilibrium state in which the kinetic correlations vanish.

In other words, though each collision between particles creates an interparticle correlation, a statistically averaged correlation between the occupancies of the states vanishes in equilibrium due to a subtle balance of the mean fluxes into the pairs of states and out of them. Displacement of the system from equilibrium (i.e., setting up of a nonequilibrium distribution) disbalances the fluxes. For the nonequilibrium distribution, the total correlating flux (4.3) does not disappear, and the statistically averaged correlation φ_{pp_1} is created. Kinematics of electron motion as well as interaction of electrons enter the problem only through the probabilities W — that is, remain the same in nonequilibrium just like in equilibrium. So, the correlation described above is of purely *kinetic* nature, in a certain sense being insensitive to the strength of the (small) interaction constant. The additional equal-time correlation in a nonequilibrium gas arises even though the individual collisions between particles are uncorrelated (cf. Chapter 5).

4.2 EXPRESSION FOR SPECTRAL INTENSITY OF CURRENT FLUCTUATIONS

In Chapter 3 we obtained a general expression (3.19) for the spectral intensity of distribution function fluctuations in the steady state, which for rare electron–electron collisions leads to expression (3.24) for the spectral intensity of the current fluctuations. When electron–electron collisions are essential, we should insert into expression (3.19) the equal-time correlation function

$$\overline{\delta F_\mathbf{p} \delta F_{\mathbf{p}_1}} = \overline{F}_\mathbf{p} \delta_{\mathbf{pp}_1} + \varphi_{\mathbf{pp}_1} \tag{4.8}$$

[see Eq. (3.9)] and use Eq. (4.6) for φ. We have

$$(I_\mathbf{p} + I_{\mathbf{p}_1})\overline{\delta F_\mathbf{p} \delta F_{\mathbf{p}_1}} = (I_\mathbf{p} + I_{\mathbf{p}_1})(\overline{F}_\mathbf{p} \delta_{\mathbf{pp}_1} + \varphi_{\mathbf{pp}_1})$$
$$= (I_\mathbf{p} + I_{\mathbf{p}_1})\overline{F}_\mathbf{p} \delta_{\mathbf{pp}_1} - I^{ee}_{\mathbf{pp}_1}\{\overline{F}, \overline{F}\}, \tag{4.9}$$

so that the final expression for the spectral intensity of distribution function fluctuations is [83]

$$(\delta F_\mathbf{p} \delta F_{\mathbf{p}_1})_\omega = \frac{1}{(-i\omega + I_\mathbf{p})(i\omega + I_{\mathbf{p}_1})}[(I_\mathbf{p} + I_{\mathbf{p}_1})\overline{F}_\mathbf{p} \delta_{\mathbf{pp}_1} - I^{ee}_{\mathbf{pp}_1}\{\overline{F}, \overline{F}\}]. \tag{4.10}$$

We recall that $\overline{F}_\mathbf{p}$ is the steady distribution function satisfying the Boltzmann equation (2.12), $I_\mathbf{p}$ is the relaxation operator (2.18), and $I^{ee}_{\mathbf{pp}_1}$ is the extra source term ("correlating term") given by Eq. (4.3). *These results, in principle, complete the theory of classical fluctuations in the nonequilibrium weakly interacting many-particle system* — in the same sense in which the Boltzmann equation (with one- and two-particle collision terms and streaming term) exhausts the theory of transport in such a system.

As we have seen [Eq. (4.7)], in the thermal equilibrium state the extra source vanishes. Hence, for fluctuations around the thermal equilibrium state, one has

$$(\delta F_\mathbf{p} \delta F_{\mathbf{p}_1})^{eq}_\omega = \frac{1}{(-i\omega + I^{eq}_\mathbf{p})(i\omega + I^{eq}_{\mathbf{p}_1})}(I^{eq}_\mathbf{p} + I^{eq}_{\mathbf{p}_1})\overline{F}^{eq}_\mathbf{p} \delta_{\mathbf{pp}_1}. \tag{4.11}$$

Here I^{eq} is the linearized collision operator for the thermal equilibrium state given by Eq. (3.37):

$$I_{\mathbf{p}}^{\text{eq}} \equiv I_{\mathbf{p}}^{\text{th}} + I_{\mathbf{p}}^{\text{ee}}(\overline{F}^{\text{eq}}).$$

Expression (4.10) leads to the following expression for the spectral intensity of current density fluctuations as defined by Eq. (3.20):

$$(\delta j_\alpha \delta j_\beta)_\omega = (e/\mathcal{V}_0)^2 \sum_{\mathbf{pp}_1} v_{\mathbf{p}\alpha} v_{\mathbf{p}_1\beta} \frac{1}{(-i\omega + I_{\mathbf{p}})(i\omega + I_{\mathbf{p}_1})}$$
$$\times [(I_{\mathbf{p}} + I_{\mathbf{p}_1})\overline{F}_{\mathbf{p}} \delta_{\mathbf{pp}_1} - I_{\mathbf{pp}_1}^{\text{ee}} \{\overline{F}, \overline{F}\}]. \quad (4.12)$$

This is the most general expression that is valid, provided that the Boltzmann equation (2.12) holds.

4.3 CRITERIA FOR APPLICABILITY OF KINETIC THEORY OF FLUCTUATIONS

Let us now discuss the limits of applicability of the Boltzmann equation for description of nonequilibrium systems (see, e.g., reference 43). First of all, the "collisional kinetics" works only if the duration of a collision, Δt^{dur}, is much shorter than the time between successive collisions, τ:

$$\Delta t^{\text{dur}}/\tau \ll 1. \quad (4.13)$$

In quantum mechanics, the "duration of a collision" is estimated as $\hbar/\varepsilon_{\mathbf{p}}$, — that is, as the time of formation of the quantum state characterized by momentum $\mathbf{p}$ and kinetic energy $\varepsilon_{\mathbf{p}}$. In the classical mechanics, Δt^{dur} is estimated as $r^{\text{sc}}/v_{\mathbf{p}}$, where $v_{\mathbf{p}}$ is the particle velocity and r^{sc} characterizes dimensions of the scatterer.

We suppose also that an applied electric field does not change remarkably the electron quasi-momentum *during a collision*:

$$eE\Delta t^{\text{dur}}/p \ll 1. \quad (4.14)$$

Naturally enough, the Boltzmann equation (2.1) is able to predict evolution of the system only on a time scale that ignores the duration of the collision:

$$\Delta t^{\text{dur}}/\Delta t \ll 1, \quad (4.15)$$

where Δt is the time interval of interest. The characteristic time of change of the distribution function due to collisions, or *mean free time*, τ, is given by the reciprocal to the order of magnitude of the collision operator acting on an *arbitrary* distribution function.

One can memorize criteria (4.13)–(4.15) as the requirement for the order of magnitude of the operators in each term in the Boltzmann equation—namely, $\partial/\partial t$, $e\mathbf{E}\partial/\partial \mathbf{p}$, and collision operators—to be small as compared with $1/\Delta t^{\mathrm{dur}}$.

The limits of applicability of the presented-above kinetic theory of fluctuations are outlined by the same criteria (4.13)–(4.15). In particular, the spectral intensities of fluctuations are given by Eqs. (4.10) and (4.12) only for not too high frequencies

$$\omega \Delta t^{\mathrm{dur}} \ll 1. \tag{4.16}$$

There are many methods of derivation of the Boltzmann equation, leading off with intuitive arguments going back to Ludwig Boltzmann and up to modern methods starting with the classical or quantum statistics (see reference 43). Similarly, there are several ways to arrive at the concept of distribution function fluctuations and to obtain the equations governing them (see references 37, 38, 42, 71, 72, 79, 83, and 85–100). We have discussed the matter relying upon rather intuitive arguments, in hope that such an approach will make the presentation more accessible.

In fact, using the small parameters $\hbar/\varepsilon_{\mathbf{p}}\tau$, $\hbar e E/\varepsilon_{\mathbf{p}} p$, $\hbar\omega/\varepsilon_{\mathbf{p}}$, and $1/N$, one can derive the Boltzmann equation and the equations of fluctuational kinetics from the first principles of quantum kinetics (see references 37, 79, 87, 91, and 98). No assumptions other than those necessary for the adequate description of the state itself are needed for the computation of classical fluctuations around a nonequilibrium state. In particular, the classical fluctuations from the nonequilibrium steady state described by the Boltzmann equation are computable without any assumptions other than those needed for the validity of that equation itself. The adequate description of fluctuations emerges parallel with the Boltzmann equation for the mean distribution function, as *two* first terms in the expansion in a power series in the reciprocal to the total number of particles. Moreover, no information about a system other than that needed to write explicitly the Boltzmann equation itself enters the expressions for the classical spectral intensities of occupation number fluctuations. That is, regression of fluctuations is governed by the time-dependent linearized Boltzmann equation, while a source of fluctuations is expressible in terms of transition probabilities and the stationary one-particle distribution function (the solution of the time-independent Boltzmann equation).

CHAPTER 5

BOLTZMANN–LANGEVIN EQUATION

5.1 LANGEVIN APPROACH

Boltzmann–Langevin Equation

We have obtained expression (4.10) for the spectral intensity of one-particle-state occupancy fluctuations using Eqs. (3.5) and (4.6) that correlation functions satisfy. However, expression (4.10) is interpretable also in a different way. The structure of Eq. (4.10) suggests that fluctuations around a nonequilibrium state can be treated using the Langevin [101] method as well. Indeed, one can obtain the generalized Langevin equation for a distribution function fluctuation $\delta F_\mathbf{p}(t)$ by adding the *random force* (rather, the *random flux in the quasi-momentum space*) to the linearized Boltzmann equation:

$$\left(\frac{\partial}{\partial t} + I_\mathbf{p}\right)\delta F_\mathbf{p}(t) = \delta J_\mathbf{p}(t). \tag{5.1}$$

This equation is referred to as the *Boltzmann–Langevin equation* or the *linearized Boltzmann equation with fluctuations* [102, 103], with the random force $\delta J_\mathbf{p}(t)$ playing the role of the "source of fluctuations." Such a source is called a Langevin source.

A comparison with expression (4.10)—the expression for the spectral intensity of occupancy fluctuations derived in Chapter 4—immediately yields the following properties of the random-force term in the Boltzmann–Langevin equation: The random fluxes $\delta J_\mathbf{p}(t)$ should be assumed to have the zero mean value,

$\overline{\delta J_{\mathbf{p}}(t)} = 0$, and to satisfy the correlation formula:

$$\overline{\delta J_{\mathbf{p}}(t)\delta J_{\mathbf{p}_1}(t_1)} = \delta(t-t_1)[(I_{\mathbf{p}}+I_{\mathbf{p}_1})\overline{F}_{\mathbf{p}}\delta_{\mathbf{pp}_1} - I^{ee}_{\mathbf{pp}_1}\{\overline{F},\overline{F}\}], \quad (5.2)$$

thus having no memory of themselves. Then the Boltzmann–Langevin equation (5.1) leads to the expression

$$(\delta F_{\mathbf{p}}\delta F_{\mathbf{p}_1})_\omega = \frac{1}{(-i\omega + I_{\mathbf{p}})(i\omega + I_{\mathbf{p}_1})}(\delta J_{\mathbf{p}}\delta J_{\mathbf{p}_1})_\omega \quad (5.3)$$

with the "white" spectral intensity of random fluxes,

$$(\delta J_{\mathbf{p}}\delta J_{\mathbf{p}_1})_\omega = (I_{\mathbf{p}}+I_{\mathbf{p}_1})\overline{F}_{\mathbf{p}}\delta_{\mathbf{pp}_1} - I^{ee}_{\mathbf{pp}_1}\{\overline{F},\overline{F}\}. \quad (5.4)$$

So, Eqs. (5.1) and (5.2) prove to be equivalent to expression (4.10).

Expressions for Spectral Intensities of Langevin Sources

The external field **E** explicitly enters expression (5.4) for spectral intensities of Langevin sources through the field terms contained in the operators $I_{\mathbf{p}}$ and $I_{\mathbf{p}_1}$. However, making use of the stationary Boltzmann equation (2.12) and of the obvious identity

$$\left(e\mathbf{E}\frac{\partial}{\partial\mathbf{p}} + e\mathbf{E}\frac{\partial}{\partial\mathbf{p}_1}\right)\overline{F}_{\mathbf{p}}\delta_{\mathbf{pp}_1} = \delta_{\mathbf{pp}_1}e\mathbf{E}\frac{\partial}{\partial\mathbf{p}}\overline{F}_{\mathbf{p}}, \quad (5.5)$$

one can exclude the field terms so that in fact the spectral intensities of Langevin sources are expressible only in terms of the distribution function $\overline{F}$ and the transition probabilities:

$$(\delta J_{\mathbf{p}}\delta J_{\mathbf{p}_1})_\omega = (\delta J_{\mathbf{p}}\delta J_{\mathbf{p}_1})^I_\omega + (\delta J_{\mathbf{p}}\delta J_{\mathbf{p}_1})^{II}_\omega, \quad (5.6)$$

where [40, 42, 86, 104]

$$(\delta J_{\mathbf{p}}\delta J_{\mathbf{p}_1})^I_\omega = \delta_{\mathbf{pp}_1}\sum_{\mathbf{k}}(W^{\mathbf{p}}_{\mathbf{k}}\overline{F}_{\mathbf{p}} + W^{\mathbf{k}}_{\mathbf{p}}\overline{F}_{\mathbf{k}}) - W^{\mathbf{p}_1}_{\mathbf{p}}\overline{F}_{\mathbf{p}_1} - W^{\mathbf{p}}_{\mathbf{p}_1}\overline{F}_{\mathbf{p}}, \quad (5.7)$$

and [38, 42, 86]

$$(\delta J_{\mathbf{p}}\delta J_{\mathbf{p}_1})^{II}_\omega = \delta_{\mathbf{pp}_1}\sum_{\mathbf{p'kk'}}(W^{\mathbf{pp'}}_{\mathbf{kk'}}\overline{F}_{\mathbf{p}}\overline{F}_{\mathbf{p'}} + W^{\mathbf{kk'}}_{\mathbf{pp'}}\overline{F}_{\mathbf{k}}\overline{F}_{\mathbf{k'}})$$

$$+ \sum_{\mathbf{kk'}}(W^{\mathbf{pp}_1}_{\mathbf{kk'}}\overline{F}_{\mathbf{p}}\overline{F}_{\mathbf{p}_1} + W^{\mathbf{kk'}}_{\mathbf{pp}_1}\overline{F}_{\mathbf{k}}\overline{F}_{\mathbf{k'}})$$

$$- 2\sum_{\mathbf{kk'}}(W^{\mathbf{k'p}_1}_{\mathbf{kp}}\overline{F}_{\mathbf{k'}}\overline{F}_{\mathbf{p}_1} + W^{\mathbf{kp}}_{\mathbf{k'p}_1}\overline{F}_{\mathbf{k}}\overline{F}_{\mathbf{p}}). \quad (5.8)$$

Taking advantage of Stueckelberg's property (2.10), expression (5.8) can be rewritten as

$$(\delta J_{\mathbf{p}}\delta J_{\mathbf{p}_1})_\omega^{II} = \delta_{\mathbf{pp}_1}\sum_{\mathbf{p'kk'}}(W_{\mathbf{pp'}}^{\mathbf{kk'}}(\overline{F}_{\mathbf{p}}\overline{F}_{\mathbf{p'}} + \overline{F}_{\mathbf{k}}\overline{F}_{\mathbf{k'}}) + \sum_{\mathbf{kk'}}W_{\mathbf{pp}_1}^{\mathbf{kk'}}(\overline{F}_{\mathbf{p}}\overline{F}_{\mathbf{p}_1} + \overline{F}_{\mathbf{k}}\overline{F}_{\mathbf{k'}})$$
$$-2\sum_{\mathbf{kk'}}(W_{\mathbf{kp}}^{\mathbf{k'p}_1}\overline{F}_{\mathbf{k'}}\overline{F}_{\mathbf{p}_1} + W_{\mathbf{k'p}_1}^{\mathbf{kp}}\overline{F}_{\mathbf{k}}\overline{F}_{\mathbf{p}}). \qquad (5.9)$$

Expressions (5.7) and (5.8) can be rewritten also in the following equivalent form [90]:

$$(\delta J_{\mathbf{p}}\delta J_{\mathbf{p}_1})_\omega^{I} = \sum_{\mathbf{rs}} W_{\mathbf{r}}^{\mathbf{s}}\overline{F}_{\mathbf{s}}(\delta_{\mathbf{rp}} - \delta_{\mathbf{sp}})(\delta_{\mathbf{rp}_1} - \delta_{\mathbf{sp}_1}), \qquad (5.10)$$

and

$$(\delta J_{\mathbf{p}}\delta J_{\mathbf{p}_1})_\omega^{II} = \frac{1}{2}\sum_{\mathbf{rr'ss'}} W_{\mathbf{rr'}}^{\mathbf{ss'}}\overline{F}_{\mathbf{s}}\overline{F}_{\mathbf{s'}}(\delta_{\mathbf{rp}} + \delta_{\mathbf{r'p}} - \delta_{\mathbf{sp}} - \delta_{\mathbf{s'p}})$$
$$\times (\delta_{\mathbf{rp}_1} + \delta_{\mathbf{r'p}_1} - \delta_{\mathbf{sp}_1} - \delta_{\mathbf{s'p}_1}), \qquad (5.11)$$

probably more transparent due to their symmetric form.

Langevin Sources at Equilibrium

For fluctuations around the equilibrium state, the expression for the spectral intensities of Langevin sources as given by Eq. (5.4) gets simplified (Kadomtsev [39]):

$$(\delta J_{\mathbf{p}}\delta J_{\mathbf{p}_1})_\omega^{eq} = (I_{\mathbf{p}}^{eq} + I_{\mathbf{p}_1}^{eq})\overline{F}_{\mathbf{p}}^{eq}\delta_{\mathbf{pp}_1} \qquad (5.12)$$

[see Eq. (4.7)]. Here I^{eq} is the relaxation operator for processes near the equilibrium state, given by expression (3.37):

$$I_{\mathbf{p}}^{eq} \equiv I_{\mathbf{p}}^{th} + I_{\mathbf{p}}^{ee}(\overline{F}^{eq}).$$

The explicit form of Eq. (5.12) is

$$(\delta J_{\mathbf{p}}\delta J_{\mathbf{p}_1})_\omega^{eq} = (\delta J_{\mathbf{p}}\delta J_{\mathbf{p}_1})_\omega^{eq\ I} + (\delta J_{\mathbf{p}}\delta J_{\mathbf{p}_1})_\omega^{eq\ II}, \qquad (5.13)$$

where

$$(\delta J_{\mathbf{p}}\delta J_{\mathbf{p}_1})_\omega^{eq\ I} = 2\delta_{\mathbf{pp}_1}\overline{F}_{\mathbf{p}}^{eq}\sum_{\mathbf{k}} W_{\mathbf{k}}^{\mathbf{p}} - W_{\mathbf{p}}^{\mathbf{p}_1}\overline{F}_{\mathbf{p}_1}^{eq} - W_{\mathbf{p}_1}^{\mathbf{p}}\overline{F}_{\mathbf{p}}^{eq} \qquad (5.14)$$

and

$$(\delta J_{\mathbf{p}}\delta J_{\mathbf{p}_1})_\omega^{eq\ II} = 2\delta_{\mathbf{pp}_1}\sum_{\mathbf{p'}}\overline{F}_{\mathbf{p}}^{eq}\overline{F}_{\mathbf{p'}}^{eq}\sum_{\mathbf{kk'}} W_{\mathbf{kk'}}^{\mathbf{pp'}} + 2\overline{F}_{\mathbf{p}}^{eq}\overline{F}_{\mathbf{p}_1}^{eq}\sum_{\mathbf{kk'}} W_{\mathbf{kk'}}^{\mathbf{pp}_1}$$
$$-2\sum_{\mathbf{kk'}}(W_{\mathbf{kp}}^{\mathbf{k'p}_1}\overline{F}_{\mathbf{k'}}^{eq}\overline{F}_{\mathbf{p}_1}^{eq} + W_{\mathbf{k'p}_1}^{\mathbf{kp}}\overline{F}_{\mathbf{k}}^{eq}\overline{F}_{\mathbf{p}}^{eq}). \qquad (5.15)$$

Equations (5.14) and (5.15) are compatible with Eqs. (5.7) and (5.9) thanks to the exponential dependence of the equilibrium distribution on electron energy, $\overline{F}_\mathbf{p}^{eq} \propto \exp(-\varepsilon_\mathbf{p}/T_0)$, and the fact that the transition probabilities in Eq. (5.9) contain δ-functions ensuring conservation of energy and quasi-momentum in a single collision (see Section 2.1).

5.2 DERIVATION OF THE BOLTZMANN–LANGEVIN EQUATION

The above-presented generalization of the Langevin formalism leading to the expressions for spectral intensities of occupancy (distribution function) fluctuations equivalent to those following from the equations for correlation functions was achieved by Kogan and Shulman [42], who exploited the idea that fluctuations of occupation numbers are caused in fact by *fluctuations in collision rate*. They used the presumption that collisions are independent events — the presumption which lies at the basis of the Boltzmann equation for the average distribution. The presumption enabled them to determine correlation properties of the Langevin random fluxes directly exploiting the Poisson statistics of collisions.

Let us present here the main points of their method. The change in fluctuation of the occupancy of a given state **p** in the quasi-momentum space (more precisely, of a group of states around **p**) is caused by fluctuation of *fluxes in the quasi-momentum space*. The variation in the flux consists of two parts. The first one exists inasmuch as the fluctuation of occupancies, $\delta F_\mathbf{p}(t)$, exists. It is determined by action of the relaxation operator $I_\mathbf{p}$ on the fluctuation $\delta F_\mathbf{p}(t)$. The second contribution reflects the statistical nature of scattering events. In general, it would not be equal to zero even if, by accident, at some instant the fluctuation of occupancy vanishes: Indeed, the number of collision events per unit time fluctuates due to the stochastic character of interaction which lies at the foundations of the kinetic description of the system. The Boltzmann–Langevin equation (5.1) is none other than the stochastic equation for the occupancy which takes into account *both* causes of change of the latter. The relaxation operator on the left-hand side of the equation ensures the trend of evolution of the existing fluctuational deviation evolving under the action of the external field and the collisional "friction." The right-hand side describes the random birth of such deviations due to stochastic character of individual collisions, each of them transferring electrons from one point of the quasi-momentum space into another (random jumps in the quasi-momentum space).

The random flux into the state (the group of states) — the right-hand side of Eq. (5.1) — consists, in turn, of two parts:

$$\delta J_\mathbf{p}(t) = \delta J_\mathbf{p}^I(t) + \delta J_\mathbf{p}^{II}(t). \tag{5.16}$$

Here

$$\delta J_\mathbf{p}^I(t) = \sum_{\mathbf{p}'}(\delta J_\mathbf{p}^{\mathbf{p}'}(t) - \delta J_{\mathbf{p}'}^{\mathbf{p}}(t)) \tag{5.17}$$

is the random flux into the state **p** due to collisions with the thermal bath, while

$$\delta J_{\mathbf{p}}^{\mathbf{p}'}(t) = J_{\mathbf{p}}^{\mathbf{p}'}(t) - \overline{J_{\mathbf{p}}^{\mathbf{p}'}} \tag{5.18}$$

is the elementary random flux from **p′** to **p**. Correspondingly, for pair collisions

$$\delta J_{\mathbf{p}}^{\mathrm{II}}(t) = \frac{1}{2} \sum_{\mathbf{p}_1 \mathbf{p}' \mathbf{p}'_1} (\delta J_{\mathbf{p}\mathbf{p}_1}^{\mathbf{p}'\mathbf{p}'_1}(t) - \delta J_{\mathbf{p}'\mathbf{p}'_1}^{\mathbf{p}\mathbf{p}_1}(t)), \tag{5.19}$$

where

$$\delta J_{\mathbf{p}\mathbf{p}_1}^{\mathbf{p}'\mathbf{p}'_1}(t) = J_{\mathbf{p}\mathbf{p}_1}^{\mathbf{p}'\mathbf{p}'_1}(t) - \overline{J_{\mathbf{p}\mathbf{p}_1}^{\mathbf{p}'\mathbf{p}'_1}} \tag{5.20}$$

is the fluctuation in the elementary flux from the pair of states **p′**, **p′**$_1$, due to collisions between the particles occupying these states, into the pair of states **p**, **p**$_1$.

One should assume the elementary fluxes to be uncorrelated at different instants; this corresponds to the assumption that the collision is an instantaneous event. This means that we deal only with not too short time intervals and, correspondingly, not too high frequencies. In conformity with the assumption that each collision event is independent — the assumption lying at the basis of the Boltzmann equation — one should assume each elementary flux to be correlated only with itself:

$$\overline{\delta J_{\mathbf{p}}^{\mathbf{p}'}(t) \delta J_{\mathbf{p}_1}^{\mathbf{p}'_1}(t')} \propto \delta(t - t') \delta_{\mathbf{p}\mathbf{p}_1} \delta_{\mathbf{p}'\mathbf{p}'_1}.$$

What has been said means that fluctuations in the collision rate should be treated as a Poisson random process; it follows that the mean square fluctuation of the number of collisions transferring an electron from **p′** to **p** is equal to the mean number of such collisions,

$$\overline{\delta J_{\mathbf{p}}^{\mathbf{p}'}(t) \delta J_{\mathbf{p}_1}^{\mathbf{p}'_1}(t')} = \delta(t - t') \overline{J_{\mathbf{p}}^{\mathbf{p}'}} \delta_{\mathbf{p}\mathbf{p}_1} \delta_{\mathbf{p}'\mathbf{p}'_1}, \tag{5.21}$$

where the average flux from **p′** to **p** is $\overline{J_{\mathbf{p}}^{\mathbf{p}'}} = W_{\mathbf{p}}^{\mathbf{p}'} \overline{F}_{\mathbf{p}'}$. This immediately leads to the expression (5.7) for the spectral intensity of random fluxes $\delta J_{\mathbf{p}}^{\mathrm{I}}$ and $\delta J_{\mathbf{p}_1}^{\mathrm{I}}$.

The spectral intensity of fluctuations in the rate of electron–electron collisions is obtainable in the same way:

$$(\delta J_{\mathbf{p}\mathbf{p}_1}^{\mathbf{p}'\mathbf{p}'_1} \delta J_{\mathbf{r}\mathbf{r}_1}^{\mathbf{r}'\mathbf{r}'_1})_\omega = \delta_{\mathbf{p}\mathbf{p}_1, \mathbf{r}\mathbf{r}_1} \delta_{\mathbf{p}'\mathbf{p}'_1, \mathbf{r}'\mathbf{r}'_1} \overline{J_{\mathbf{p}\mathbf{p}_1}^{\mathbf{p}'\mathbf{p}'_1}}, \tag{5.22}$$

where $\delta_{\mathbf{p}\mathbf{p}_1, \mathbf{r}\mathbf{r}_1} = \delta_{\mathbf{p}\mathbf{r}} \delta_{\mathbf{p}_1 \mathbf{r}_1} + \delta_{\mathbf{p}\mathbf{r}_1} \delta_{\mathbf{p}_1 \mathbf{r}}$ and the average flux is

$$\overline{J_{\mathbf{p}\mathbf{p}_1}^{\mathbf{p}'\mathbf{p}'_1}} = 2 W_{\mathbf{p}\mathbf{p}_1}^{\mathbf{p}'\mathbf{p}'_1} \overline{F}_{\mathbf{p}'} \overline{F}_{\mathbf{p}'_1}.$$

As a result, one arrives to expression (5.8) for the spectral intensity of random fluxes $\delta J_{\mathbf{p}}^{II}$ and $\delta J_{\mathbf{p}_1}^{II}$.

In conclusion, the results following from equations for correlation functions (valid provided that the Boltzmann equation is valid) coincide with those obtained by the generalized Langevin method based on the assumption that fluctuations in the collision rate are treated as a Poisson process. Within the limits of applicability of the Boltzmann equation, the collisions should be regarded as statistically independent events, whereas the occupation numbers of different states could be correlated.

CHAPTER 6

FLUCTUATIONS AND DIFFUSION

At thermal equilibrium, three basic kinetic coefficients are interrelated: the ohmic free-electron mobility μ^{eq}, the spectral intensity of current fluctuations S_I^{eq}, and the *free-electron diffusion coefficient* D^{eq}. Indeed, the Nyquist relation (3.42) expresses the spectral intensity of current fluctuations in terms of the conductivity and the thermal equilibrium temperature, while the *Einstein relation* (see reference 64) expresses the diffusion coefficient of carriers in terms of their mobility μ^{eq} and thermal-equilibrium temperature T_0: For carrier densities far from degeneracy, we have

$$D^{eq} = k_B T_0 \mu^{eq}/e, \tag{6.1}$$

where e is the elementary charge. Thus, in the linear-response regime an independent determination of the diffusion coefficient and the noise temperature [see Eq. (1.7)], in addition to the drift velocity, does not add any particular information about the kinetic properties of the material as compared to the knowledge obtained from a determination of the ohmic mobility. Under hot-electron conditions, not only Nyquist's relation but also Einstein's relation no longer hold in general. Therefore an independent determination of conductivity, noise temperature, and diffusion coefficient in principle can provide new information.

6.1 FLUCTUATION–DIFFUSION RELATIONSHIP

Investigating, by means of the *linear* Boltzmann equation, slightly *nonuniform* processes near a uniform *nonequilibrium* state of the system of charged particles moving through a uniform gas under the action of a static uniform electric field,

Wannier [34] identified a supplementary current that is proportional to the small space gradient of the carrier density. Thus he showed that the diffusion concept can still be applicable to a nonequilibrium system. Wannier demonstrated that smoothing-out of electron density variation is describable as an *anisotropic* diffusion process: The flux of electrons, $\mathbf{j}^D$, resulting from a small gradient of the electron density n, is given by the expression

$$j_\alpha^D = -D_{\alpha\beta}\frac{\partial n}{\partial r_\beta},$$

where $D_{\alpha\beta}(\mathbf{E})$ is the *tensor* of electron diffusion coefficients. Wannier was able to define unambiguously the tensor of carrier diffusion coefficients for a nonequilibrium state (see also reference 105) as

$$D_{\alpha\beta} = \sum_{\mathbf{p}} v_{\mathbf{p}\alpha}\tilde{I}_\mathbf{p}^{-1}(v_{\mathbf{p}\beta} - V_\beta)\overline{F}_\mathbf{p}/N, \qquad (6.2)$$

where $\overline{F}_\mathbf{p}$ was the steady-state distribution of *nonintercolliding* carriers in the absence of spatial gradients, satisfying the linear spatially uniform Boltzmann equation (3.23),

$$\tilde{I}_\mathbf{p}\overline{F}_\mathbf{p} \equiv \left(e\mathbf{E}\frac{\partial}{\partial \mathbf{p}} + I_\mathbf{p}^{\text{th}}\right)\overline{F}_\mathbf{p} = 0,$$

and $\mathbf{V}$ was the steady value of the drift velocity of the electron system as a whole, given by Eq. (2.15):

$$\mathbf{V} = \frac{1}{en_0}\bar{\mathbf{j}} = \frac{1}{N}\sum_{\mathbf{p}}\mathbf{v}_\mathbf{p}\overline{F}_\mathbf{p}. \qquad (6.3)$$

A comparison of Wannier's expression (6.2) for the tensor of diffusion coefficients of nonequilibrium nonintercolliding carriers with expression (3.24) for the spectral intensity of *spatially homogeneous* (uniform, space-independent) fluctuations of current density leads to the relation originally proposed by Price [41]:

$$(\delta j_\alpha \delta j_\beta)_{\omega\tau\ll 1} = (e^2 n_0/\mathcal{V}_0)(D_{\alpha\beta} + D_{\beta\alpha}), \qquad (6.4)$$

where $(\delta j_\alpha \delta j_\beta)_{\omega\tau\ll 1}$ is the low-frequency limit of the spectral intensity of space-independent current density fluctuations given by Eq. (3.24):

$$(\delta j_\alpha \delta j_\beta)_{\omega\tau\ll 1} = \left(\frac{e}{\mathcal{V}_0}\right)^2 \left[\sum_{\mathbf{p}} v_{\mathbf{p}\alpha}(\tilde{I}_\mathbf{p})^{-1}(v_{\mathbf{p}\beta} - V_\beta)\overline{F}_\mathbf{p} \right.$$

$$\left. + \sum_{\mathbf{p}} v_{\mathbf{p}\beta}(\tilde{I}_\mathbf{p})^{-1}(v_{\mathbf{p}\alpha} - V_\alpha)\overline{F}_\mathbf{p}\right]. \qquad (6.5)$$

The Price relation (6.4) proved to be very useful in enabling one to gain information about *spatially inhomogeneous* processes in a nonequilibrium system from noise measurements in *spatially homogeneous* nonequilibrium state or, *vice versa*, to predict properties of fluctuational phenomena in nonequilibrium system from the results of investigation of the response of the system to the spatially inhomogeneous perturbation. For numerous examples of usefulness of the Price relation in the experimental investigation of noise and transport in semiconductors and semiconductor structures, see references 10 and 11. For scores of years, the diffusion of electrons moving through an ordinary gas under the action of a static uniform electric field remained in the spotlight (see reference 106). Theoretical methods and results obtained in that field often prove to be useful while investigating carrier diffusion in semiconductors, and *vice versa*.

6.2 VIOLATION OF FLUCTUATION–DIFFUSION RELATION; ADDITIONAL CORRELATION TENSOR

Being as valuable as it proved to be, the Price relation has its natural limits of applicability. In this section, the modifications caused by interelectron collisions in nonequilibrium state will be revealed.

Drift and Diffusion in System of Intercolliding Carriers

In the general case of the nonlinear Boltzmann equation (2.1), the expression for the diffusion coefficient is modified. The tensor of coefficients of proportionality in the expression for the supplementary current proportional to the small space gradient of the carrier density — i.e., the carrier diffusion tensor — was identified by Gantsevich et al. [37]:

$$D_{\alpha\beta} = \sum_{\mathbf{p}} v_{\mathbf{p}\alpha} I_{\mathbf{p}}^{-1} (v_{\mathbf{p}\beta} - V'_{\beta}) \partial \overline{F}_{\mathbf{p}}/\partial N. \qquad (6.6)$$

Here $I_{\mathbf{p}}$ is the relaxation operator given by Eq. (2.18), and $\overline{F}_{\mathbf{p}}$ is the uniform stationary distribution function — the solution of Eq. (2.12). The derivative of this function with respect to a carrier number (or carrier density), $\partial \overline{F}_{\mathbf{p}}/\partial N$, appears in the expression (6.6) for the diffusion tensor. The function $\partial \overline{F}_{\mathbf{p}}/\partial N$ coincides with the function $\overline{F}_{\mathbf{p}}/N$ standing in Eq. (6.2) only in the thermal equilibrium or if interelectron collisions are negligible. Indeed, in these cases, $\overline{F}_{\mathbf{p}}/N$ is independent of the electron density; that is, the *shape* of the electron distribution in quasi-momentum space is insensitive to changes in the density. In the general case of the *nonlinear* Boltzmann equation (2.12), the functions $\partial \overline{F}_{\mathbf{p}}/\partial N$ and $\overline{F}_{\mathbf{p}}/N$ can differ significantly. Accordingly, a quantity

$$\mathbf{V}' = \sum_{\mathbf{p}} \mathbf{v}_{\mathbf{p}} \partial \overline{F}_{\mathbf{p}}/\partial N \qquad (6.7)$$

appears in Eq. (6.6), instead of the drift velocity of the carrier system given by Eq. (6.3) and standing in Eq. (6.2). Due to the possible complicated manner of the dependence of $\overline{F}_\mathbf{p}$ on N (or n_0), the quantity $\mathbf{V}'$ does not coincide with the drift velocity of the carrier system, $\mathbf{V}$. In fact, a smooth spatially-inhomogeneous disturbance in the system of intercolliding carriers (a smooth local carrier-density perturbation) travels in the space at the "differential drift velocity" $\mathbf{V}'$ (see Section 6.3), while the *carrier system as a whole* travels at the velocity $\mathbf{V}$.

Additional Correlation Tensor

On comparing the general expressions for the spectral intensity of current density fluctuations and for the diffusion tensor, Eqs. (4.12) and (6.6), we conclude that nonlinearity of the Boltzmann equation and the additional correlation created by electron–electron collisions in the nonequilibrium state makes it impossible to express the low-frequency spectral intensity of current fluctuations through the diffusion coefficient as was done neglecting interelectron collisions in the previous section [see Eq. (6.4)]. In the general case, by comparing Eqs. (4.12) and (6.6) we conclude that [37]

$$(\delta j_\alpha \delta j_\beta)_{\omega\tau \ll 1} = (e^2 n_0 / \mathcal{V}_0)(D_{\alpha\beta} + D_{\beta\alpha} - \Delta_{\alpha\beta}), \tag{6.8}$$

where the diffusion tensor $D_{\alpha\beta}$ is given by Eq. (6.6) while the symmetric tensor $\Delta_{\alpha\beta}$ is the low-frequency limit of the tensor

$$\Delta_{\alpha\beta}(\omega) = \Delta_{\beta\alpha}(-\omega) = \frac{1}{N} \sum_{\mathbf{p}\mathbf{p}_1} v_{\mathbf{p}\alpha} v_{\mathbf{p}_1 \beta} \frac{1}{(-i\omega + I_\mathbf{p})(i\omega + I_{\mathbf{p}_1})}$$
$$\times [I^{ee}_{\mathbf{p}\mathbf{p}_1}\{\overline{F}, \overline{F}\} - (I_\mathbf{p} + I_{\mathbf{p}_1})(\overline{F}_\mathbf{p} - N\partial \overline{F}_\mathbf{p}/\partial N)\delta_{\mathbf{p}\mathbf{p}_1}], \tag{6.9}$$

which can be called the *tensor of additional correlation*. The tensor $\Delta_{\alpha\beta}$ describes the specific influence of electron–electron collisions on electric noise, being a characteristic of the degree of correlation (or nonideality) and of significance of nonlinearity, in electron density, of the distribution function of a nonequilibrium electron gas.

The additional correlation tensor *vanishes in thermal equilibrium*

$$\Delta^{eq}_{\alpha\beta}(\omega) = 0, \tag{6.10}$$

since $\partial \overline{F}^{eq}_\mathbf{p}/\partial N = \overline{F}^{eq}_\mathbf{p}/N$ and $I^{ee}_{\mathbf{p}\mathbf{p}_1}\{\overline{F}^{eq}, \overline{F}^{eq}\} = 0$ [see Eq. (4.7)]. So, in thermal equilibrium the Price relation

$$(\delta j_\alpha \delta j_\beta)^{eq}_{\omega\tau \ll 1} = (e^2 n_0 / \mathcal{V}_0)(D^{eq}_{\alpha\beta} + D^{eq}_{\alpha\beta}) \tag{6.11}$$

holds *independently* of frequency of interelectron collisions. In thermal equilibrium, the expression (6.6) for the diffusion tensor goes over to

$$D_{\alpha\beta}^{eq} = \sum_{\mathbf{p}} v_{\mathbf{p}\alpha}(I_{\mathbf{p}}^{eq})^{-1} v_{\mathbf{p}\beta} \overline{F}_{\mathbf{p}}^{eq}/N, \qquad (6.12)$$

where the relaxation operator near the equilibrium state $I_{\mathbf{p}}^{eq}$ is defined by Eq. (3.37). Let us be reminded that expression (3.39),

$$(\delta j_\alpha \delta j_\beta)_\omega^{eq} = (\delta j_\beta \delta j_\alpha)_\omega^{eq} = (e/\mathcal{V}_0)^2 2\,\mathrm{Re} \sum_{\mathbf{p}} v_{\mathbf{p}\alpha}(-i\omega + I_{\mathbf{p}}^{eq})^{-1} v_{\mathbf{p}\beta} \overline{F}_{\mathbf{p}}^{eq},$$

is valid for the spectral intensity of current density fluctuations in thermal equilibrium.

Fluctuations, Diffusion, and Conductivity

At equilibrium, the low-frequency limit of spectral intensity of current density fluctuations and the carrier diffusion coefficient are connected by relation (6.11) which is in line with two other well-known relations valid at equilibrium — the Einstein relation connecting the diffusion coefficient of carriers and the conductivity [cf. Eq. (6.1)]

$$D_{\alpha\beta}^{eq} = (k_B T_0/e^2 n_0)\sigma_{\alpha\beta}^{eq}, \qquad (6.13)$$

and the Nyquist relation (3.42)

$$(\delta j_\alpha \delta j_\beta)_\omega^{eq} = 2(T_0/\mathcal{V}_0)\,\mathrm{Re}\,\sigma_{\alpha\beta}^{eq}(\omega)$$

we have discussed in Section 3.5. In the framework of the kinetic theory, the Einstein relation (6.13) follows from comparison of expression (6.12) for diffusion coefficient of equilibrium carriers and expression (3.40) for conductivity near equilibrium.

So, the fundamental relations among fluctuation, diffusion, and electric transport characteristics hold at thermal equilibrium state:

$$(\delta j_\alpha \delta j_\beta)_{\omega\tau\ll 1}^{eq} = (2e^2 n_0/\mathcal{V}_0)D_{\alpha\beta}^{eq} = (2T_0/\mathcal{V}_0)\sigma_{\alpha\beta}^{eq}. \qquad (6.14)$$

A departure from thermal equilibrium generally breaks the tie between the diffusion coefficient, or the spectral intensity of current fluctuations, and the conductivity. Unlike this, the relation between the diffusion and the current fluctuations proves to be more persistent: It can survive in the nonequilibrium state and disappears only if electron–electron collisions play an essential role in the kinetics of the nonequilibrium (hot electron) state.

In Chapters 12–14 we shall present some results of analytical calculation and of Monte Carlo simulation — taking into account electron–electron collisions — of the noise, diffusion, and correlation characteristics of hot electrons, and we shall analyze for real situations the degree of violation of the Price relation and aftereffects of the violation, provided that it takes place.

6.3 DRIFT–DIFFUSION EQUATION; GENERALIZED CHAPMAN–ENSKOG PROCEDURE

Formulation of Problem

In the preceding section, we presented expression (6.6) for the tensor of carrier diffusion coefficients without derivation. Now we are going to expound such a derivation. To do it, we should derive, from the kinetic equation, the drift–diffusion equation for transport near a nonequilibrium state. The derivation will be based on the ideas of the well-known Chapman–Enskog method, enabling us to obtain the hydrodynamic equations from the kinetic equation in ordinary gases (see reference 107).

Up to now we dealt with uniform (homogeneous) distributions of carriers governed by the space-independent Boltzmann equation (2.1). In the general case the Boltzmann equation contains the *convective* term $\mathbf{v_p}\partial \overline{F}_\mathbf{p}(\mathbf{r}, t)/\partial \mathbf{r}$, describing a change of the space- and time-dependent mean distribution function, $\overline{F}_\mathbf{p}(\mathbf{r}, t)$, due to free motion of carriers in the space between the collisions:

$$\left(\frac{\partial}{\partial t} + \mathbf{v_p}\frac{\partial}{\partial \mathbf{r}}\right)\overline{F}_\mathbf{p}(\mathbf{r}, t) = -e\mathbf{E}\frac{\partial}{\partial \mathbf{p}}\overline{F}_\mathbf{p}(\mathbf{r}, t) - I_\mathbf{p}^{\text{th}}\overline{F}_\mathbf{p}(\mathbf{r}, t) - I_\mathbf{p}^{\text{ee}}\{\overline{F}(\mathbf{r}, t), \overline{F}(\mathbf{r}, t)\}. \tag{6.15}$$

We suppose the electric field to remain uniform; that is, for simplicity we ignore the field perturbation induced by redistribution of charge carriers in space; the arising self-consistent field can be easily taken into account by using the Poisson equation (for details see references 37 and 70). We also suppose the thermostat to be uniform. As a result, the steady-state distribution is space-independent, being given, as it was in the previous sections, by the solution of Eq. (2.12):

$$e\mathbf{E}\frac{\partial}{\partial \mathbf{p}}\overline{F}_\mathbf{p} + I_\mathbf{p}^{\text{th}}\overline{F}_\mathbf{p} + I_\mathbf{p}^{\text{ee}}\{\overline{F}, \overline{F}\} = 0.$$

We shall deal only with situations where deviations from the uniform steady-state distribution are *small*,

$$\overline{F}_\mathbf{p}(\mathbf{r}, t) = \overline{F}_\mathbf{p} + \Delta \overline{F}_\mathbf{p}(\mathbf{r}, t), \qquad \Delta \overline{F}_\mathbf{p}(\mathbf{r}, t) \ll \overline{F}_\mathbf{p}. \tag{6.16}$$

Inserting Eq. (6.16) into the Boltzmann equation (6.15) and linearizing the latter with respect to $\Delta \overline{F}_\mathbf{p}(\mathbf{r}, t)$, we arrive at the equation governing evolution of a small space-dependent deviation of distribution of electrons from the uniform steady distribution:

$$\left(\frac{\partial}{\partial t} + \mathbf{v_p}\frac{\partial}{\partial \mathbf{r}} + I_\mathbf{p}\right)\Delta \overline{F}_\mathbf{p}(\mathbf{r}, t) = 0, \tag{6.17}$$

where $I_\mathbf{p}$ is the linearized operator of the uniform Boltzmann equation, given by Eq. (2.18) and referred to by us as the *relaxation operator*:

$$I_\mathbf{p} \equiv e\mathbf{E}\partial/\partial \mathbf{p} + I_\mathbf{p}^{th} + I_\mathbf{p}^{ee}(\overline{F}).$$

Our task is to obtain from Eq. (6.17), in the case of a *smooth* spatial disturbance of carrier distribution (small gradients of distribution), a macroscopic drift–diffusion equation, as well as expressions for coefficients entering the equation. The method is to apply to Eq. (6.17) an iterative, in small spatial gradients and corresponding time derivatives, procedure leading to an equation for smoothly changing *density* of carriers. The method works because the evolution of the nearly uniform *initial* distribution $\overline{F}_\mathbf{p}(\mathbf{r}; t = 0)$ passes through two stages. During the first stage, the relatively fast *relaxation in the quasi-momentum space* prevails so that at the end of this stage the *shape* of the quasi-momentum distribution approaches that imposed by the applied field and collisions; that is, the distribution becomes close to the function $\overline{F}_\mathbf{p}$ normalized to *local* density of carriers. In other words, an arbitrary smooth initial distribution $\overline{F}_\mathbf{p}(t = 0; \mathbf{r})$ after the first stage of evolution approximately factorizes:

$$\overline{F}_\mathbf{p}(\mathbf{r}, t) \approx n(\mathbf{r}, t)\overline{F}_\mathbf{p}/n_0, \qquad t \gg \tau, \tag{6.18}$$

where we denote as $n(\mathbf{r}, t)$ the local density of carriers:

$$n(\mathbf{r}, t) = \sum_\mathbf{p} \overline{F}_\mathbf{p}(\mathbf{r}, t)/V_0, \tag{6.19}$$

$n_0 = N/V_0$ is the stationary density of carriers, and the inverse of the relaxation time τ gives the order of magnitude of the relaxation operator.

This first stage of evolution is short enough, with τ being the time of relaxation of the nonstationary *uniform* distribution of carriers to the stationary one. Contrary, the second stage of evolution — the approaching to the uniform steady distribution by carrier density in space — will be shown to be a comparatively slow process. We are going to demonstrate that the slow stage of evolution from the distribution (6.18) to the uniform stationary distribution $\overline{F}_\mathbf{p}$ is governed by the *drift–diffusion equation*, entered by the coefficients: the "differential" drift velocity given by Eq. (6.7) and the tensor of diffusion coefficients (6.6). The differential drift velocity (6.7) and the tensor of diffusion coefficients (6.6) *are functionals of the uniform steady-state distribution function* $\overline{F}_\mathbf{p}$; naturally enough, the coefficients in the drift–diffusion equation depend on the corresponding steady state, with the diffusion near an equilibrium state being a special case of a rather wide class of drift–diffusion processes.

Inverse Relaxation Operator

The evolution of a small initial space-dependent deviation of distribution of electrons from the uniform steady distribution, $\Delta \overline{F}_\mathbf{p}(\mathbf{r}; t = 0)$, is governed by the linearized Boltzmann equation (6.17). For detailed analysis of this process we need a suitable technique (see references 37 and 70).

Let us consider the *linear inhomogeneous* equation

$$I_\mathbf{p} x_\mathbf{p} = y_\mathbf{p}. \tag{6.20}$$

From definition (2.18) of the relaxation operator $I_\mathbf{p}$, we have for any physically reasonable distribution in the quasi-momentum space, $x_\mathbf{p}$,

$$\sum_\mathbf{p} I_\mathbf{p} x_\mathbf{p} = 0. \tag{6.21}$$

Indeed, under the action of collisions and of the electric field the *number* of carriers is not changed — carriers are only *redistributed* in the quasi-momentum space. In a formal way, property (6.21) — for any reasonable function $x_\mathbf{p}$ — follows from the obvious properties of the field term, $e\mathbf{E}\partial/\partial\mathbf{p}$, and collision terms, Eqs. (2.2) and (2.20), entering the definition (2.18) of the relaxation operator $I_\mathbf{p}$.

Property (6.21) means that Eq. (6.20) is soluble in the manifold of "physically reasonable" functions only if the right-hand side of the equation — the function $y_\mathbf{p}$ — satisfies the condition

$$\sum_\mathbf{p} y_\mathbf{p} = 0. \tag{6.22}$$

We shall refer to property (6.22) as *zero-sum property* of the function $y_\mathbf{p}$. It follows from what was said that, if we would like to introduce the inverse of the relaxation operator $I_\mathbf{p}$, the latter could be allowed *to act only on functions having the zero-sum property* (6.22).

However, this restriction is not sufficient to construct the solution of Eq. (6.20) — that is, to define the inverse operator $I_\mathbf{p}^{-1}$. Indeed, differentiating the steady-state Boltzmann equation (2.12) with respect to the total number of free electrons N, we come to Eq. (2.25):

$$I_\mathbf{p} \partial \overline{F}_\mathbf{p}/\partial N = 0,$$

where the function $\partial \overline{F}_\mathbf{p}/\partial N$ is the solution of the *homogeneous* equation

$$I_\mathbf{p} z_\mathbf{p} = 0. \tag{6.23}$$

We deal with the steady-state distribution function normalized to N [see Eq. (2.14)], so the function $\partial \overline{F}_\mathbf{p}/\partial N$ is normalized to unity [see Eq. (2.26)]:

$$\sum_\mathbf{p} \partial \overline{F}_\mathbf{p}/\partial N = 1.$$

In other words, the function $\partial \overline{F}_\mathbf{p}/\partial N$ is the normalized-to-unity eigenfunction of the relaxation operator $I_\mathbf{p}$ corresponding to its zero eigenvalue. Because the homogeneous equation (6.23) has a solution, the solution of the *inhomogeneous* equation (6.20) apparently *is not unique*: If $x_\mathbf{p}^{(0)}$ is a solution of the *inhomogeneous* equation (6.20), then the function

$$x_\mathbf{p}^{(1)} = x_\mathbf{p}^{(0)} + \text{const}\, \partial \overline{F}_\mathbf{p}/\partial N \tag{6.24}$$

is a solution as well. We conclude that, for the present, the *inverse* of the operator standing on the left-hand side of Eq. (6.20) — of the relaxation operator $I_\mathbf{p}$ — is not uniquely defined.

The right-hand side of the inhomogeneous equation (6.20) — the function $y_\mathbf{p}$ — has zero-sum property (6.22). To define the inverse of the operator $I_\mathbf{p}$ uniquely, let us select from a number of solutions of Eq. (6.20) the one which has the zero-sum property. In other words, we require the function generated by action of the inverse operator on the function $y_\mathbf{p}$, the latter having the zero-sum property by definition, also to have the zero-sum property. Let us designate this function as

$$I_\mathbf{p}^{-1} y_\mathbf{p}; \tag{6.25}$$

that is, let us use the symbol $I_\mathbf{p}^{-1} y_\mathbf{p}$ for the particular solution $x_\mathbf{p}^{(1)}$ of Eq. (6.20) which satisfies the condition $\sum_\mathbf{p} x_\mathbf{p}^{(1)} = 0$:

$$\sum_\mathbf{p} I_\mathbf{p}^{-1} y_\mathbf{p} = 0. \tag{6.26}$$

The solution satisfying the condition (6.26) can always be constructed from any particular solution $x_\mathbf{p}^{(0)}$ by proper choosing of the constant in Eq. (6.24): const $= -\sum_\mathbf{p} x_\mathbf{p}^{(0)}$.

Now the *general* solution of Eq. (6.20) can be written as

$$x_\mathbf{p} = (\partial \overline{F}_\mathbf{p}/\partial N) \sum_{\mathbf{p}'} x_{\mathbf{p}'} + I_\mathbf{p}^{-1} y_\mathbf{p}, \tag{6.27}$$

where the *inverse of relaxation operator*, $I_\mathbf{p}^{-1}$, is allowed to act only on the functions having the zero-sum property and generates functions with the same property. These requirements guarantee existence and uniqueness of the solution of Eq. (6.20) denoted as $I_\mathbf{p}^{-1} y_\mathbf{p}$.

As we have seen, the relaxation operator $I_\mathbf{p}$ generates functions having the zero-sum property [see Eq. (6.21)]. It is convenient to demand that the relaxation operator $I_\mathbf{p}$ would also act only on the functions having the zero-sum property. This requirement can always be fulfilled because of the identity following from Eq. (2.25):

$$I_\mathbf{p} x_\mathbf{p} \equiv I_\mathbf{p} \left(x_\mathbf{p} - (\partial \overline{F}_\mathbf{p}/\partial N) \sum_{\mathbf{p}'} x_{\mathbf{p}'} \right). \tag{6.28}$$

DRIFT–DIFFUSION EQUATION; GENERALIZED CHAPMAN–ENSKOG PROCEDURE   59

According to this convention, we should rewrite Eq. (6.20) as follows:

$$I_\mathbf{p}\left(x_\mathbf{p} - (\partial \overline{F}_\mathbf{p}/\partial N)\sum_{\mathbf{p}'} x_{\mathbf{p}'}\right) = y_\mathbf{p}. \tag{6.29}$$

Now formally the general solution of Eq. (6.20), given by Eq. (6.27), is obtainable from Eq. (6.20), rewritten in the form (6.29), by simple multiplication by the symbol $I_\mathbf{p}^{-1}$. It means that the accepted "zero-sum rule" enables us to manipulate the operators $I_\mathbf{p}$ and $I_\mathbf{p}^{-1}$ quite mechanically ($I_\mathbf{p} I_\mathbf{p}^{-1} = I_\mathbf{p}^{-1} I_\mathbf{p} = 1$).

The above-described procedure is none other than the adaptation to our needs of the method well known in the theory of linear integral equations and linear operators (see references 108 and 109).

From Boltzmann Equation to Drift–Diffusion Equation

Now we are ready to deal with Boltzmann equation (6.17) in the case of *smooth* spatial disturbance of carrier distribution (in other words, in the limit of small spatial gradients). The summation of Eq. (6.17) over $\mathbf{p}$ leads, after division by the volume $\mathcal{V}_0$, to the *continuity equation*

$$\frac{\partial}{\partial t}\Delta n(\mathbf{r},t) + \frac{\partial}{\partial \mathbf{r}}\Delta \mathbf{j}^{(e)}(\mathbf{r},t) = 0, \tag{6.30}$$

where we denote as $\Delta n(\mathbf{r}, t)$ the local disturbance of density of carriers:

$$\Delta n \equiv \Delta n(\mathbf{r},t) = \sum_\mathbf{p} \Delta \overline{F}_\mathbf{p}(\mathbf{r},t)/\mathcal{V}_0, \tag{6.31}$$

and we denote as $\Delta \mathbf{j}^{(e)}(\mathbf{r}, t)$ the local disturbance of carrier flux density:

$$\Delta \mathbf{j}^{(e)} \equiv \Delta \mathbf{j}^{(e)}(\mathbf{r},t) = \sum_\mathbf{p} \mathbf{v}_\mathbf{p} \Delta \overline{F}_\mathbf{p}(\mathbf{r},t)/\mathcal{V}_0. \tag{6.32}$$

Multiplying the continuity equation (6.30) by $\mathcal{V}_0 \partial \overline{F}_\mathbf{p}/\partial N$ and subtracting the result from Eq. (6.17), we get the equation for the disturbance of carrier distribution, $\Delta \overline{F}_\mathbf{p}(\mathbf{r}, t)$, prepared for the action by the inverse operator $I_\mathbf{p}^{-1}$:

$$\frac{\partial}{\partial t}\left(\Delta \overline{F}_\mathbf{p} - \Delta n \frac{\partial \overline{F}_\mathbf{p}}{\partial N}\mathcal{V}_0\right) + \frac{\partial}{\partial \mathbf{r}}\left(\mathbf{v}_\mathbf{p}\Delta \overline{F}_\mathbf{p} - \Delta \mathbf{j}^{(e)}\frac{\partial \overline{F}_\mathbf{p}}{\partial N}\mathcal{V}_0\right)$$
$$+ I_\mathbf{p}\left(\Delta \overline{F}_\mathbf{p} - \Delta n \frac{\partial \overline{F}_\mathbf{p}}{\partial N}\mathcal{V}_0\right) = 0. \tag{6.33}$$

The last term in Eq. (6.33) is rewritten in accordance with the convention (6.28). Each of the expressions put in brackets has the zero-sum property. This allows us to deal with each of these expressions separately while acting on Eq. (6.33) by the operator $I_\mathbf{p}^{-1}$:

$$\Delta \overline{F}_\mathbf{p} = \Delta n \frac{\partial \overline{F}_\mathbf{p}}{\partial N} V_0 - \frac{\partial}{\partial t} I_\mathbf{p}^{-1} \left(\Delta \overline{F}_\mathbf{p} - \Delta n \frac{\partial \overline{F}_\mathbf{p}}{\partial N} V_0 \right)$$
$$- \frac{\partial}{\partial \mathbf{r}} I_\mathbf{p}^{-1} \left(\mathbf{v}_\mathbf{p} \Delta \overline{F}_\mathbf{p} - \Delta \mathbf{j}^{(e)} \frac{\partial \overline{F}_\mathbf{p}}{\partial N} V_0 \right). \quad (6.34)$$

This equation—still equivalent to Eq. (6.17)—is much better prepared for iterative procedure than Eq. (6.17). Indeed, the first term on the right-hand side of Eq. (6.34) is factorized in the manner mentioned at the beginning of the section [cf. Eq. (6.18)]. The remaining terms apparently are small, provided that the space and time derivatives are small enough. Of course, these terms cannot be neglected if we want to obtain the equation governing evolution of distribution in time and space, but they *can be treated as corrections* to the zero-order term

$$\Delta \overline{F}_\mathbf{p}(\mathbf{r}, t) \approx \Delta n(\mathbf{r}, t) V_0 \partial \overline{F}_\mathbf{p}/\partial N. \quad (6.35)$$

We arrived at the central idea of the Chapman–Enskog-type iterative procedure, which should lead us from the kinetic equation to the drift–diffusion equation. Indeed, multiplying Eq. (6.34) by the component $v_{\mathbf{p}\alpha}$ of the carrier velocity vector $\mathbf{v}_\mathbf{p}$ and dividing by the volume V_0, we obtain after summation over $\mathbf{p}$ the following expression for the corresponding component of the electron flux density vector $\Delta \mathbf{j}^{(e)}$:

$$\Delta j_\alpha^{(e)} = V'_\alpha \Delta n - \frac{\partial}{\partial t} \sum_\mathbf{p} v_{\mathbf{p}\alpha} I_\mathbf{p}^{-1} \left(\Delta \overline{F}_\mathbf{p}/V_0 - \Delta n \frac{\partial \overline{F}_\mathbf{p}}{\partial N} \right)$$
$$- \frac{\partial}{\partial r_\beta} \sum_\mathbf{p} v_{\mathbf{p}\alpha} I_\mathbf{p}^{-1} \left(v_{\mathbf{p}\beta} \Delta \overline{F}_\mathbf{p}/V_0 - \Delta j_\beta^{(e)} \frac{\partial \overline{F}_\mathbf{p}}{\partial N} \right). \quad (6.36)$$

The already defined by Eq. (6.7) quantity $\mathbf{V}'$ appeared as the coefficient in the first term on the right-hand side of Eq. (6.36). As one sees from the structure of Eq. (6.36), this means that a profile of a smooth spatially inhomogeneous disturbance in the system of carriers [a smooth local carrier-density perturbation $\Delta n(\mathbf{r}, t)$] travels in the space at the velocity $\mathbf{V}'$, which thus can be referred to as *differential drift velocity*. Due to the possible complicated manner of the dependence of $\overline{F}_\mathbf{p}$ on N (or n_0) in cases where collisions among carriers play a role, the quantity $\mathbf{V}'$ in general does not coincide with the drift velocity $\mathbf{V}$ of the carrier system as a whole, given by Eq. (6.3).

Equations (6.34) and (6.36) can be iterated, repeatedly inserting Eq. (6.34) into Eq. (6.36) and *vice versa*. As a result, the carrier flux density is expressible as a series of terms containing higher and higher space–time derivatives, each standing side by side with higher and higher powers of the operator $I_{\mathbf{p}}^{-1}$. The order of magnitude of action of this operator on the momentum-dependent function is given by the relaxation time τ, thus the series in fact is an expansion in powers of the ratio of the relaxation time τ to the characteristic time of evolution of the nonuniform system, and of the ratio of the "mean free path" $l \equiv v\tau$ to the characteristic distance.

During the first stage of evolution, these ratios are by no means small. But after the time interval exceeding the relaxation time τ, the fast stage — the relaxation in the quasi-momentum space — practically ends, with the quasi-momentum distribution being forced to take the shape of the steady-state one. The further evolution of the distribution is much slower, provided that the distribution in space is smooth enough from the very beginning — that is, provided that the ratio of the "mean free path" $l \equiv v\tau$ to the characteristic scale of inhomogeneity in the system of carriers is small.

At the second stage of evolution, one should wait for a long enough time (large in comparison with τ) so that the distribution and the corresponding observables change remarkably. In other words, at the second stage of evolution, terms with space–time derivatives in Eq. (6.36) are small enough so that one can retain only the first terms of the expansion, obtaining the approximate expression for the carrier flux density:

$$\Delta j_\alpha^{(e)} = V'_\alpha \Delta n - D_{\alpha\beta} \partial \Delta n / \partial r_\beta, \tag{6.37}$$

where the coefficient in front of the gradient of carrier density — the tensor of diffusivity — is given by Eq. (6.6). For simplicity, from the very beginning [see Eq. (6.15)] we have ignored the electric field perturbation induced by redistribution of charge carriers in space. Correspondingly, in Eq. (6.15) and expression (6.37) the more or less trivial terms describing the action of the electric field perturbation are lacking. For details see references 37 and 70 and Chapter 19.

In conclusion, let us stress once more that the Chapman–Enskog-type solution of the Boltzmann equation describes a state of an electron gas with two well-separated characteristic length scales. One, denoted by l, is the mean-free path of an electron while the other is the length scale L over which the macroscopic ("hydrodynamic") variable $n(\mathbf{r}, t)$ varies. It is assumed in the Chapman–Enskog-type solution that $l \ll L$. It is this separation of microscopic and macroscopic length scales that permits a derivation of the drift–diffusion equation, as well as of the hydrodynamic equations in the theory of the ordinary gas, from the corresponding Boltzmann transport equation.

CHAPTER 7

FEATURES OF HOT-ELECTRON FLUCTUATION SPECTRA

In the previous chapters, the theory of fluctuations near a nonequilibrium state was presented in a rather general form. In this chapter we present results of explicit calculation of the spectral intensity of spatially homogeneous current fluctuations for analytically tractable models which are of special physical interest. This should give the reader an idea of real hot-electron fluctuation spectra and of possibilities that the fluctuation spectroscopy presents. In this chapter, the features of nonequilibrium fluctuation spectra of *bulk* semiconductors will be outlined. Peculiarities of electric noise spectra of two-dimensional structures will be described in Chapters 16 and 17.

In this chapter, we do not take into consideration influence of interelectron collisions. Consequently, the models described here could be directly applicable only to semiconductors with low enough density of free carriers. The effects conditioned by frequent interelectron collisions will be described in Chapter 12.

The formal expressions for the spectral intensities of current fluctuations obtained in Chapter 3 in the framework of the kinetic theory of fluctuations [see Eq. (3.24)] are complete, with their structure being transparent and rather simple. However, their direct use presupposes solution of the inhomogeneous integro-differential equations (calculation of the corresponding Green functions) that is possible analytically and/or numerically in few interesting cases but by no means in a general way. However, a *qualitative analysis* is possible on the basis of these expressions and equations, revealing characteristic features of noise, especially in comparison with response, and enabling interpretation and even prognostication of experimental results. Such a qualitative analysis (a quantitative one when available) is the task of this chapter.

7.1 CURRENT-FLUCTUATION SPECTRUM AT EQUILIBRIUM

In thermal equilibrium state, a spatially homogeneous (or averaged over a volume) current fluctuation in a given direction is proportional to an "instantaneous" value of the corresponding component of fluctuation of drift velocity of the free-carrier system [cf. Eq. (3.4)], with the average value in equilibrium state vanishing. The decay of such a fluctuation is determined by relaxation of quasi-momenta of carriers due to their interaction with the thermal bath. The relaxation is governed by the operator $I_\mathbf{p}^{th}$ [see Eq. (2.2)] and depends on details of interaction. In some cases the action of the operator $I_\mathbf{p}^{th}$ turns into multiplication by a scalar function of the carrier's quasi-momentum, $\tau^{-1}(\mathbf{p})$, or even more, of its energy only, $\tau^{-1}(\varepsilon_\mathbf{p})$. In the latter case, if, moreover, the carrier's energy depends only on the absolute value of the quasi-momentum $\mathbf{p}$, $\varepsilon_\mathbf{p} = \varepsilon(p)$, only the diagonal elements of the tensorial expression (3.39) for the spectral intensity of current fluctuations differ from zero:

$$(\delta j_\alpha \delta j_\beta)_\omega^{eq} = (\delta j_\alpha^2)_\omega^{eq} \delta_{\alpha\beta}, \tag{7.1}$$

where $\delta_{\alpha\beta}$ is the Kroenecker symbol. For a simple parabolic dependence of energy on quasi-momentum, $\varepsilon(p) = p^2/2m$, it follows from Eq. (3.39) that

$$(\delta j_\alpha^2)_\omega^{eq} = \frac{4e^2 n_0 \int_0^\infty \tau(\varepsilon) \varepsilon^{3/2} e^{-\varepsilon/T_0} d\varepsilon/(1 + \omega^2 \tau^2(\varepsilon))}{3m \mathcal{V}_0 \int_0^\infty \varepsilon^{1/2} e^{-\varepsilon/T_0} d\varepsilon}. \tag{7.2}$$

Naturally, the expression on the right-hand side of Eq. (7.2) does not depend on α: In the isotropic medium at equilibrium the intensity of current fluctuations is direction-independent. In the low-frequency limit

$$(\delta j_\alpha^2)_{\omega\tau \ll 1}^{eq} = \frac{4e^2 n_0 \int_0^\infty \tau(\varepsilon) \varepsilon^{3/2} e^{-\varepsilon/T_0} d\varepsilon}{3m \mathcal{V}_0 \int_0^\infty \varepsilon^{1/2} e^{-\varepsilon/T_0} d\varepsilon} \equiv \frac{4e^2 n_0}{3m \mathcal{V}_0} \overline{\varepsilon \tau} \tag{7.3}$$

(here the bar denotes averaging, with distribution function, over energies).

Under the oversimplifying assumption that the relaxation time is energy-independent, $\tau(\varepsilon_\mathbf{p}) = \tau$, expression (7.2) reduces into the Drude-type formula

$$(\delta j_\alpha^2)_\omega^{eq} = \frac{(\delta j_\alpha^2)_{\omega\tau \ll 1}^{eq}}{1 + \omega^2 \tau^2}; \tag{7.4}$$

the frequency dependence of the spectral intensity of current fluctuations has a simple Lorentzian form. In the low-frequency limit we have

$$(\delta j_\alpha^2)_{\omega\tau \ll 1}^{eq} = \frac{2e^2 n_0 T_0 \tau}{m \mathcal{V}_0}. \tag{7.5}$$

We remind the reader that here and above temperature is measured in energy units.

Of course, expressions (7.4) and (7.5) are in complete agreement with the Nyquist theorem discussed in Section 3.4, since the conductivity tensor (3.40) for the model in question takes the form

$$\sigma_{\alpha\alpha}^{eq}(\omega) = \frac{e^2 n_0 \tau}{m(1 - i\omega\tau)}. \tag{7.6}$$

Expression (7.6) is known as the Drude formula.

7.2 CONVECTIVE NOISE

Contribution of Energy Fluctuations into Hot-Electron Noise

Other types of fluctuations taking place in the system of carriers, such as energy fluctuations, fluctuations of free-carrier density due to generation–recombination processes, and so on, in equilibrium cannot manifest themselves through the current fluctuations. Contrary to this, in a current-carrying state, fluctuations of energy, free-carrier density, and so on, can be revealed through the fluctuations of current in the steady-current direction. For example, due to the carrier mobility dependence on the carrier energy ε_p, a fluctuation of the energy of the system of carriers causes a fluctuation of the current in the current-carrying state. In fact, the two types of fluctuations, namely, the fluctuations of the drift-velocity component described in the preceding section and the energy fluctuations, determine the spectrum of current fluctuations in a one-valley current-carrying semiconductor, provided that the scattering of carriers is *quasi-elastic*.

The term "quasi-elastic scattering," or "nearly elastic scattering," is used to refer to a situation where the carrier energy is changed only slightly during a collision, while its quasi-momentum can be changed remarkably. A well-known example is the electron scattering by acoustic phonons at not-too-low lattice temperatures. A fundamental property of the quasi-elastic scattering is a rather slow relaxation of the carrier energy so that two time constants characterize the relaxation of the distribution function as well as that of the current. These are the mean time of quasi-momentum relaxation, τ, and the mean time of energy relaxation, τ^{en}, the latter being large compared to the first:

$$\tau \ll \tau^{en} \tag{7.7}$$

(characteristic values of τ can be of the order of 10^{-12} s while τ^{en} is of the order of $10^{-10} - 10^{-11}$ s). As a result, the drift velocity in a current-carrying state remains small as compared to (possibly, remarkably enhanced in the presence of an external electric field) "chaotic-motion velocity" $\bar{v} = (\overline{v^2})^{1/2}$:

$$V \ll \bar{v}. \tag{7.8}$$

In the case of quasi-elastic scattering, the energy fluctuation possesses an "inertia": It tends to "survive" during the time interval of the order of τ^{en}, while the quasi-momentum fluctuation dies within the time τ. In the current-carrying state, the *energy fluctuation*, through a dependence of the mobility on energy, *is accompanied by the current fluctuation in the direction of the steady current*, decaying in the same way as the energy fluctuation. Indeed, due to dependence of the electron mobility on its energy, the energy variation results in the current variation. As a result, the spectral intensity of *longitudinal* current fluctuations (current fluctuations in the direction of the steady current) contains a term caused by energy fluctuations, also called the convective noise [41, 110]. Frequency dependence of this term, being determined by τ^{en}, differs remarkably from that of the velocity-component-fluctuation term of the type of Eq. (7.4). Under the oversimplifying assumption that the energy (but not quasi-momentum) relaxation time is energy-independent, the energy-fluctuation-caused contribution into the spectral intensity of longitudinal current fluctuations takes a Lorentzian form:

$$\frac{C_\parallel}{1+\omega^2(\tau^{en})^2}. \tag{7.9}$$

The coefficient $C_\parallel$ — the low-frequency limit ($\omega\tau^{en} \ll 1$) of the contribution in question — can be roughly estimated as follows:

$$C_\parallel \sim \frac{e^2 n_0}{V_0}\left(\frac{\partial V}{\partial \bar{\varepsilon}}\right)^2 \tau^{en}\overline{\delta\varepsilon^2}, \tag{7.10}$$

where V is the carrier drift velocity, $\bar{\varepsilon}$ is the mean energy of a carrier, and $\overline{\delta\varepsilon^2}$ is the mean square of the carrier energy fluctuation.

Thus, the spectral intensity of longitudinal current fluctuations in a current-carrying state consists of two terms, each of them having, roughly speaking, a Lorentzian form:

$$(\delta j_\parallel^2)_\omega = \frac{(\delta j^2)_{\omega\tau\ll 1}}{1+\omega^2\tau^2} + \frac{C_\parallel}{1+\omega^2(\tau^{en})^2}, \tag{7.11}$$

where the first term in the case of an isotropic medium does not depend on the direction of measurement — that is, is the same for longitudinal current fluctuations and transverse current fluctuations (fluctuations of current in the direction perpendicular to the steady current). For brevity, sometimes we shall refer to this isotropic contribution as hot-electron "thermal" noise.

Since, as we have already mentioned, the energy-fluctuation-caused term appears only in the steady-current direction, in the transverse direction we see that only the direction-independent term remains, proportional to the spectral intensity of short-lived fluctuations of the corresponding component of drift velocity of the carrier system — that is, the hot-electron "thermal" contribution. In the explicit form we have

$$(\delta j_\perp^2)_\omega = \frac{4e^2 n_0 \int_0^\infty \tau(\varepsilon)\varepsilon^{3/2} F(\varepsilon)\,d\varepsilon/(1+\omega^2\tau^2(\varepsilon))}{3mV_0 \int_0^\infty \varepsilon^{1/2} F(\varepsilon)\,d\varepsilon}, \qquad (7.12)$$

where $F(\varepsilon)$ is the distribution of carrier energy in the current-carrying state for quasi-elastic scattering by thermal bath,

$$F(\varepsilon) \propto \exp\left[-\int_0^\varepsilon \frac{d\varepsilon'}{T_0 + 2e^2 E^2 \tau(\varepsilon')\tau^{\text{en}}(\varepsilon')/3m}\right], \qquad (7.13)$$

referred to as the Davydov distribution [111].

In the low-frequency limit ($\omega\tau \ll 1$) for the direction-independent ("thermal") contribution we have

$$(\delta j^2)_{\omega\tau\ll 1} = (\delta j_\perp^2)_{\omega\tau\ll 1} = \frac{4e^2 n_0 \int_0^\infty \tau(\varepsilon)\varepsilon^{3/2} F(\varepsilon)\,d\varepsilon}{3mV_0 \int_0^\infty \varepsilon^{1/2} F(\varepsilon)\,d\varepsilon}. \qquad (7.14)$$

It can be shown that in the case of quasi-elastic scattering, one has $V^2/\overline{v^2} \sim \tau/\tau^{\text{en}}$, so we can roughly rewrite the order-of-magnitude estimate of the coefficient $C_\parallel$ given by Eq. (7.10) in the following way:

$$C_\parallel \sim \frac{e^2 n_0}{V_0}\overline{v^2}\tau. \qquad (7.15)$$

The low-frequency limit of the energy-fluctuation contribution into the current fluctuations, Eq. (7.15), and that of the "thermal" contribution, Eq. (7.14) [cf. Eq. (7.5)], *are of the same order of magnitude*; consequently, both types of fluctuations should necessarily be taken into account.

Sign of Convective Contribution

Caused by direct exchange of energy between the carriers and the lattice, the contribution to the current fluctuation spectrum estimated by Eqs. (7.9) and (7.10) is, of course, positive. Indeed, this in fact is an excess noise source arising in the current-carrying state, of the same sort as generation–recombination or intervalley noise (see below). However, this is not the only stochastic process changing the carrier energy. Another is an energy gain from the electric field acting on the already existing current fluctuation, $\delta\varepsilon_E \sim \delta\mathbf{j}\cdot\mathbf{E}$. Its contribution to the spectral intensity of the current fluctuations may be either *positive* or *negative*. Moreover, the latter contribution, being of the same order of magnitude as the *direct* energy-fluctuation contribution (7.9) and (7.15), can be shown [42,

110, 112] to exceed it in magnitude. Therefore the sign of the coefficient $C_{\|}$ entering Eq. (7.11) is conditioned by the sign of the contribution caused by the power absorbed from the external supply in the presence of spontaneous current fluctuations. In turn, the latter sign can be shown to correlate with the sign of the derivative of carrier mobility with respect to the carrier energy.

Indeed, in the case of quasi-elastic scattering, the *small-signal AC conductivity* [see Eq. (2.32)] also has the structure of the type of Eq. (7.11):

$$\text{Re}\{\sigma_{\|}(\omega)\} = \frac{\tilde{\sigma}}{1 + (\omega\tau)^2} + \frac{G_{\|}}{1 + (\omega\tau^{\text{en}})^2}, \quad (7.16)$$

where the sign of $G_{\|}$ determines the type of the current–voltage characteristic (sublinear or superlinear). It has been shown [42, 112] that the sign of $C_{\|}$ coincides with the sign of $G_{\|}$. In other words, there exists a correlation between the sign of the convective noise contribution and the type of the current–voltage characteristic of the semiconductor. In semiconductors with a sublinear current–voltage characteristic, the current fluctuations are partially suppressed due to a decrease of the mobility with increasing energy. In this case the fluctuations of the absorbed power *reduce* the total spectral intensity of current fluctuations, as shown in Fig. 7.1.

The electric-field dependence of the intensity of the current fluctuation spectrum is determined by the dependences of the quasi-momentum relaxation time and the energy relaxation time on energy. For example, for acoustic-phonon scattering ($\tau(\varepsilon) \propto \tau^{\text{en}}(\varepsilon) \propto \varepsilon^{-1/2}$) the current–voltage characteristic is sublinear. At sufficiently high electric fields, where

$$E \gg E_0 \equiv \left(\frac{3mT_0}{2e^2\tau\tau^{\text{en}}}\right)^{1/2} \quad (7.17)$$

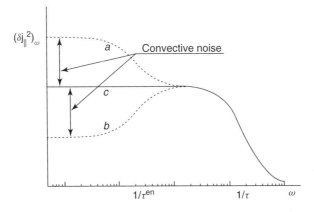

Figure 7.1. Schematic frequency dependence of the spectral intensity of hot-electron current fluctuations in the directions parallel (a,b) and transverse (c) to the steady current in a one-valley semiconductor for nearly elastic electron scattering in the cases of a superlinear (a) and a sublinear (b) current–voltage characteristic.

(*hot-electron* situation proper, $\bar{\varepsilon} = e^2 E^2 \tau \tau^{en}/m \gg 3T_0/2$), one has

$$\sigma_\| |_{\omega \tau^{en} \ll 1} = \tfrac{1}{2} \sigma_\perp |_{\omega \tau \ll 1} \propto E^{-1/2}. \tag{7.18}$$

For that particular type of scattering, the longitudinal current fluctuations are partially suppressed by the energy fluctuations. In the hot-electron situation [110], we have

$$(\delta j_\|^2)_{\omega \tau^{en} \ll 1} = 0.49 (\delta j_\perp^2)_{\omega \tau \ll 1} \propto E^{1/2}. \tag{7.19}$$

So, for the scattering on acoustic phonons the anisotropy of the current fluctuations and that of the AC small-signal conductivity, being well-pronounced, is nearly the same.

Anisotropy of Noise Temperature

Different (in a general case) degree of anisotropy of the spectral intensity of current fluctuations [given by Eq. (7.11)] and of AC small-signal conductivity [given by Eq. (7.16)] leads to anisotropic character of the noise temperature as defined in Section 1.5:

$$T_{n,\alpha}(\omega, \mathbf{E}) = \frac{\mathcal{V}_0 (\delta j_\alpha^2)_\omega}{2 \operatorname{Re} \sigma_{\alpha\alpha}(\omega)}. \tag{7.20}$$

On the other hand, in the special case of scattering on acoustic phonons, in the region of electric fields defined by Eq. (7.17), as it follows from Eqs. (7.18) and (7.19), the noise temperature (increasing *linearly* with increasing field) is nearly isotropic *at all frequencies*:

$$T_{n\|}(\omega) \approx T_{n\perp}(\omega) \propto E. \tag{7.21}$$

One should note that the near-equality of hot-carrier noise temperatures $T_{n\|}$ and $T_{n\perp}$ in the case of acoustic-phonon scattering is more readily an exception, being conditioned by the identical dependence of the momentum- and energy-relaxation times on energy. When $\tau(\varepsilon) \propto \varepsilon^s$ and $\tau^{en}(\varepsilon) \propto \varepsilon^t$ but $s \neq t$, the anisotropy of the noise temperature at frequencies $\omega \tau^{en} \lesssim 1$ is well-pronounced [113].

In any case, the noise temperature, the spectral intensity of current fluctuations, and the AC small-signal conductivity become isotropic at frequencies $\omega \tau^{en} \gg 1$, where $(\delta j^2)_\omega$ and $\sigma(\omega)$ have a nearly Lorentzian shape of the type $(1 + (\omega \tau)^2)^{-1}$, while T_n is independent of frequency up to very high (quantum) frequencies.

More details concerning the convective noise caused by energy fluctuations in the case of quasi-elastic scattering are presented in references 70 and 113. The convective noise is a good example illustrating how the mechanisms of relaxation reflect themselves in the hot-electron noise spectrum pattern. The current, or voltage, fluctuation spectral intensities prove to be quite sensitive to the details of the scattering mechanisms, with the features being revealed by the biasing current. The spectrum is "rich": The noise in the microwave frequency region is "colored" (not "white").

Case of Isotropic Spectrum

The convective contribution into hot-electron noise is absent in a very special case of energy-independent quasi-momentum relaxation time, $\tau(\varepsilon) \equiv \tau$, ensuring a linear current–voltage characteristic of a sample. Then the second term in Eq. (7.16) vanishes ($G_\| = 0$), and

$$\text{Re}\{\sigma_\|(\omega)\} = \frac{e^2 n_0 \tau}{m(1 + \omega^2 \tau^2)}. \tag{7.22}$$

The differential and chord conductivities coincide:

$$\sigma_\| \equiv \sigma_\|(\omega)|_{\omega\tau \ll 1} = \sigma_\perp(\omega)|_{\omega\tau \ll 1} = \bar{j}/E.$$

Longitudinal and transverse spectral intensities of current fluctuations also coincide [in Eq. (7.11), $C_\| = 0$, since in Eq. (7.10) $\partial V/\partial \bar{\varepsilon}$ vanishes]. The isotropic spectral intensity of current fluctuations is given by the expression

$$(\delta j_{\perp,\|}^2)_\omega = \frac{4e^2 n_0 \tau \int_0^\infty \varepsilon^{3/2} F(\varepsilon)\,d\varepsilon}{3m\mathcal{V}_0(1 + \omega^2\tau^2) \int_0^\infty \varepsilon^{1/2} F(\varepsilon)\,d\varepsilon}, \tag{7.23}$$

where the energy distribution function now is

$$F(\varepsilon) \propto \exp\left[-\int_0^\varepsilon \frac{d\varepsilon'}{T_0 + 2e^2 E^2 \tau \tau^{\text{en}}(\varepsilon')/3m}\right], \tag{7.24}$$

Correspondingly, the noise temperature (7.20) also becomes isotropic and, moreover, frequency-independent in all the region of classical frequencies:

$$T_{n\perp,\|} = \frac{2}{3} \frac{\int_0^\infty \varepsilon^{3/2} F(\varepsilon)\,d\varepsilon}{\int_0^\infty \varepsilon^{1/2} F(\varepsilon)\,d\varepsilon}. \tag{7.25}$$

In this case, the noise temperature directly gives the degree of electron system heating, while the current–voltage characteristic is unsensitive to electron heating.

7.3 FLUCTUATION PECULIARITIES UNDER OPTICAL-PHONON EMISSION

The scattering of carriers is nearly elastic only in special cases. For example, in semiconductors with comparatively strong coupling of free carriers with optical phonons (such as p-Ge, n-InSb) at low temperatures $T_0 \ll \hbar\omega^{\text{opt}}$, where $\hbar\omega^{\text{opt}}$ is the optical-phonon energy, quasi-elastic scattering by acoustic phonons and impurities prevails at all essential carrier energies only in the range of fields

$$E < E_c \equiv \left(\frac{2m\hbar\omega^{\text{opt}}}{e^2 \tau \tau^{\text{en}}}\right)^{1/2}. \tag{7.26}$$

In the range of fields $E < E_c$, optical-phonon emission is relatively rare and its contribution to the carrier energy balance is small enough so that the situation investigated in Section 7.2 occurs.

At higher fields,

$$E > E_c \equiv \left(\frac{2m\hbar\omega^{\text{opt}}}{e^2 \tau \tau^{\text{en}}}\right)^{1/2}, \tag{7.27}$$

the contribution of emission of optical phonons by heated carriers becomes essential. Let us consider the case where the constant of coupling of carriers with optical phonons is large enough for the probability of optical-phonon emission to be higher than any other scattering probability in the energy range $\varepsilon > \hbar\omega^{\text{opt}}$. Moreover, we suppose that the carrier having reached the energy $\varepsilon = \hbar\omega^{\text{opt}}$ will emit an optical phonon almost immediately and then occur near the state $\varepsilon \approx 0$. So, in the "active" region $\varepsilon > \hbar\omega^{\text{opt}}$ the carriers are practically absent. In the "passive" region ($\varepsilon < \hbar\omega^{\text{opt}}$), the elastic impurity scattering and quasi-elastic acoustic-phonon scattering are important, while the number of thermally available optical phonons is low, so that the optical-phonon absorption probability is negligible. This is the Rabinovich [114] model. In this model, the distribution function is almost isotropic in the passive energy region ($\varepsilon < \hbar\omega^{\text{opt}}$) and is near zero in the active region $\varepsilon > \hbar\omega^{\text{opt}}$.

Intensive optical-phonon emission by hot carriers in the field region $E > E_c$ results in important and interesting new features of the noise spectrum at a low lattice temperature: saturated and even reduced spectral intensity of current fluctuations with increasing electric field E, and so on (see reference 115). The current–voltage characteristic, the spectral behavior of current fluctuations and noise temperature are different in pure and doped semiconductors. Let us discuss these two cases separately.

Pure Semiconductors

In pure semiconductors, the first ohmic region (low fields, $E < E_0 \equiv (2mT_0/e^2\tau\tau^{\text{en}})^{1/2}$) is followed by the nonlinear region ($E_0 < E < E_c$) described in Section 7.2, and then by the *second ohmic region* ($E > E_c$) caused by the optical-phonon emission and carrier transitions from the active to the passive region, where only acoustic-phonon scattering is important. The AC small-signal conductivities, $\sigma_{\|,\perp}(\omega)$, start with the first ohmic value σ^{eq}, behave as $E^{-1/2}$ in the field region $E_0 \ll E < E_c$, and reach the second ohmic value $\frac{9}{8}(\pi T_0/\hbar\omega^{\text{opt}})^{1/2}\sigma^{\text{eq}}$ in the field region $E > E_c$. The latter region is the subject of our interest at the moment [115].

The spectral intensity of current fluctuations increases as $E^{1/2}$ in the hot-carrier region $E_0 \ll E < E_c$ (Section 7.2) and saturates in the region $E > E_c$, becoming isotropic (the convective contribution to the current fluctuations disappears). Both $(\delta j_{\|}^2)_{\omega\tau^{\text{en}}\ll 1}$ and $(\delta j_{\perp}^2)_{\omega\tau\ll 1}$ reach the value $\frac{9}{32}(\pi\hbar\omega^{\text{opt}}/T_0)^{1/2}(\delta j^2)_{\omega\tau\ll 1}^{\text{eq}}$. Correspondingly, the noise temperature increases as a linear function of E in the region $E_0 \ll E < E_c$ and saturates in the region $E > E_c$. The saturation value is isotropic, $T_{\text{n}\|,\perp} = \hbar\omega^{\text{opt}}/4$, while the calculated mean energy of the carrier is $\bar{\varepsilon} = 9\hbar\omega^{\text{opt}}/25$ [116]. In this case the relation $\bar{\varepsilon} = 3T_{\text{n}}/2$ is almost true.

In the high-frequency ($\omega\tau \gg 1$) region, there is an ω^{-2} behavior of $(\delta j^2)_{\omega}$ as well as of $\sigma(\omega)$, with the noise temperature being frequency independent: $T_{\text{n}} = 2\hbar\omega^{\text{opt}}/9$.

Doped Semiconductors

The case of doped (compensated) semiconductors when carrier quasi-momentum scattering in the passive region is controlled by ionized impurities was also studied in reference 115. First of all, in contrast to pure materials, the current–voltage characteristic in the field region $E_0 \ll E < E_c$ in this case is superlinear, so that, in the low-frequency case, $\sigma_{\|} > \sigma_{\perp}$. Correspondingly, $(\delta j_{\|}^2) > (\delta j_{\perp}^2)$. At higher fields $E > E_c$, both $(\delta j_{\|}^2)_{\omega\tau^{\text{en}}\ll 1}$ and $(\delta j_{\perp}^2)_{\omega\tau\ll 1}$ are nonmonotonic functions of the electric field. Each of them passes through a maximum and then decreases as E^{-1} with increasing E. In the low-frequency limit there is an essential anisotropy in the spectral intensity of current fluctuations: for $E_c < E$,

$$(\delta j_{\|}^2)_{\omega\tau^{\text{en}}\ll 1} = K(\delta j_{\perp}^2)_{\omega\tau\ll 1}, \qquad K = 8\left(\frac{2}{3} - \frac{1}{\pi^2}\right) \approx 4.5. \qquad (7.28)$$

The decrease of the spectral intensity of current fluctuations with increasing field strength is conditioned by carrier accumulation at low energies due to the pumping of carriers from the active region to the vicinity of the state $\varepsilon = 0$ and comparatively intensive ionized-impurity scattering of "cool" electrons.

The transverse small-signal AC conductivity also passes through a maximum, and the transverse noise temperature $T_{\text{n}\perp}$ at $E > E_c$ is independent of field and sufficiently small: $T_{\text{n}\perp} = \hbar\omega^{\text{opt}}/8$. The mean carrier energy $\bar{\varepsilon}$ in this "cooling" regime decreases as E^{-1} with increasing electric field strength; $\bar{\varepsilon}$ and $T_{\text{n}\perp}$ have different functional dependencies on E. Notice that here, as well as in pure semiconductors, the saturated value of $T_{\text{n}\perp}$ does not depend on the lattice temperature T_0. In the case of a doped semiconductor, under condition $\hbar\omega^{\text{opt}} \gg T_0 > \hbar\omega^{\text{opt}}/8$, the transverse noise temperature $T_{\text{n}\perp}$ even may be less than T_0.

In this cooling region, the current–voltage characteristic tends to saturate and the longitudinal DC differential conductivity $\sigma_\parallel$ is small and can even be negative, so the behavior of the low-frequency longitudinal noise temperature $T_{n\parallel}$ depends strongly on the detailed assumptions about the character of carrier penetration into the active region.

In any case, $(\delta j_\parallel^2)_\omega$ approaches $(\delta j_\perp^2)_\omega$ and $\sigma_\parallel(\omega)$ approaches $\sigma_\perp(\omega)$ when ω increases in the range of frequencies $\omega > (\tau^{en})^{-1}$. It means that in the frequency interval $(\tau^{en})^{-1} \ll \omega \ll \tau^{-1}$ the isotropic frequency-independent and low-noise case is realized.

At high frequencies $\omega\tau \gg 1$ there is a nontrivial frequency dependence of $(\delta j^2)_\omega$, $\sigma(\omega)$, and $T_n(\omega)$; that is, the decrease of isotropic $(\delta j^2)_\omega$ and $\sigma(\omega)$ for $\omega\tau \gg 1$ does not obey the law ω^{-2}:

$$(\delta j^2)_\omega \propto E\omega^{-2}\ln(\omega^2/E^2), \qquad \sigma(\omega) \propto E\omega^{-4/3} \qquad (7.29)$$

and hence

$$T_n(\omega) \propto \omega^{-2/3}\ln(\omega^2/E^2). \qquad (7.30)$$

A saturation of T_n is not reached with increase of frequency ω, as in the ordinary case, but T_n becomes essentially less than the lattice temperature T_0. These dependencies are obtained by integration of the Lorentzian-shape contributions $(1 + \omega^2\tau^2(\varepsilon))^{-1}$ over energy ε, with the integral being dominated by the small values of ε due to the dependence $\tau(\varepsilon) \propto \varepsilon^{3/2}$.

7.4 FLUCTUATIONS IN STREAMING-MOTION REGIME

In the framework of the Rabinovich model described in the previous section, carriers perform a drift–diffusion type of motion in *energy* space toward the passive-region boundary $\varepsilon = \hbar\omega^{opt}$ and, after reaching it, almost immediately return to the point $\varepsilon \approx 0$. The model works, provided that the carrier suffers many nearly elastic collisions before reaching the passive-region boundary.

The character of carrier motion changes at fields $E \sim E^- \equiv p_0/e\tau$, where $p_0 \equiv (2m\hbar\omega^{opt})^{1/2}$ is the value of the carrier quasi-momentum at the boundary of the passive region. At $E > E^-$, the scattering in the passive region becomes ineffective: Being accelerated by the electric field, the carrier passes the passive region rapidly enough to have a good chance of avoiding scattering up to the passive-region boundary $\varepsilon = \hbar\omega^{opt}$. On the other hand, the carrier penetration into the active region $\varepsilon > \hbar\omega^{opt}$ remains negligibly small up to much higher fields $E^+ \gg E^-$, depending on the characteristic time of the optical-phonon emission. As a result, in the field region between E^- and E^+ the motion of a carrier consists of nearly "ballistic" passages: The carrier, having reached the energy $\varepsilon = \hbar\omega^{opt}$, almost immediately emits an optical phonon, occurs near the state $\varepsilon = 0$, is accelerated again, reaches the passive-region boundary $\varepsilon = \hbar\omega^{opt}$ in the "time of flight" $\tau_E = p_0/eE$, and repeats thereafter the same streaming motion almost without scattering in the passive region. The frequency of the cyclic motion is $\omega_E = 2\pi/\tau_E = 2\pi eE/(2m\hbar\omega^{\text{opt}})^{1/2}$ (~35 GHz at 60 V/cm in n-type GaAs). Moreover, even if scattered in the passive region, the carrier, having reached the boundary and emitted an optical phonon, returns to $\varepsilon \approx 0$. As a result, at fields $E^- < E < E^+$ almost all the electrons are concentrated on the "main" trajectory, with the background of the other trajectories being rather low (Fig. 7.2). The carrier distribution function is "needle-shaped." The net effect of the external field is the ordering of the carrier motion randomly spread at equilibrium [117–119] (see also references 120–122).

The presence, in quasi-momentum space, of a region with a pronounced dynamics of the carrier motion gives rise to a number of interesting phenomena. In high-frequency (microwave) characteristics the resonance phenomena are to take place. Actually, the spectral intensity of current fluctuations was shown [123, 124] to have peaks at frequencies that are multiples of the "flight" frequency ω_E. Indeed, the expressions (3.24) in the zeroth-order approximation with respect to the small parameters E^-/E and E/E^+—that is, neglecting both the scattering in the passive region and the penetration into the active region—yield the following expressions:

$$(\delta j_\perp^2)_\omega = 0, \tag{7.31}$$

$$(\delta j_\parallel^2)_\omega = (e^2 n_0/\mathcal{V}_0) \sum_{l=1}^{\infty} |v(l)|^2 [\delta(\omega - l\omega_E) + \delta(\omega + l\omega_E)], \tag{7.32}$$

where $v(l) \propto l^{-1}$ is the lth Fourier component of the carrier velocity in its periodic motion. In other words, the carrier executing the cyclic motion in the passive region is "noisy" at the "flight" frequency and its multiples. The current fluctuation intensity of the harmonic decreases with the number of harmonic as l^{-2} (for a parabolic band).

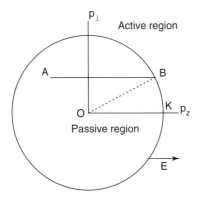

Figure 7.2. Streaming motion of carriers in an electric field **E** in quasi-momentum space: OK is the main trajectory; AB is an untypical trajectory. Optical phonon emission is not possible while the electron is in the passive region.

The scattering in the passive region and the carrier penetration into the active region cause the broadening of the resonance peaks. The shape of the peaks is determined by a competition between these two weak processes. Having taken into account merely the broadening caused by the scattering in the passive region, the half-width of each peak is, of course, of the order of $1/\tau$. The peaks are well-pronounced, provided that the distance between them, $\omega_E = 2\pi eE/(2m\hbar\omega^{\text{opt}})^{1/2}$, is large compared with the half-width $1/\tau$ — that is, provided that $E > E^-$. Scattering in the passive region leads also to appearance of the weak background in the spectrum of the current fluctuations. The background is due to the non-vanishing carrier distribution function in the passive region outside the main trajectory.

The phase of the "carrier density fluctuation wave" in quasi-momentum space, characterized by the wavelength p_0/l, can be destroyed also due to the carrier penetration into the active region. The penetration is estimated as $p_0(E/E^+)^{2/3}$. As a result, the half-width of the lth peak is estimated as

$$\Gamma_l \approx 0.6\omega_E l^2 (E/E^+)^{2/3}; \tag{7.33}$$

that is, broadening of the peak strongly depends on its number [124].

In contrast to the current fluctuation spectral intensity $(\delta j_\parallel^2)_\omega$, the longitudinal AC small-signal conductivity $\sigma_\parallel(\omega)$ calculated in the zeroth-order approximation with respect to the small parameters E^-/E and E/E^+ (i.e., neglecting both the scattering in the passive region and the penetration into the active region) is equal to zero. Indeed, in zeroth-order approximation with respect to weak mechanisms (scattering and penetration) the average current in the streaming model is constant, so that the longitudinal small-signal conductivity arises only when weak mechanisms are taken into account. As was shown [125–127], the longitudinal AC small-signal conductivity differs from zero essentially only at frequencies that are multiples of the "flight" frequency ω_E. The weak alternating electric field of the frequency close to one of the resonance frequencies is able, in the presence of the weak scattering in the passive region and/or the weak penetration into the active region, to group, or to bunch, in some peculiar way, carriers moving along the main trajectory. Again, the shape of the resonance line is determined by competition between scattering in the passive region and penetration into the active region.

As a result, the spectrum of the longitudinal current fluctuations and the frequency dependence of the longitudinal AC small-signal conductivity prove to be similar. Both quantities differ from zero substantially only near the resonances, where both have Lorentzian shapes, with the Lorentz denominators being the same in both cases. Therefore, in the vicinity of each resonance the noise temperature, $T_{n\parallel}^{(l)}$, is proportional to either $\hbar\omega^{\text{opt}}(E^+/E)^{2/3}$ or $\hbar\omega^{\text{opt}}E/E^-$; that is, it is large. It should be noted that large values of the noise temperature in the vicinity of resonance does not mean large values of the carrier average energy which cannot exceed $\hbar\omega^{\text{opt}}$ in the streaming model.

7.5 INTERVALLEY NOISE

It is well known [41] that the differential mobility of a *many-valley* semiconductor is not equal to the sum of the corresponding components of the differential mobilities in the individual valleys weighted by the relative numbers of electrons in them. In general the overall differential mobility is not given by this sum over the valleys even when *intervalley scattering* is not strong enough (compared with *intravalley scattering*) to have an appreciable influence on the electron distribution *in* the valley. Indeed, the applied electric field perturbs the intervalley transition rates, provided that the probability for a carrier to make a transition to some other valley depends on its position in the valley — that is, on its energy. Hence the electric field perturbs the *relative numbers* of carriers in the valleys. Consequently, there is a contribution to the differential conductivity ascribed to the modulation of the relative numbers.

The noise in a many-valley semiconductor in a current-carrying state also is not equal to the sum of the contributions from the individual valleys. Due to intervalley transitions, fluctuations of the numbers of carriers in the valleys take place. Provided that the components, in a given direction, of the drift velocities in the different valleys differ, the carrier number fluctuation is accompanied by a current fluctuation. The "intervalley noise" [128] arising in such a way can even exceed the intravalley noise, with the former manifesting itself as a main source of an excess noise in many-valley semiconductors.

We introduce the *intervalley relaxation time* $\tau_{j' \leftarrow j}^{\text{inter}}$ and assume it to be large compared with the characteristic *intravalley relaxation time* $\tau_{<j>}^{\text{intra}}$ (parentheses around the subscript or superscript will indicate that the quantity to which it is attached applies to the hypothetical system in which there is no intervalley scattering but which is otherwise unchanged from the actual one). We consider the case typified by the conduction bands of germanium and silicon, which have a number of so-called *equivalent* valleys (or equivalent pairs of valleys). In that case, $\tau_{j' \leftarrow j}$ depends only on j but not on j' (for the Boltzmann statistics):

$$\tau_{j' \leftarrow j}^{\text{inter}} \equiv \tau_j^{\text{inter}} \quad (7.34)$$

(for details, see, e.g., Section 3.5 in reference 75). Then the spectral intensity of low-frequency ($\omega \tau_j^{\text{inter}} \ll 1$) current fluctuations is given by the following expression:

$$(\delta j_\alpha \delta j_\beta)_\omega = \sum_j [(\delta j_\alpha \delta j_\beta)_{\omega \tau_{<j>}^{\text{intra}} \ll 1}^{(j)} + (2e^2/\mathcal{V}_0) n_j \Delta V_\alpha^{(j)} \Delta V_\beta^{(j)} \tau_j^{\text{inter}}], \quad (7.35)$$

where $(\delta j_\alpha \delta j_\beta)^{(j)}$ is the spectral intensity of the "intravalley" current fluctuations in the jth valley, $\mathbf{V}^{(j)}$ is the drift velocity of carriers in the jth valley,

$$\Delta \mathbf{V}^{(j)} = \mathbf{V}^{(j)} - \mathbf{V}, \quad \mathbf{V} = \sum_j \mathbf{V}^{(j)} n_j / n_0, \quad (7.36)$$

and n_j is the mean density of carriers in the jth valley.

One can see that the ratio of the intensities of the intervalley noise to the intravalley one is of the order of

$$\Delta V^2 \tau^{\text{inter}} / \overline{v^2} \tau^{\text{intra}}. \qquad (7.37)$$

While obtaining Eq. (7.37), the inequality

$$\tau^{\text{inter}} \gg \tau^{\text{intra}} \qquad (7.38)$$

was used. Estimate (7.37) shows that, if the components of the drift velocities in different valleys differ substantially, the intervalley noise can prevail. In the case of equivalent valleys the intervalley noise is quite sensitive to the geometry of the experiment (i.e., to the crystallographic orientation of the applied DC field and the direction of current fluctuation measurement).

7.6 HISTORICAL COMMENTS ON THE THEORY OF FLUCTUATIONS

Let us insert here a short survey of the history of modern theory of fluctuations. The prominent achievements until the middle of the twentieth century were, as we have already noted: (i) the Einstein relation (1905; see reference 64) between (a) the diffusion coefficient determining the rate at which a gradient of the charge carrier density in a homogeneous medium dissolves and (b) the mobility of carriers; and (ii) the Nyquist relation (1928; see reference 3 and, e.g., references 6, 68 and 69) between the spectral intensity of current fluctuations in a conductor and the real part of its conductivity, as well as a generalization of the Nyquist theorem achieved by Callen and Welton (1951; see reference 67 and also, e.g., references 6 and 68) and referred to as the fluctuation–dissipation theorem. According to these theorems, the spectral intensity of fluctuations in a physical system at a given frequency is expressible in terms of the dissipative part of the response of the system to some external perturbation. Accordingly, calculation, or measurement, of the linear response of the system at a given frequency provides the knowledge of the spectral intensity of fluctuations of the corresponding variables at the same frequency.

These fundamental relations have been established for systems in thermal equilibrium. Since, in the second half of the twentieth century, investigation and utilization of nonequilibrium states became very important in physics and electronics, a fundamental problem of investigation of fluctuations in nonequilibrium systems arose. Progress in the field was stimulated by the development of physics of nonequilibrium phenomena in semiconductors. Experimental as well as most of theoretical results in the field concern just the noise of nonequilibrium carriers in semiconductors, where the nonequilibrium distribution of charge carriers can be realized with comparative ease under the action of an external electric field.

The solution of the problem of investigation of fluctuations in nonequilibrium systems implied, first of all, answers to the following questions: (i) Do the fluctuation–dissipation relations remain valid in a nonequilibrium state and, if not, what is the recipe of calculation of the spectral intensity of fluctuations in a conductor characterized by the nonequilibrium distribution of charge carriers? (ii) Does the Einstein relation still work in a nonequilibrium state (of course, provided that the diffusion coefficient can be introduced to describe spatially inhomogeneous transport near the far-from-equilibrium spatially homogeneous state)? (iii) What, if any, new features of sources of fluctuations reveal themselves in a nonequilibrium state as compared with thermal equilibrium? In Chapters 3–7 we presented the answers to these questions. Now, let us enumerate the main stages of the historical development of the theory of fluctuations in nonequilibrium states.

Importance of the problem of nonequilibrium fluctuations as a special problem of physical kinetics has been realized by Leontovich [80] as early as in 1935. In 1956–1957, Hashitsume [76] and Kadomtsev [39] have introduced a concept of fluctuations of the distribution function of gas particles in momentum space. In 1960, Lax [40] aimed at constructing the theory of fluctuations in the nonequilibrium state by using a minimal number of assumptions additional to those needed for the adequate description of the state itself. He demonstrated that it is possible for a class of systems having the Markoffian property. The quantitative theory of hot-electron fluctuations in semiconductors for the simple model of quasi-elastic scattering of electrons by phonons was developed in 1965 by Price [41] and Gurevich and Katilius [110] (an important conclusion that the spectral intensity of longitudinal current fluctuations in the high electric field is frequency-dependent at frequencies of the order of inverse energy relaxation time was made by Gurevich [129] and Kogan [130] even earlier). In 1967, Kogan and Shulman [131] at the cost of some specific conjecture concerning correlation properties of the Langevin sources in the balance equations were able to calculate the spectrum of hot-electron fluctuations in the case of frequent electron–electron collisions shaping the energy distribution (the effective electron temperature case).

The results obtained in reference 131 were shown to contradict what followed from the Price fluctuation–diffusion relation. At that time it seemed that the simple connection noted by Price between the diffusion coefficient and fluctuations is none other than a fundamental property taking in the nonequilibrium state the place of the Nyquist theorem. Hence, the contradiction between the fluctuation–diffusion relation and the result of the explicit calculation performed in reference 131 was taken as a challenge pointing out the necessity of a deeper approach to the problem. A development of a self-consistent kinetic theory of fluctuations in a nonequilibrium state based on the first principles of quantum mechanics and statistical physics was called for.

Such a theory was developed and published in 1969 by Gantsevich, Gurevich, and Katilius [37, 79]. In references 37 and 79, the diagram technique was adopted for investigation of the correlation functions of fluctuations in a nonequilibrium electron gas, taking into account electron–electron collisions. The equations for the correlation functions (sometimes referred to as equations of fluctuation kinetics) were obtained without any assumptions other than those needed for validity of the Boltzmann equation itself. At the same time, Kogan and Shulman [42] developed the Langevin treatment of fluctuations of occupation numbers in a nonequilibrium gas with two-particle collisions based only on the statistical properties of collisions indispensable for the kinetic description of the gas. In references 83 and 84 the equivalence of the results obtained by diagram and Langevin methods was demonstrated. Thus the main equations of kinetic theory of fluctuations were established.

From these equations, important physical consequences were immediately obtained. The new physical phenomenon—the existence, in a nonequilibrium state, of the interparticle correlation appearing due to collisions between particles—was first reported in reference 79. Shulman [132] was the first to calculate the contribution of the interelectron correlation to the spectrum of current fluctuations. He demonstrated that the contribution of the interelectron correlation at high electric fields can be appreciable. The above-mentioned contradiction between the results of reference 131 and the relation proposed by Price was explained: The limits of validity of the Price relation were established. It was shown that the expressions for the spectral intensities of current fluctuations obtained by Kogan and Shulman [131] had in fact even a wider region of applicability than the authors claimed.

Thus, it was shown that, due to the existence of interparticle correlation, the problem of calculation of fluctuation characteristics of the nonequilibrium systems differ in essence from the problem of calculation of the reaction to an external perturbation. It was proved that, in a sense, a gas in which interparticle collisions are essential is less ideal in a nonequilibrium state than in thermal equilibrium. This circumstance shed light on the very nature of the thermal equilibrium state of a gas as one in which, contrary to a nonequilibrium state, occupation numbers of one-particle states are uncorrelated. If a nonequilibrium system is allowed to relax to equilibrium (external forces are switched off), the correlation decays as the system evolves toward the equilibrium, and it vanishes when the equilibrium is reached. One can say that a tendency of minimization of interparticle correlations in the gas matches the tendency to evolve toward the equilibrium.

The appearance of the self-consistent theory of fluctuations in a nonequilibrium state immediately drew the attention of those working in the field of physical kinetics, provoking further investigation (the references are given in Chapters 4 and 19). Much attention was attached to the most nontrivial result of the theory—that is, to the occurrence of correlation in a nonequilibrium gas with pair collisions. This result was confirmed, reobtaining it by various methods [38, 71, 72, 86, 89–92, 95, 133, 134]. Moreover, it was shown [95] that the nonequilibrium interparticle correlation forms the basis of a *nonexponential decay of response* which is usually referred to as a *long-time tail*, and of the *logarithmic terms* in the density expansion of transport coefficients.

The development of the kinetic theory laid down solid theoretical foundations for further theoretical investigations of hot-electron noise in semiconductors. Spectra of longitudinal and transverse current fluctuations were calculated for the needle-shaped distribution [124], for the Davydov distributions [113], and for the carrier energy scattering by optical phonons [115]. Noise spectra in the case of frequent electron–electron collisions were investigated in detail, paying special attention to the possibility of violation of the Price relation in semiconductors [135, 136].

By using the same methods, fluctuations in the electron and magnon systems of a *ferromagnetic semiconductor* in high electric fields were investigated [137], and the expressions for the correlation functions of the magnon temperature and the longitudinal magnetic moment fluctuations were obtained.

On the basis of the kinetic theory of fluctuations, the theory of collision-controlled electromagnetic wave scattering by nonequilibrium charge carriers in semiconductors, including many-valley ones, was developed [37, 63, 138–144] (see Chapter 19). It exerted a noticeable influence upon the development of experimental investigation of light scattering from solid-state plasma (see references 145–147). The contemporary theory of light scattering from nonequilibrium gas of neutral molecules is also based on the equations of fluctuation kinetics (see reference 148 and references therein). The applications of the theory of fluctuations in nonequilibrium state to fluctuation phenomena in plasma were reviewed in reference 149.

"Quasi-hydrodynamic" equations for comparatively large-scale and low-frequency fluctuations, valid in the case of frequent electron–electron collisions (the effective electron temperature case), were derived containing "quasi-hydrodynamic" Langevin random forces [150]). The investigation of peculiarities of spectra of electron-density and electron-temperature fluctuations in semiconductors, as well as of their cross-correlations, was performed on the basis of these equations [143, 144].

The kinetic equations with fluctuations including generation-recombination processes were derived from the first principles [87]. It was demonstrated how, in the limit of frequent electron scattering in the conduction band, the fluctuational drift-diffusion equations with generation-recombination terms emerge from the kinetic description.

This progress in the theoretical research made it possible to seek and find concrete answers to the pressing questions put by the accelerating experimental investigation of nonequilibrium fluctuation phenomena in semiconductors (e.g., what degree of anisotropy of noise the experimentalist should expect while applying the electric field to the semiconductor, and what are the possible reasons for such an anisotropy; what is the influence on noise of variation of free carrier concentration; up to what carrier concentration the diffusion coefficient of hot carriers can be safely evaluated from noise measurements, i.e., at what concentration the Price relation is expected to be violated in practice; what is the influence of the departure from equilibrium on space-dependent carrier-density fluctuations in solid-state plasma, etc.). A new field of research — physics of fluctuations from a nonequilibrium state of charge carriers — appeared in close interaction between theoretical and experimental studies of solids. The theory helped experimentalists to realize that physical quantities tightly interrelated at thermal equilibrium — namely, the carrier mobility, the carrier diffusion coefficient, and noise — become relatively independent under nonequilibrium conditions. Independent measurements of each of them was understood as a fundamental problem important for the diagnostics of non-equilibrium solid-state plasma. This stimulated the development of adequate Monte Carlo methods for detailed computation sufficient to interpret the available experimental data on noise. The close contact of analytic investigations, Monte Carlo simulation, and experimental work was fruitful for interpretation of experimental data. It helped to determine the electron–phonon coupling and other constants from noise measurements in the cases where these constants were known with great uncertainty or were unknown. In particular, these studies were useful for determination of field-dependent longitudinal and transverse carrier diffusion coefficients, energy and intervalley relaxation time constants, and carrier scattering mechanisms in the most popular semiconductors. Information obtained from noise experiments proved to be nontrivial; the "noise spectroscopy" became a powerful tool for diagnostics of nonequilibrium solid-state plasma.

The methods and main results of experimental research of nonequilibrium noise in semiconductors will be described in detail in the subsequent chapters.

CHAPTER 8

EXPERIMENTAL TECHNIQUES

Failure of the universal fluctuation–dissipation relation at nonequilibrium conditions and practical needs stimulate experimental investigation of hot-electron noise in semiconductors and semiconductor structures. The aim of this chapter is to give basic and up-to-date information about the tools and methods needed for such an investigation. Our main concern is noise due to velocity fluctuations caused by chaotic motion of mobile electrons; this chaotic motion is enhanced by an applied high electric field. The sources of hot-electron noise are associated with the kinetic processes taking place inside the conduction band; the corresponding relaxation times are in the picosecond and subpicosecond range (see Chapter 7). Therefore, the velocity-fluctuation spectrum extends to microwave frequencies; this frequency range is suited for experimental investigation of hot-electron fluctuations. At microwave frequencies, other sources of noise, such as $1/f$ fluctuations and generation–recombination noise, do not mask the hot-electron noise. Microwave techniques for hot-electron noise measurements in semiconductors subjected to high electric fields are presented in this chapter.

Noise spectroscopy, unlike the usual optical spectroscopy, deals with relaxation — that is, aperiodic processes. Different electronic processes cause steps in noise intensity as a function of frequency near the corresponding cutoff frequencies. Each step in the spectrum has a simple Lorentzian form, provided that the decay of fluctuations is exponential. The superposition of noise sources, the cross-correlation effects, possible existence of the processes leading to damped oscillations, and so on, make the noise spectra more complicated. Investigations of noise characteristics as a function of frequency, of current, of lattice temperature, and so on, help to resolve different sources of hot-electron noise. In order to extract information on a particular electronic process and the associated noise source, the measured noise spectra are analyzed using the noise theory outlined in Chapters 3–7.

8.1 EARLY EXPERIMENTS ON HOT-ELECTRON NOISE

Experimental research of hot-electron noise in semiconductors was initiated by Erlbach and Gunn in 1962 [151]. They measured transverse noise temperature in a semiconductor (n-type Ge) at 420-MHz frequency and demonstrated that the applied electric field enhanced the kinetic motion of mobile electrons in the direction transverse to the steady DC current. Similar results were obtained for hot electrons in n-type GaAs subjected to a high electric field [152].

Systematic studies of hot-electron fluctuations became possible in 1966 when the X-band waveguide-type pulsed technique for hot-electron noise measurements was developed [153]. Avoiding possible contribution of generation–recombination noise, Bareikis, Vaitkevičiūtė, and Požela [154] measured the longitudinal noise temperature of hot electrons and hot holes in n-type and p-type germanium at 10-GHz frequency. The results on the longitudinal noise temperature, together with those on the transverse noise temperature [151], gave the first experimental evidence for the field-induced anisotropy of hot-electron velocity fluctuations [154]. The experiment was performed on samples containing a relatively low density of carriers, and the results were interpreted in terms of the Price fluctuation–diffusion relation with the correct conclusion [154] that the diffusion coefficient of hot carriers was anisotropic as well. Some time later, Wagner, Davis, and Hurst [155] observed the anisotropy of electron diffusion in ordinary gases at high electric fields (see Chapter 1 in reference 106). Thus, microwave noise experiments demonstrated the possibility to obtain results on field-dependent longitudinal and transverse diffusion coefficients for majority carriers in uniform samples without introducing carrier density gradients, and this technique [154, 156] later was applied to investigate hot-electron diffusion in main semiconductors used in electronics (see references 10 and 11).

The detailed investigation of noise and diffusion anisotropy in p-type Ge was performed by Bareikis, Matulionienė, and Požela [156, 157]. The resolved difference between the longitudinal and transverse noise temperatures was interpreted as manifestation of the convective fluctuations (cf. Chapter 7).

The early experimental results on hot-electron noise and diffusion in n-type and p-type Ge [151, 154, 156, 157], as well as those on n-type Si [158] and n-type GaAs [152], later were confirmed by different authors using other techniques. Nougier and Rolland measured noise temperature in p-type Ge [159] (for details of the experimental technique see reference 160). Time-of-flight technique was applied to measure the longitudinal diffusion coefficient in n-type GaAs by Ruch and Kino [161] and in p-type Ge by Reggiani, Canali, Nava, and Alberigi–Quaranta [162] (see also reference 163). The fluctuations in one-valley semiconductors at low temperatures (p-type Ge and n-type InSb at 10 K) were observed and interpreted in terms of streaming motion of carriers [164–167] (the model is described in Section 7.4). Magnetic-field dependence of the noise temperature was evidenced experimentally and associated with hot-electron fluctuations [168, 169] (this dependence is absent at thermal equilibrium). Intervalley noise and diffusion were thoroughly investigated in semiconductors with equivalent valleys (n-type Ge, n-type Si) [163, 170–180] and nonequivalent valleys (n-type GaAs, n-type InP) [161, 181–184].

8.2 E-SPECTROSCOPY AND ω-SPECTROSCOPY OF NOISE

It is well known that frequency-dependent spectra of noise (ω-spectroscopy) can be used to determine intensities and time constants of different sources of hot-electron noise and the associated fast and ultrafast processes of dissipation. Figure 8.1 illustrates the spectra for a single relaxation process assuming the same integral intensity of the spectrum but different values of the relaxation time τ_m. The source of noise is cut off at frequency $\omega\tau_m \sim 1$. As one can see from Fig. 8.1, a wide range of frequencies is needed for a reliable estimation of the relaxation time. The estimation procedure would be even more complicated if several relaxation processes acted simultaneously.

The relaxation time can be estimated in another way. Since the average energy is field-dependent and the relaxation time is energy-dependent, sweeping of the field leads to a continuous variation of the relaxation time. Let us keep the frequency fixed and vary the relaxation time (by the applied electric field): The results are illustrated in Fig. 8.2. In this plot, the maximum forms in the vicinity of the condition $\omega\tau_m \sim 1$. Thus the maximum can be used to obtain information otherwise available from the step of the Lorentzian at $\omega\tau_m \sim 1$ present in the traditional frequency-dependent spectra determined at a fixed electric field.

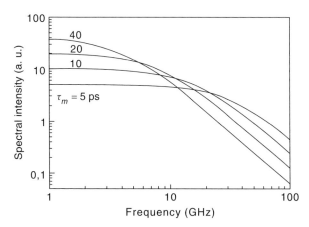

Figure 8.1. Velocity fluctuation spectra for a single relaxation process assuming different values of the relaxation time τ_m and the same integral intensity. The source of noise is cut off at frequency $f \sim 1/(2\pi\tau_m)$.

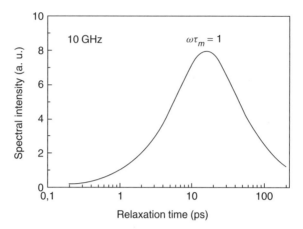

Figure 8.2. The idea of **E**-spectroscopy of hot-electron noise is illustrated for a single relaxation process with the electric-field-dependent relaxation time: $\tau_m(E)$ (cf. Fig. 8.1). Sweeping of the field at a fixed frequency forms the maximum of the spectral intensity near $\omega\tau_m \sim 1$.

Consequently, the **E**-spectroscopy of hot-electron noise can be introduced alongside with the ω-spectroscopy. Sometimes, especially at microwave frequencies, the **E**-spectroscopy is preferable over the ω-spectroscopy from technical point of view. Achievements of the **E**-spectroscopy of hot-electron noise have been reviewed successively by Nougier [10, 185], Bareikis et al. [9, 11, 75, 186, 187], and Pozhela [188]. Some of the results are discussed in Chapter 9.

An electronic process of dissipation and the associated source of noise can be eliminated, provided that the electron transit time is made essentially shorter than the corresponding relaxation time; as a result, the hot electrons leave the sample without having participated in the process and without having contributed to the particular source of noise. Once the source of noise is cut off, a better access to the remaining sources becomes possible. Variation of the sample length L and the electric field **E** is the basic idea behind the L–**E** spectroscopy. The L–**E** spectroscopy of hot-electron noise, alongside with the ω-spectroscopy, is nowadays becoming a modern tool of hot-electron investigations, and we shall discuss the results in Chapter 11.

8.3 MEASURABLE QUANTITIES AND DEFINITIONS

This section gives a brief review of methods for investigating microwave noise in a two-terminal sample: a bulk semiconductor sample, a thin channel, a short n^+–n–n^+ structure, a 2DEG channel, a diode, and so on. The investigation is mainly concerned with the hot-electron noise—the noise component associated with the electron velocity fluctuations in a semiconductor subjected to a uniform electric field of sufficiently high strength. In a given experimental configuration, two independent parameters characterize the noise properties: the available noise power and the spectral intensity of current fluctuations.

Noise Temperature and Intensity of Current Fluctuations

The available noise power $\Delta P_n(\omega, \mathbf{E})$ is related to the equivalent noise temperature $T_{n\alpha}(\omega, \mathbf{E})$ [cf. Eq. (1.1)]:

$$\Delta P_{n\alpha}(\omega, \mathbf{E}) = k_B T_{n\alpha}(\omega, \mathbf{E}) \Delta f. \tag{8.1}$$

$\Delta P_{n\alpha}(\omega, \mathbf{E})$ is the maximum noise power dissipated by the biased sample on the matched load (a coaxial line, a waveguide, etc.) in the frequency interval Δf at the angular frequency ω; the subscript α indicates direction of current fluctuations in respect to the direction of the steady DC current caused by the applied DC electric field $\mathbf{E}$.

According to Eq. (1.7), the spectral intensity of current fluctuations in the direction α is expressible as

$$S_{I\alpha}(\omega, \mathbf{E}) = 4k_B T_{n\alpha}(\omega, \mathbf{E}) \, \text{Re}\{Z_\alpha^{-1}(\omega, \mathbf{E})\}. \tag{8.2}$$

where $\text{Re}\{Z_\alpha^{-1}(\omega, \mathbf{E})\}$ is the real part of small-signal admittance at angular frequency ω, measured in the direction α (the parameter of the response to a weak AC electric field applied in the same direction). Two directions are of exceptional importance: $\alpha = \|$ stands for the direction parallel to the steady current, and $\alpha = \perp$ corresponds to the transverse direction. Thus, measurements of the available noise power and the real part of the small-signal AC admittance in the direction of interest provide us with the experimental data on $S_{I\alpha}(\omega, \mathbf{E})$.

The relation between S_I and $(\delta j^2)_\omega$ is as follows (see Section 3.3):

$$S_{I\alpha}(\omega, \mathbf{E}) = 2Q^2 (\delta j_\alpha^2)_\omega, \tag{8.3}$$

where Q is the cross-sectional area.

Ramo–Shockley Theorem

It is a convention to deal with the short-circuit current fluctuations unless otherwise is mentioned. Under these conditions, the Ramo–Shockley theorem relates two fluctuating quantities: the instantaneous value of the short-circuit current and the instantaneous value of the velocity of the moving charge (e.g., the velocity of the mass center of all mobile electrons — this quantity is usually available from microscopic simulation using Monte Carlo technique). Let the charge q move between two planar parallel electrodes located at a distance L apart. The motion of the charge with velocity v_x perpendicular to the electrodes induces the short-circuit current $I(t) = qv_x/L$, where the charge $q = en_0QL$. Thus, the spectral intensities of fluctuations of the mass center velocity S_V and of the current

S_I are interrelated:

$$S_I(\omega) = \frac{e^2 n_0 Q}{L} S_V(\omega), \tag{8.4}$$

where n_0 is the electron density and e is the elementary charge.

Diffusion Coefficient

The steady-state diffusion coefficient of hot electrons can be obtained from the experimental data on the small-signal conductivity $\sigma(E)$ and the noise temperature $T_n(E)$, provided that the interelectron collisions are not important and the Price relation [41] is valid (see Chapter 6). Indeed,

$$\frac{D(E)}{D(0)} = \frac{T_n(E)\sigma(E)}{T_0\sigma(0)}. \tag{8.5}$$

Equation (8.5) has suggested a convenient way to measure $D(\mathbf{E})$ without introducing electron density gradient [154]. The microwave noise technique has been successfully applied to investigate hot-carrier diffusion in Ge [154, 156, 159, 177, 189], Si [177, 178, 190, 191], InSb [166, 167], GaAs [181, 183, 184, 192, 193], InP [182, 184, 192, 194], AlGaAs [195], and other semiconductors.

8.4 WAVEGUIDE-TYPE SHORT-TIME-DOMAIN GATED RADIOMETER

Commercially available spectral analyzers for a wide range of frequencies up to and including millimeter waves operate in a continuous-wave (cw) mode, supporting standard cw measurements of noise in semiconductor devices in many laboratories. However, investigation of hot-electron effects at high electric fields require pulsed rather than cw modes of operation. Pulsed measurements help to avoid thermal walkout due to the consumed power, while the Joule effect in a cw mode leads to an increase in the lattice temperature masking hot-electron effects. Unfortunately, spectral analyzers for pulsed measurements are not commercially available yet, and radiometric techniques operating at fixed frequencies are commonly used to obtain data specific to hot electrons. The noise power in the chosen frequency band is selected by a filter, amplified, and fed into the radiometer for noise power measurement.

Pulsed measurements of noise impose several special requirements. The noise power is to be measured when the electric field is on: One deals with a low noise signal in the presence of the high pulsed voltage, with the latter penetrating into the noise measuring circuit and disturbing the sensitive amplifier unless the radiometric circuit is safely decoupled from that used to heat the electrons. The decoupling is easier achieved at microwave frequencies. This frequency range is also favorable from another point of view: Flicker and generation–recombination noise sources are cut off at microwave frequencies and do not interfere with hot-electron noise measurements. Noise measurements at high electric fields meet another problem: The electric field changes the sample impedance and introduces a mismatch of the sample and the load. The mismatch must be eliminated by changing the load impedance.

The idea of pulsed radiometric measurements is as follows. The noise power emitted by hot electrons into the waveguide is measured with a sensitive gated radiometer opened for a short time during the voltage pulse (Fig. 8.3). The voltage pulses are repeated in a low-duty cycle giving enough time for heat dissipation. All the problems have been solved by developing a waveguide-type short-time-domain gated modulation radiometer [153, 196]. Low transmission losses inherent to waveguide technique, a parametric narrow-band low-noise high-gain microwave amplifier available at X-band frequencies, and efficient filtering-out of the parasitic signals (unable to propagate in the waveguide) make a waveguide-type radiometer a valuable instrument for research of hot-electron fluctuations (see references 9, 11, and 12).

In a modulation radiometer, the hot-electron noise power is compared with that of a standard noise source (Fig. 8.4). For this the gated radiometer is opened twice during the modulation cycle: first to sample the microwave noise signal of the hot-electrons, and next to sample the standard noise signal. The sampled noise signals are amplified, squared, and integrated, and the signal difference is taken. The best accuracy is reached if the difference is zero; this means that the both sources of noise emit equal powers (Fig. 8.4). Low frequency fluctuations of the amplifier and its "zero" drift are completely eliminated, while the microwave power loss and excess noise in the radiometric circuit are identical for the both sources.

Figure 8.5 presents a schematic view of the setup for hot-electron noise measurements at microwave frequencies (see references 12 and 75). The experimental procedure for determining the noise temperature T_n consists of two steps. The first step includes matching the waveguide impedance to that of the sample at each bias. For this, switch 7 through port 7a is connected to the microwave generator 4, and the transformer 8 is tuned to reach the minimum standing microwave

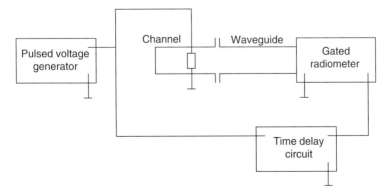

Figure 8.3. The idea of gated radiometric measurements: A biased channel emits the noise power into the waveguide, and the gated radiometer is opened to register the hot-electron noise power at a chosen moment of time.

88 EXPERIMENTAL TECHNIQUES

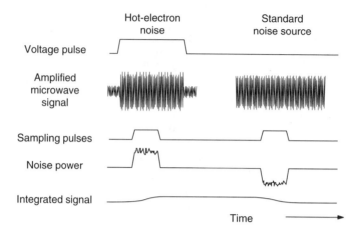

Figure 8.4. Sequence of operations in a gated modulation-type short-time-domain radiometer. The voltage pulse generator agitates hot electrons, their noise is amplified and sampled, and the noise power is integrated and compared to that of the standard noise source.

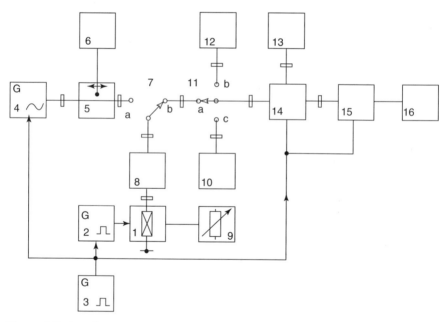

Figure 8.5. A schematic diagram of the X-band waveguide-type setup for gated measurements of hot-electron noise temperature [12]: 1—the investigated channel in the waveguide; 2—the pulsed voltage generator; 3—the master generator; 4—the microwave generator; 5—the microwave line; 6—the SWR indicator; 7, 11—the microwave switches; 8—the impedance transformer; 9—the resistance bridge; 10, 12, and 13—the reference noise generators; 14—the modulator; 15—the gated modulation radiometer; 16—the indicator.

ratio in the microwave line 5. The resistance bridge 9 allows one to control the sample resistance at each bias (the typical duration of the bias pulse ranges from 1 to 5 μs, and the repetition frequency is 125 Hz).

The second step includes measurement of the noise temperature of the sample at a chosen bias using the best matching data (transformer readings) available from the first step. Switches 7 and 11 are connected to ports 7b and 11a, respectively. The noise power emitted by the sample 1 and that of the reference noise generator 13 are periodically fed into the input of the gated radiometer 15 which is opened twice during the period of modulation: first to connect the biased sample 1, and for the second time to connect the reference noise generator 13 to the radiometer 15. The difference of the signals is used to determine the noise power emitted by the sample. The reference standard noise sources 10 and 12 help to control the zero level (switch 11 in port 11c) and the gain of the radiometer amplifiers (port 11b). The parameters of the X-band radiometer with 10^{-7} s gating time are as follows: The power sensitivity is 10^{-15} W, the systematic error is ±0.25 dB, and the noise temperature is up to $100\,T_0$.

The emitted noise power is not equal to the available noise power, unless the sample is perfectly matched to the waveguide. This requirement is met with certain difficulty. Since the sample conductance depends on **E**, the optimal transmission is not obtained over a wide range of electric field. The transmission can be taken into account while determining the equivalent noise temperature T_n [75]:

$$T_n = T'_n/\gamma, \qquad (8.6)$$

where T'_n is the measured noise temperature, and the transmission coefficient γ depends on the sample differential resistance as illustrated in Fig. 8.6. The latter two quantities are available from additional microwave measurements.

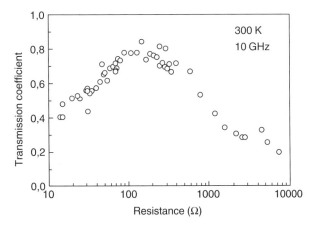

Figure 8.6. Dependence of the transmission coefficient on the real part of impedance for a quantum well channel QW-2498 [313].

Thermal Walkout Effect

Consumed DC power introduces a thermal walkout due to the Joule effect. The walkout effect can be investigated by measuring the noise power after certain delay following the bias voltage onset. The noise temperature is found to depend on the delay time [197, 198]. Provided that the excess hot-electron temperature remains almost independent of the lattice temperature, the walkout effect and the hot-electron effect are additive: The measured noise temperature increases as the lattice temperature increases. Figure 8.7 illustrates the additive effect [198]). The applied electric field causes a fast increase in the noise temperature followed by a slow walkout effect (Fig. 8.7): The 40 K temperature is gained during the voltage pulse in addition to the 170 K excess temperature corresponding to the initial fast increase, with the latter being due to the hot electrons. When the voltage is switched off, the hot-electron noise switches off almost immediately (typical hot-electron energy relaxation time is in the picosecond range). This fast drop is nearly the same in magnitude as the initial fast increase (Fig. 8.7). The remaining "tail" can be attributed to the lattice temperature; the relaxation is caused by the heat dissipation to the environment. The excess lattice temperature immediately after the voltage pulse can be estimated through extrapolation of the exponential "tail" (dashed line in Fig. 8.7). The increase in the lattice temperature seems not to cause a noticeable change of the hot-electron temperature in this particular case. However, the thermal walkout effect can be more complicated provided the hot-electron noise depends on the lattice temperature. For example, an increase in the lattice temperature can cause a decrease of the excess hot-electron noise temperature [197].

Samples

A typical sample for hot-electron noise measurements in a bulk semiconductor is a nonlinear resistor prepared in the shape of a rectangular parallelepiped with two

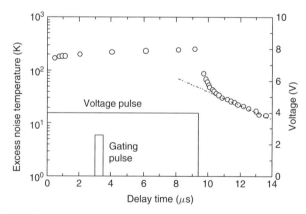

Figure 8.7. Dependence of the excess noise temperature (O) on the delay time for an AlGaAs sample [198].

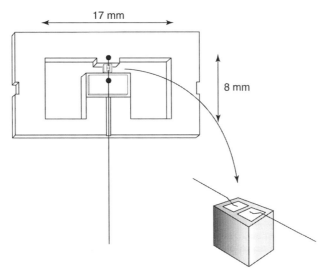

Figure 8.8. An epitaxial sample in a waveguide frame of a special shape for longitudinal noise temperature measurements.

ohmic electrodes on its bases. Since shot noise can be important at microwave frequencies, especially if the current is controlled by a barrier, perfect ohmic contacts, uniform doping, and relatively high density of majority carriers are prerequisites in order to avoid interference of shot noise during the experiments on hot-electron velocity fluctuations. The longitudinal current fluctuations are measured in the direction of the steady DC current. The transverse noise voltage is usually coupled capacitatively to the transmission line connected with the radiometer.

Investigations of thin epitaxial channels are usually performed on test samples cut from a transmission-line-model structure containing coplanar electrodes. A sample is mounted into a special frame of X-band waveguide (Fig. 8.8). For measurements of longitudinal noise the sample is oriented so that the electrons drift normal to the wide walls of the waveguide, and H_{10} mode of the AC field is excited by the electron drift-velocity fluctuations. The test samples with side arms (for the Hall effect measurements) can be used to investigate fluctuations in the direction transverse to the steady DC current.

On-wafer microwave noise measurements can be performed using microprobes. Each microprobe consists of a central wire and two side wires attached for screening. Microprobes are connected to coaxial lines or waveguides and are manipulated to contact the sample electrodes on the wafer.

8.5 SMALL-SIGNAL RESPONSE

According to Eq. (8.2) the AC small-signal conductivity is needed for determination of the spectral intensity of current fluctuations at the microwave frequency ω

of interest. The technique described in reference 199 can be modified to perform the measurements on the setup illustrated in Fig. 8.5 without changing the sample holder. The experimental procedure is as follows. First, the standing wave ratio $K(\omega, E)$ is measured for the biased sample. Next, the electric field is switched off, and the previous value of the standing wave ratio is restored by applying external magnetic field B:

$$K(\omega, E = 0, B) = K(\omega, E, B = 0). \tag{8.7}$$

Now, the DC low-field conductivity $\sigma(\omega = 0, E = 0, B)$ is measured at a very low DC electric field in presence of the magnetic field B in the standard way.

At microwaves a strong inequality

$$\omega\tau \ll 1 \tag{8.8}$$

holds, where τ is the momentum relaxation time entering the zero-field mobility and conductivity. At equilibrium (and at very low electric field) under condition (8.8), one has

$$\sigma(\omega = 0, E = 0, B) = \sigma(\omega, E = 0, B), \tag{8.9}$$

where $\sigma(\omega, E = 0, B)$ is the zero-field AC conductivity in the magnetic field.

When the standing wave ratios are equal, Eq. (8.7), the microwave small-signal conductivities are equal as well; with Eq. (8.9) in mind, one has

$$\sigma(\omega, E = 0, B) = \sigma(\omega, E, B = 0) = \sigma(\omega = 0, E = 0, B). \tag{8.10}$$

In this way, the standard measurement of the DC conductivity at a weak electric field $\sigma(\omega = 0, E = 0, B)$ provides us with the required small-signal conductivity at the microwave frequency under strong electric field: $\sigma(\omega, E, B = 0)$.

8.6 COAXIAL-TYPE GATED SETUP FOR NOISE MEASUREMENTS

Experimental determination of the available noise power without perfect matching has been repeatedly considered. An excellent solution [200] is illustrated in Fig. 8.9, where a schematic view of the computer-controlled setup is given [198]. Appropriate filters are used to perform measurements at the chosen frequencies: 50, 220, 320, 460, 650, 850 MHz, and 4 GHz. The power of the standard noise generator is directed through the circulator and the commutator to the biased semiconductor device under test (DUT). Because of the device mismatch, the generator noise power is partially reflected from the device. The reflected power, together with the sample noise power, is forwarded through the circulator into the input circuit of the measuring device.

The noise power at the input of the preamplifier is given by

$$P = \rho P_g + (1 - \rho) P_n + P_c, \tag{8.11}$$

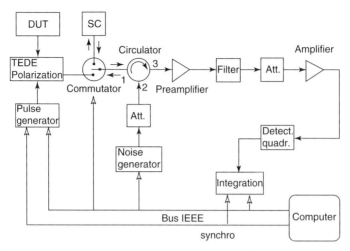

Figure 8.9. A schematic diagram of the computer-controlled coaxial setup for pulsed hot-electron noise temperature measurements in the frequency range from 220 MHz to 10 GHz covered with seven sets of circulators and filters [198, 200]. The circulator performs summing of the noise power emitted by the device under test (DUT) with the power of a standard noise source, and the resultant AC signal is filtered, amplified, detected, integrated, and registered. Four independent measurements are performed at the given frequency and bias (see text for details).

where P_n is the available noise power of the biased sample, ρ is its squared reflection coefficient, P_g is the noise power of the standard noise generator, and P_c is the noise power generated by the circuit elements.

The registered signal X is proportional to the input power P:

$$X = \lambda P. \tag{8.12}$$

The transfer function λ takes into account (a) amplification of the power and (b) loss and generation of the power in the filter, in the attenuator, and in other elements of the radiometric circuit. Let us express different noise contributions in terms of noise temperatures:

$$X = \lambda[\rho T_g + (1-\rho)T_n + T_c]. \tag{8.13}$$

Equation (8.13) contains four unknown variables: λ, ρ, T_n, and T_c. All of them can be determined through four independent measurements.

During the *first* step the standard noise generator is switched off, and its noise temperature equals the ambient temperature $T_g = T_0$. The sample is disconnected, and the short-circuit element (SC in Fig. 8.9) is connected; the reflection coefficient of the latter equals unity ($\rho = 1$). In this situation, the output signal is

$$X_1 = \lambda(T_0 + T_c). \tag{8.14}$$

During the *second* measurement the standard noise generator is switched on, its noise temperature T_g is known, and the sample remains short-circuited ($\rho = 1$). Now, the output signal is given by

$$X_2 = \lambda(T_g + T_c). \tag{8.15}$$

For the *third* measurement the standard noise generator is switched off ($T_g = T_0$), and the short-circuit element is taken away. The biased sample is characterized by its noise temperature T_n and its reflection coefficient. The resultant output signal is

$$X_3 = \lambda[\rho T_0 + (1 - \rho)T_n + T_c]. \tag{8.16}$$

For the *fourth* measurement the short-circuit element is taken away, the standard noise generator is switched on, and the biased sample feeds its noise into the circuit. All sources of noise cause the output signal

$$X_4 = \lambda[\rho T_g + (1 - \rho)T_n + T_c]. \tag{8.17}$$

The computer-controlled circuit performs the required measurements of the quantities X_1, X_2, X_3, and X_4 and determines the four unknown variables including the noise temperature of the sample T_n. Additional measurements are performed to check the noise temperature of the standard noise generator, if necessary.

The techniques based on the setups presented in Figs. 8.5 and 8.9 give comparable results on hot-electron noise temperature available under similar conditions. Figure 8.10 compares the results on experimental investigation of field-dependent noise temperature obtained for the same sample on the both setups. A satisfactory fit is obtained.

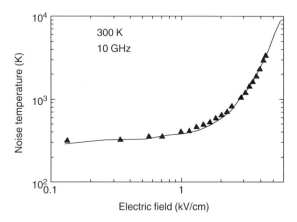

Figure 8.10. Dependence of hot-electron longitudinal noise temperature on electric field for a lattice-matched InGaAs 2DEG channel 13Q (for the 2DEG data see Chapter 17). Solid curve stands for the results measured by the waveguide-type gated modulation radiometer (Fig. 8.5), symbols correspond to the results for the same sample obtained on the pulsed coaxial-type setup (Fig. 8.9).

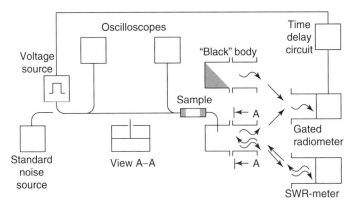

Figure 8.11. A schematic diagram for the combined rectangular-waveguide–coaxial radiometric setup for nanosecond-time-domain measurements of hot-electron noise temperature at an X-band microwave frequency [201].

8.7 NANOSECOND-TIME-DOMAIN GATED RADIOMETRIC SETUP

Nanosecond-time-domain measurements are indispensable at very high applied electric fields. Figure 8.11 presents a schematic view of the setup combining elements of coaxial nanosecond and of waveguide techniques [201]. A nanosecond/microwave sample holder was designed to support short-time-domain pulsed measurements of hot-electron noise power at an X-band frequency. The sample under test is placed into a coaxial cable of the nanosecond circuit, and a T-shaped antenna is used for coupling the sample circuit to the waveguide. The electrons are heated up by voltage pulses (pulse duration is 100 ns) in order to reduce the channel overheat. The incident/reflected nanosecond pulse waveforms were fed into the oscilloscopes and analyzed to obtain current dependence on voltage. The coupling and transmission coefficients and waveguide losses were determined using the noise standard. The delay of the gating ensured the noise power measurements before, during, and after the voltage pulse. This was sufficient to control the channel overheat and keep it low enough. Matching of the channel circuit to the waveguide was controlled by the standing-wave-ratio-meter (SWR-meter). Its readings were voltage-dependent, and they were taken into account while determining the noise temperature. The noise power radiated by the channel and detected by the gated radiometer was compared to that of the "black-body" radiation source kept at known temperature. This technique was used [201] to perform measurements of the hot-electron noise in the direction of applied electric field.

CHAPTER 9

HOT-ELECTRON MICROWAVE NOISE IN ELEMENTARY SEMICONDUCTORS

This chapter presents experimental results on hot-electron noise and related fluctuation phenomena in elementary semiconductors. A detailed information on the spectral properties and the associated processes of dissipation is available from noise measurements supported by interpretation of the experimental data in terms of realistic semiconductor models developed for lightly doped semiconductors containing a low/intermediate density of electrons. This choice gives us a possibility to consider the experimental data in terms of the fluctuation theory formulated in Chapter 3, thereby avoiding complications due to interelectron collisions. While analytical models (see Chapter 7) provide us with perfect illustrations of the kinetic theory with a deep insight into physics of hot-electron noise, numerical techniques are useful to extract quantitative information on the dominant kinetic processes. The experimental results are illustrated and interpreted using Monte Carlo simulation data where available.

9.1 TRANSVERSE NOISE TEMPERATURE

The hot-electron noise temperature $T_n(\mathbf{E})$ is an anisotropic quantity: The directions parallel and transverse to the steady current are not equivalent even in the simplest case of spherical bands and isotropic scattering mechanisms. This is illustrated by Fig. 9.1, which presents the noise temperatures measured for p-type germanium at 9.6-GHz frequency and 80 K lattice temperature [156]. The samples had the resistivity $\rho(300\ \text{K}) = (20\text{--}40)$ ohm·cm, and the DC electric field was applied along the $\langle 110 \rangle$ crystallographic direction. The longitudinal noise temperature $T_{n\parallel}(E)$ was measured in the same direction, and the transverse noise temperature $T_{n\perp}(E)$ was measured in the $\langle 1\bar{1}0 \rangle$ direction.

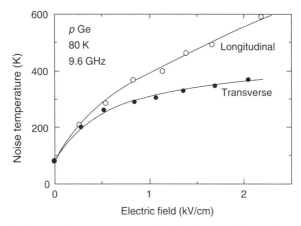

Figure 9.1. Equivalent noise temperature of hot holes at 9.6-GHz frequency in longitudinal (○) and transverse (●) directions to the steady current for p-type germanium, $\rho(300\ \text{K}) = 20$ ohm·cm, measured at 80 K lattice temperature [156]. Solid curves are guides to the eye.

In the transverse direction, the noise is caused by the isotropic part of the electron (hole) velocity distribution. After Eqs. (7.12) and (7.20), at frequencies $\omega\tau \ll 1$ one has

$$T_{n\perp} = \frac{2}{3}\frac{\overline{\varepsilon\tau}}{k_B \overline{\tau}}, \tag{9.1}$$

where $\overline{\tau}$ is the momentum relaxation time entering the small-signal AC response mobility, $\mu_\perp = (e/m)\overline{\tau}$. In the case of one-valley conduction controlled by acoustic phonon scattering, one has $\overline{\varepsilon\tau} = 1.144\overline{\varepsilon}\,\overline{\tau}$ [41]. A stronger dependence of the relaxation time on energy results from ionized impurity scattering, and the factor deviates from unity more than 14%. On the other hand, the impurity scattering acting together with acoustic phonon scattering in lightly doped semiconductors flattens the dependence $\tau(\varepsilon)$, and the approximate equality $\overline{\varepsilon\tau} \approx \overline{\varepsilon}\,\overline{\tau}$ holds within 5%, as obtained from Monte Carlo simulation for the models developed to interpret the experimental data on the transverse noise temperature for p-type germanium and n-type InSb [75]. Of course, $\overline{\varepsilon\tau} = \overline{\varepsilon}\,\overline{\tau}$, provided that the relaxation time is independent of electron energy (see Section 7.2). Consequently, the electron kinetic energy $\overline{\varepsilon}$ and its dependence on electric field can be estimated from measurements of the transverse noise temperature $T_{n\perp}$.

The velocity fluctuations, determined by the kinetic energy, is the only source of noise in the transverse direction in an isotropic semiconductor. For brevity, let us call it hot-electron "thermal" noise (or hot-hole "thermal" noise). The transverse noise is "white" in the frequency range $\omega\overline{\tau} < 1$. The frequency dependence $T_{n\perp}(\omega)$ appears at frequencies $\omega\overline{\tau} \sim 1$, provided that $\overline{\varepsilon\tau}/(\overline{\varepsilon}\,\overline{\tau})$ is not equal to unity [see Eqs. (7.12) and (7.20)]. Because of small values of $\overline{\tau}$, the corresponding frequencies usually fall into the millimeter and submillimeter wave range. As mentioned, $T_{n\perp}$ remains independent of frequency in the range $\omega\overline{\tau} \ll 1$.

Let us remind the reader that some models (discussed in Chapter 7) predict electron cooling by the applied DC electric field. For example, in the case of dominant optical-phonon scattering, the transverse noise temperature can be lower than the equilibrium one: $T_{n\perp} < T_0$ (see discussion in Section 7.3). In this particular case the reduced transverse noise temperature is an indication that the chaotic motion in the transverse direction is partially quenched by the almost periodic streaming motion of hot electrons in the longitudinal direction. In compensated semiconductors containing high density of ionized impurities, the cooling of electrons by the electric field due to optical phonon emission has been observed in the direction of current as well (see reference 202 and references therein).

9.2 ENERGY RELAXATION TIME ESTIMATED BY NOISE TECHNIQUE

Dissipation of the consumed power can be considered using the data on hot-electron kinetic energy available from experimental data on the transverse noise temperature $T_{n\perp}$. Let us treat the energy dissipation in terms of the energy relaxation time. The effective energy relaxation time $\bar{\tau}^{en}$ is introduced in the following way:

$$\bar{\tau}^{en}(\mathbf{E}) = \frac{(\bar{\varepsilon} - \varepsilon_0)}{e\mathbf{E} \cdot \mathbf{v}^{dr}}. \tag{9.2}$$

where $\mathbf{v}^{dr}$ is the steady drift velocity, $e\mathbf{E} \cdot \mathbf{v}^{dr}$ is the mean consumed power per electron, and $(\bar{\varepsilon} - \varepsilon_0)/\bar{\tau}^{en}(\mathbf{E})$ is the energy dissipation rate.

If we assume that $\overline{\varepsilon \tau} = \bar{\varepsilon}\,\bar{\tau}$, Eqs. (9.2) and (9.1) lead to

$$\bar{\tau}^{en}(\mathbf{E}) = \frac{3}{2} \frac{k_B(T_{n\perp} - T_0)}{e\mathbf{E} \cdot \mathbf{v}^{dr}}. \tag{9.3}$$

Equation (9.3) suggests a technique to estimate $\bar{\tau}^{en}$ from the experimental data on the transverse noise temperature $T_{n\perp}$ and the drift velocity $\mathbf{v}^{dr}$.

Figure 9.2 presents the effective energy relaxation time determined according to Eq. (9.3) [168, 203]. On can see that the energy relaxation time is quite long. It is considerably longer than the momentum relaxation time $\bar{\tau}$ available from the mobility data. The strong inequality $\bar{\tau}^{en} \gg \bar{\tau}$ gives an experimental evidence that quasi-elastic scattering is important at low electric fields. In the considered case the lattice temperature is within the range $\hbar\omega^{acoust} \ll k_B T_0 \ll \hbar\omega^{opt}$ (here $\hbar\omega^{acoust}$ and $\hbar\omega^{opt}$ are the energies of the acoustic and optical phonons involved in the energy dissipation), and the energy loss on acoustic phonons dominates at low electric fields. The transition from the acoustic-phonon-controlled to the optical-phonon-controlled scattering takes place upon carrier heating (Fig. 9.2) [see also Eqs. (7.26) and (7.27)].

The experimental data on the energy relaxation time obtained using the noise technique are in a good agreement with those available from other experiments [204] (see also reference 205). The noise technique seems to be especially important if the momentum relaxation time is *independent* of the electron energy (see

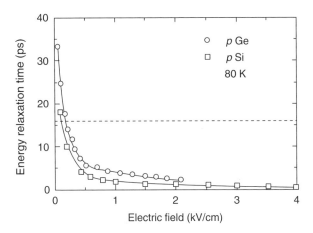

Figure 9.2. Hot-hole energy relaxation time $\bar{\tau}^{en}$ in germanium [168] and in silicon [203] at 80 K lattice temperature estimated from the transverse noise temperature measured in the $\langle 110 \rangle$ direction for electric field $\mathbf{E} \parallel \langle 1\bar{1}0 \rangle$: ○ represents p-type Ge, $\rho(300\ K) = 20$ ohm·cm; □ represents p-type Si, $\rho(300\ K) = 30$ ohm·cm. Dashed line stands for ω^{-1} at 10-GHz frequency. Solid curves are guides to the eye.

Section 7.2); under this condition, no deviation from Ohm's law takes place, and techniques based on signal mixing fail. Nevertheless, the noise technique [168] remains applicable.

The observed strong dependence of the energy relaxation time on the electric field can be exploited for the E-spectroscopy of noise (see Section 8.2). The broken line in Fig. 9.2 gives the value ω^{-1} at 9.6 GHz. One can see that the condition $\omega \bar{\tau}^{en} \sim 1$ is satisfied for p-type Ge at $T_0 = 80$ K in the microwave X-band at electric fields $E \sim 150$ V/cm. A similar behavior is observed in p-type silicon at 80 K lattice temperature, as well as in some other lightly doped semiconductors.

9.3 CONVECTIVE NOISE IN p-TYPE GERMANIUM

Now let us concentrate our main attention on the longitudinal fluctuations, longitudinal noise, and other longitudinal quantities. The spectral intensity of the longitudinal current fluctuations $S_{I\parallel}$ in the presence of external electric field contains the energy-fluctuation-born term called the convective noise (see Section 7.2). The convective noise is most important at frequencies $\omega \bar{\tau}^{en} < 1$; its appearance is due to quasi-elastic scattering — this type of scattering assumes that a collision changes the direction of the electron motion remarkably with little effect on the absolute value of the electron velocity. As a result, many collisions are needed for the energy relaxation, and two time constants take part in the relaxation processes: the momentum relaxation time, τ, and the energy relaxation time, τ^{en}, with the latter being large compared to the first. The well-known example is the dominant scattering of electrons and holes by acoustic phonons at not-too-low lattice temperatures at low and moderate electric fields. In the case of scattering by acoustic phonons in a spherical parabolic band, the theory predicts a negative convective contribution to longitudinal current fluctuations (see Section 7.2).

The phenomenon of convective noise received the first experimental confirmation in p-type germanium [156, 157]. The noise temperature $T_n(\omega, \mathbf{E})$ and the small-signal conductivity $\sigma(\omega, \mathbf{E})$ were measured at 9.6-GHz frequency for the samples of lightly acceptor-doped germanium: $\rho(300\ K) = 20$–40 ohm·cm. The choice of the frequency eliminated the generation–recombination noise. The electric field was applied along $\langle 110 \rangle$ crystallographic direction. The longitudinal quantities were measured in the direction of the field, and the transverse ones were measured in $\langle 1\bar{1}0 \rangle$ direction. The spectral intensity of current fluctuations $S_I(\omega, \mathbf{E})$ was obtained according to Eq. (8.2).

As predicted by the theory, under the condition $\omega \bar{\tau}^{en} < 1$ for a semiconductor with a sublinear current–voltage characteristic, the spectral intensity of the transverse current fluctuations exceeds the spectral intensity of the longitudinal current fluctuations. The experimental results for p-type Ge presented in Fig. 9.3 confirm the theoretical prediction. The anisotropy of the spectral intensities of current fluctuations at not-too-high electric fields is mainly decided by the energy fluctuations (cf. Section 7.2).

It is well known that the valence band of germanium consists of two subbands containing heavy and light holes in case of p-type doping. Therefore, in the current direction, the noise consists of the convective noise and the interband transfer noise in addition to the field-enhanced hot-hole "thermal" noise. The estimations show that the interband noise is smaller by an order of magnitude as compared to the intraband noise [75]. The contribution of light holes to the noise

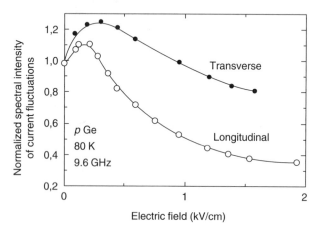

Figure 9.3. Normalized spectral intensity $S_I(E)/S_I(0)$ of longitudinal (○) and transverse (●) current fluctuations at 9.6-GHz frequency in p-type germanium, $\rho(300\ K) = 20$ ohm·cm, at 80 K lattice temperature for electric field $\mathbf{E} \parallel \langle 110 \rangle$ [156]. Solid curves are guides to the eye.

temperature is also less than 10%. Therefore, in p-type Ge, the intraband noise of heavy holes prevails over the other sources efficient at microwave frequencies. The assumption of a spherically symmetrical band is also acceptable. For these reasons, p-type Ge is a suitable semiconductor for observation of the effect of the convective noise caused by the energy fluctuations. Of course, one should be cautious using the formulae derived for the acoustic-phonon-controlled scattering (see Section 7.2), because the gradual transition from the quasi-elastic scattering to the inelastic scattering takes place (cf. Fig. 9.2) in the field range where the experimental data are available (see Fig. 9.3).

In particular, the observed decrease of the effective energy relaxation time upon hole heating (see Fig. 9.2) causes a nonmonotonous dependence of the spectral intensity of current fluctuations on the field (Fig. 9.3). The experiment is performed at a fixed frequency ($\omega^{-1} \approx 17$ ps; see dashed line in Fig. 9.2). As mentioned, the condition $\omega \bar{\tau}^{en} < 1$ is satisfied at high electric fields, but the opposite inequality holds at low fields. As a result, the maximum of $S_I(\mathbf{E})$ forms in the field range including the condition $\omega \bar{\tau}^{en} = 1$. The maximum of similar origin is observed in p-type silicon at 80 K lattice temperature (Fig. 9.4). The energy relaxation time in n-type InSb is essentially shorter than the corresponding values for p-type Ge and p-type Si, and the maximum (if any) in n-type InSb is weak (see Fig. 9.5).

The maximum-containing dependence on electric field can be used in the E-spectroscopy of fluctuations (see Section 8.1) for estimation of the relaxation time. The following result is obtained at 80 K lattice temperature for p-type Ge $\rho(300 \text{ K}) = 20$ ohm·cm : $\bar{\tau}^{en} \sim 17$ ps at $E \sim 150$ V/cm. This result is in satisfactory agreement with that available from Fig. 9.2. It is noteworthy that these

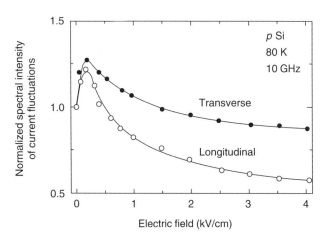

Figure 9.4. Normalized spectral intensity $S_I(E)/S_I(0)$ of longitudinal (○) and transverse (●) current fluctuations at 10-GHz frequency in p-type silicon, $\rho(300 \text{ K}) = 30$ ohm·cm, at 80 K lattice temperature for electric field $\mathbf{E} \parallel \langle 110 \rangle$ [186]. Solid curves are guides to the eye.

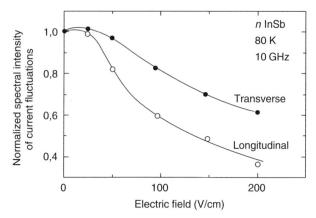

Figure 9.5. Normalized spectral intensity $S_I(E)/S_I(0)$ of longitudinal (O) and transverse (●) current fluctuations at 10-GHz for n-InSb ($n_0 = 1.2 \times 10^{14}$ cm^{-3}) for electric field $\mathbf{E} \parallel \langle 110 \rangle$ at 80 K lattice temperature [166]. Solid curves are guides for the eye.

two independent experimental techniques based on the transverse noise temperature (Fig. 9.2) and on the cutoff frequency of the convective contribution to the longitudinal current fluctuations (Fig. 9.3) lead to essentially the same result for the effective relaxation time.

No maximum of the spectral intensity of current fluctuations at 10-GHz frequency is observed in p-type germanium at room temperature (Fig. 9.6). In this particular case, the energy relaxation time satisfies the condition $\omega \bar{\tau}^{en} < 1$ in the whole range of electric fields. The experimental results are in good agreement with the Monte Carlo simulation data [206] obtained for the model, taking into

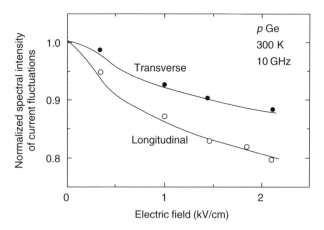

Figure 9.6. Normalized spectral intensity $S_I(E)/S_I(0)$ of longitudinal (O) and transverse (●) current fluctuations at 10-GHz frequency for p-type germanium, $\rho(300\ \text{K}) = 20$ ohm·cm, for electric field $\mathbf{E} \parallel \langle 110 \rangle$ at 300 K lattice temperature [186]. Solid curves are calculated by the Monte Carlo method [206].

account acoustic and nonpolar optical scattering in a single heavy-hole spherical parabolic band. Such a simple model is acceptable for microscopic interpretation of numerous experimental data on transport properties of p-type Ge (see reference 73).

9.4 RESONANCE DUE TO STREAMING MOTION

Inelastic scattering of hot electrons by optical phonons is the main energy loss mechanism at elevated electron energies. This scattering mechanism supports a resonance-type spectrum of velocity fluctuations at moderate electric fields at low lattice temperatures (see Section 7.4). One-valley semiconductors suit for demonstration of the resonant hot-electron noise behavior caused by the streaming motion terminated by the optical phonon emission.

The spectral intensity $S_{v_\parallel}(\omega, \mathbf{E})$ of the longitudinal velocity fluctuations obtained through Monte Carlo simulation shows the resonance peak [164, 166, 167; see also reference 75]. The resonant behavior occurs at the frequency equal to the reciprocal time of the acceleration of a hole with the zero initial energy until the optical-phonon energy:

$$\tau_E = \frac{(2m\hbar\omega^{\mathrm{opt}})^{1/2}}{eE}. \tag{9.4}$$

The resonance is pronounced in the field range determined by the condition that $\tau^{\mathrm{opt}} \ll \tau_E \ll \tau$, where τ^{opt} is the time for spontaneous optical phonon emission by an electron present in the active region $\varepsilon_\mathbf{p} > \hbar\omega^{\mathrm{opt}}$ (see Fig. 7.2). According to Eq. (9.4), the acceleration time decreases as the field strength E is increased, and the resonance peak of $S_{v_\parallel}(\omega, \mathbf{E})$ shifts to higher frequencies. Figure 9.7 illustrates

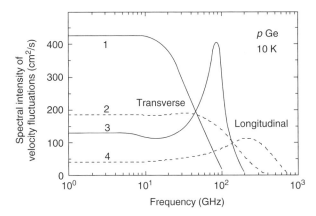

Figure 9.7. Calculated spectra of transverse (1, 2) and longitudinal (3, 4) hot-hole velocity fluctuations for the p-type Ge model, taking into account only scattering on optical phonons at 10 K lattice temperature [75]. Electric field: $E = 500$ V/cm (solid curves), $E = 2000$ V/cm (dashed curves).

the results of simulation for p-type Ge [164]. The shift of the resonance frequency with the field is accompanied by a stronger damping at high fields: Supposing that $\tau_E \sim \tau^{opt}$, the carriers penetrate into the active region, and fluctuations of the acceleration time of individual carriers come into play. On the other hand, the quasi-elastic scattering by acoustic phonons and impurities tend to wash out the resonance (Fig. 9.8) at low electric fields corresponding to $\tau_E \sim \tau$, where τ is the momentum relaxation time determined by the quasi-elastic scattering mechanisms at energies $\varepsilon_p < \hbar\omega^{opt}$. Provided that the scattering by ionized impurities were negligible ($N^{ion} = 0$), the resonance would be observed in the microwave and millimeter-wave frequency range.

The resonant behavior of fluctuations can be demonstrated using the E-spectroscopy of noise, thereby measuring the field-dependence of the noise at a fixed frequency. Since the acceleration time [see Eq. (9.4)] depends on the applied electric field, the resonance condition is met at a certain field strength — the associated maximum of spectral intensity develops at $\omega\tau_E \sim 1$ (see Fig. 8.2). The experiments were performed at 10-GHz frequency for p-type Ge [164] and n-type InSb [166,167] at 10 K lattice temperature. Figure 9.9 shows the field-dependent spectral intensity of current fluctuations for p-type germanium. The transverse component exceeds the longitudinal one in the field range $E > 140$ V/cm, but the opposite inequality $S_{I\|} > S_{I\perp}$ holds in the field range below 140 V/cm. The maximum of intensity of the longitudinal fluctuations is resolved in the vicinity of the condition $\omega\tau_E = 1$. A similar behavior has been reported for n-InSb as well [166, 167].

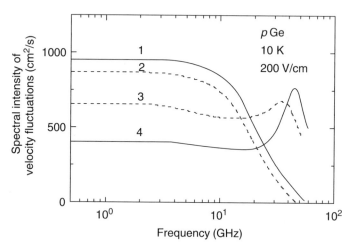

Figure 9.8. Calculated spectra of transverse (1, 2) and longitudinal (3, 4) hot-hole velocity fluctuations at 10 K lattice temperature for p-type Ge model at $E = 200$ V/cm. Broken curves take into account ionized impurity scattering ($N^{ion} = 8 \times 10^{13}$ cm^{-3}); solid curves refer to $N^{ion} = 0$ [75].

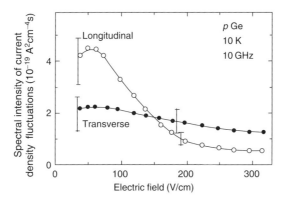

Figure 9.9. Experimental dependence of spectral intensity $S_j(E)$ of transverse (●) and longitudinal (○) current density fluctuations on electric field $\mathbf{E} \parallel \langle 110 \rangle$ for p-type Ge, $\rho(300\,\mathrm{K}) = 20$ ohm·cm, at 10-GHz frequency and 10 K lattice temperature. Solid curves are guides to the eye. The error bars indicate the uncertainty of the measured points [164, 165].

9.5 INTERVALLEY FLUCTUATIONS IN Si AND Ge

Two-Valley Model

The contributions of hot-electron "thermal," convective, and intervalley noise depend on frequency: The cutoff frequencies are determined by the corresponding relaxation times. Assuming that the intravalley processes are fast as compared to the intervalley processes, the spectral intensity of current fluctuations can be written as the sum of Lorentzians. According to Eqs. (7.35) and (7.11), after the discussion on the effective relaxation time (see Sections 9.1 and 9.2), for two groups of electrons ($i = 1, 2$) one obtains an approximate expression:

$$S_I(\omega, \mathbf{E}) = \sum_i \left(\frac{S_i^{\text{"therm"}}(\mathbf{E})}{1 + (\omega \bar{\tau}_i)^2} + \frac{S_i^{\text{en}}(\mathbf{E})}{1 + (\omega \bar{\tau}_i^{\text{en}})^2} \right) + \frac{S^{\text{inter}}(\mathbf{E})}{1 + (\omega \bar{\tau}^{\text{inter}})^2}, \quad (9.5)$$

where $\bar{\tau} \ll \bar{\tau}^{\text{en}} \ll \bar{\tau}^{\text{inter}}$ are the effective momentum, energy, and intervalley relaxation times, respectively; $S_i^{\text{"therm"}}(\mathbf{E})$, and $S_i^{\text{en}}(\mathbf{E})$ are the low-frequency limits for the spectral intensities of the hot-electron "thermal" and the convective noise sources in the ith valley, correspondingly. According to the Ramo–Shockley theorem, the spectral intensity of current fluctuations S_I is proportional to that of velocity fluctuations S_V. The low-frequency limit of the spectral intensity of velocity fluctuations due to the intervalley transitions is given by

$$S_V^{\text{inter}}(\mathbf{E}) = 4 \frac{n_1 n_2}{n_0^2} (v_1^{\text{dr}} - v_2^{\text{dr}})^2 \bar{\tau}^{\text{inter}}, \quad (9.6)$$

where n_0 is the electron density, and n_i and v_i^{dr} are the steady electron density and the steady drift velocity component in the ith valley, respectively.

Consequently, the hot-electron noise in a many-valley semiconductor is not equal to the sum of the corresponding intravalley contributions weighted by the partial numbers of electrons. The intervalley noise given by Eq. (9.6) is always positive, and it vanishes in several cases: in the absence of the intervalley transfer (either $n_1 = 0$, or $n_2 = 0$), for equivalent valleys ($v_1^{dr} = v_2^{dr}$), and at thermal equilibrium ($v_1^{dr} = 0$ and $v_2^{dr} = 0$). The nonzero drift velocity is necessary for the intervalley contribution to appear, but electron heating is not necessary in general. For example, at a relatively high lattice temperature, the transitions between nonequivalent valleys can lead to the intervalley noise revealed at low electric fields without electron heating. It is common to observe the intervalley noise in the direction of steady current. In an isotropic semiconductor with spherically symmetric valleys, the intervalley noise in the transverse direction is zero. However, the transverse component of the intervalley noise appears in anisotropic semiconductors and in isotropic semiconductors containing anisotropic valleys, with silicon being a suitable example.

Conduction-Band Structure and Intervalley Noise

The bottom of conduction band of germanium and silicon has several equivalent valleys containing equal densities of the electron gas in equilibrium. The effect of applied electric field is twofold. A low field introduces difference in the drift velocities in the ellipsoidal valleys oriented at different angles to the field direction. Thus, the fluctuations of valley occupancies, present at thermal equilibrium, cause the associated fluctuations of the current around its steady value. At a high electric field the hot-electron repopulation due to intervalley transitions takes place, and the associated noise — hot-electron intervalley noise — appears. The intervalley noise is anisotropic with respect to the electric field direction and to the crystallographic orientation [172]. Intervalley noise is an example of so-called partition noise.

Germanium has four valleys oriented along $\langle 111 \rangle$ axes. The field applied in $\langle 100 \rangle$ direction finds all valleys at the same angle in respect to the field. Consequently, the steady drift velocity of the electrons is the same in each valley. In agreement with Eq. (9.6), there is no intervalley contribution to longitudinal noise in this configuration. However, in $\mathbf{E} \parallel \langle 111 \rangle$ configuration, one valley has the long axis oriented along the field, while the three remaining valleys have their long axes at an angle to the field — the field-induced anisotropy appears.

Silicon has six equivalent valleys located at $\langle 100 \rangle$ axes. The electric field directed along the $\langle 100 \rangle$ axis finds the valleys in nonequivalent positions: The long axes of two valleys ($\langle 100 \rangle$ and $\langle \overline{1}00 \rangle$) are parallel to the field, and the remaining four valleys (of $\langle 010 \rangle$ and $\langle 001 \rangle$ type) are transverse to the field. In this case, a more efficient heating of electrons takes place in the four valleys. The electrons tend to "evaporate" from the "hot" valleys and "condense" in the "cold" valleys; the hot-electron intervalley repopulation leads to a decrease in the mobility (averaged over all valleys). The intervalley noise of hot-electrons appears as well. In another configuration — $\mathbf{E} \parallel \langle 111 \rangle$ — all six valleys of silicon remain equivalent, and no intervalley noise appears in the longitudinal direction. Experiments confirm the expectations.

Figure 9.10 presents the experimental results on the longitudinal noise temperature $T_{n\|}(E)$ for n-type germanium at 80 K lattice temperature. Closed squares stand for the configuration with the electric field applied along the ⟨111⟩ direction. For comparison, the longitudinal noise temperature measured in the configuration $\mathbf{E} \parallel ⟨100⟩$ is given by open squares. In agreement with the expectations based on Eq. (9.6), the longitudinal noise temperature in the ⟨111⟩ direction is higher that that in the ⟨100⟩ direction at the fields supporting the hot-electron intervalley transfer.

The experimental data for n-type silicon (Fig. 9.11) differ from those for n-type germanium: the longitudinal noise temperature in $\mathbf{E} \parallel ⟨111⟩$ configuration

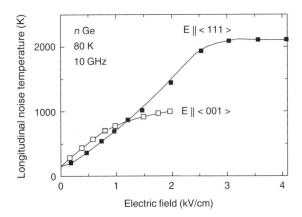

Figure 9.10. Hot-electron longitudinal noise temperature $T_{n\|}$ measured at 10-GHz frequency for n-type germanium, $\rho(300\text{ K}) = 10$ ohm·cm, in $\mathbf{E} \parallel ⟨111⟩$ and $\mathbf{E} \parallel ⟨001⟩$ configurations at 80 K lattice temperature [177]. Solid curves are guides to the eye.

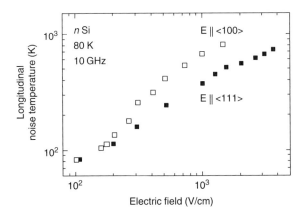

Figure 9.11. Hot-electron longitudinal noise temperature $T_{n\|}$ measured at 10-GHz frequency for n-type silicon, $\rho(300\text{ K}) = 30$ ohm·cm, in $\mathbf{E} \parallel ⟨111⟩$ and $\mathbf{E} \parallel ⟨100⟩$ configurations at 80 K lattice temperature [177].

is lower than that measured in **E** ∥ ⟨100⟩ configuration. Nevertheless, there is no contradiction to the expectations based on the band structure: The lowest valleys in the conduction band of silicon are located along ⟨100⟩ crystallographic directions; correspondingly, no intervalley noise is observed, provided that the electric field is directed along an ⟨111⟩ crystallographic axis.

9.6 STANDARD ω-SPECTROSCOPY OF HOT-ELECTRON NOISE IN SILICON

Generation–Recombination and Hot-Electron Noise

Measurements of noise spectra at low frequencies necessitate application of long pulses of voltage, and the consumed (dissipated) power limits the range of electric fields where hot-electron effects can be investigated experimentally. For n-type silicon, long pulses of electric field up to 200 V/cm could be applied at 78 K temperature without a noticeable effect of lattice heating due to the Joule effect [180]. Figure 9.12 illustrates the experimental results on spectral intensity of longitudinal current fluctuations measured in ⟨100⟩ direction over the wide range of frequencies. As expected, the $1/f$ fluctuations are observed at low frequencies. In addition to these fluctuations, two plateaus of the excess noise are resolved in the frequency range below 10 GHz. The fluctuations of electron number in the conduction band dominate at frequencies below 50 MHz, while the hot-electron velocity fluctuations prevail at microwave frequencies. Solid curve is the fitted approximation assuming 20 ns and 50 ps relaxation times for the two Lorentzians. The hot-electron noise (broken curve) is the most important contribution in the

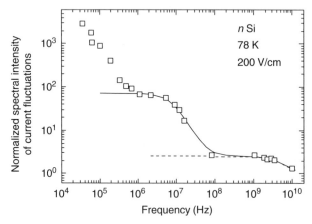

Figure 9.12. Experimental results (□) on normalized spectral intensity $S_{I_\parallel}(E)/S_I(0)$ of longitudinal current fluctuations in n-type silicon ($n_0 = 3 \times 10^{13}$ cm^{-3}) in **E** ∥ ⟨100⟩ configuration ($E = 200$ V/cm) at 78 K lattice temperature [180]. In the frequency range from 1 MHz to 10 GHz, two Lorentzian contributions (curves) with the time constants $\tau^{gr} = 20$ ns and $\tau^{inter} = 50$ ps are resolved.

frequency range $\omega\tau^{gr} \gg 1$, where $\tau^{gr} = 20$ ns is the relaxation time assumed for the generation–recombination process.

The step at $\omega\tau^{gr} \sim 1$ due to generation–recombination fluctuations shifts toward higher frequencies at higher electric fields. This behavior has been evidenced experimentally in p-type silicon at 77 K lattice temperature and interpreted in terms of the hot-hole effect on trapping and release probabilities (see reference 10). Eventually, in the range of frequencies from 100 MHz to 1 GHz, the interplay of hot-hole velocity fluctuations and hot-hole effect on generation–recombination fluctuations takes place in p-type silicon [207] (for the interplay analysis see references 208 and 209 and review papers [10, 210]).

"Thermal", Convective, and Intervalley Noise in Si

Important information on the origin of hot-electron fluctuations in n-type Si at $\omega\tau^{gr} \gg 1$ follows from comparison [180] of the longitudinal fluctuations measured for two directions of applied electric fields, $\mathbf{E} \parallel \langle 100 \rangle$ and $\mathbf{E} \parallel \langle 111 \rangle$ (open and closed squares in Fig. 9.13). According to the conduction-band structure of silicon, all valleys are oriented at the same angle to the electric field in the $\mathbf{E} \parallel \langle 111 \rangle$ configuration, and there is no intervalley noise, but the intervalley noise is activated in the other configuration $\mathbf{E} \parallel \langle 100 \rangle$ [see Eq. (9.6)].

The results of Monte Carlo simulation of the longitudinal velocity fluctuations (Fig. 9.13, solid lines [180]) give a satisfactory description of the experimental data (Fig. 9.13, symbols [180]). In the configuration corresponding to no

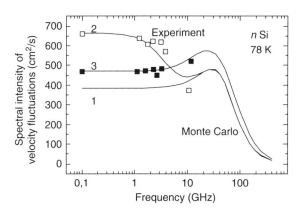

Figure 9.13. Experimental and simulated spectra of longitudinal velocity fluctuations of hot electrons in n-type silicon ($n_0 = 3 \times 10^{13}$ cm^{-3}) at 78 K lattice temperature and 200 V/cm electric field [180]. The intervalley fluctuations contribute at frequencies below ~ 10 GHz for $\mathbf{E} \parallel \langle 100 \rangle$ configuration, and the convective fluctuations contribute at frequencies below ~ 20 GHz in $\mathbf{E} \parallel \langle 111 \rangle$ and $\mathbf{E} \parallel \langle 100 \rangle$ configurations. Experimental results: □ represents $\mathbf{E} \parallel \langle 100 \rangle$, ■ represents $\mathbf{E} \parallel \langle 111 \rangle$. Results of Monte Carlo simulation: 1 represents $\mathbf{E} \parallel \langle 100 \rangle$, intervalley transitions neglected; 2 represents $\mathbf{E} \parallel \langle 100 \rangle$, intervalley transitions included; 3 represents $\mathbf{E} \parallel \langle 111 \rangle$, the intervalley transitions are included, but they cause no intervalley fluctuations because $v_1^{dr} = v_2^{dr}$, [see Eq. (96)].

110   HOT-ELECTRON MICROWAVE NOISE IN ELEMENTARY SEMICONDUCTORS

intervalley noise, the negative contribution due to convective noise appears at frequencies below $\omega \sim 1/\bar{\tau}^{en}$ (curve 3 of Fig. 9.13, cf. Fig. 7.1). The numerical results treated in terms of Eq. (9.5) lead to the following fitting parameters at $E = 200$ V/cm, $T_0 = 78$ K: the energy relaxation time $\bar{\tau}^{en} \approx 15$ ps and the momentum relaxation time $\bar{\tau} \approx 2$ ps (curve 3 in Fig. 9.13).

The convective noise and the intervalley noise compete in the $\mathbf{E} \parallel \langle 100 \rangle$ configuration. One obtains $\bar{\tau}^{en} \approx 5$ ps for the energy relaxation time ($\langle 100 \rangle \parallel \mathbf{E} = 200$ V/cm, $T_0 = 78$ K). The energy relaxation time appears to be shorter than the intervalley time $\bar{\tau}^{inter} \approx 50$ ps, and the local minimum is resolved at frequencies $(\bar{\tau}^{inter})^{-1} < \omega = 2\pi f < (\bar{\tau}^{en})^{-1}$ as evidenced by the Monte Carlo simulation data (curve 2 in Fig. 9.13).

9.7 E-SPECTROSCOPY OF HOT-ELECTRON NOISE

As mentioned, the energy relaxation time decreases as the applied electric field is increased (see Fig. 9.3). A similar behavior is observed for the intervalley relaxation time $\bar{\tau}^{inter}$ (Figs. 9.14 and 9.15). This can be used for the E-spectroscopy of noise (see Section 8.2) — the mth source of noise vanishes in the range of electric fields where $\omega \tau_m(\mathbf{E}) \gg 1$. The field-dependent interplay in n-type germanium and silicon is illustrated in Figs. 9.16 and 9.17.

Interplay of Noise Sources in E-Spectroscopy: *n*-Type Ge

Figure 9.16 presents the spectral intensity of current fluctuations normalized to its value at zero electric field, $S_{I\parallel}(\omega, \mathbf{E})/S_{I\parallel}(\omega, 0)$, measured for n-type Ge at the fixed 10 GHz frequency [177]. The experimental results correspond to two different configurations. In the $\mathbf{E} \parallel \langle 111 \rangle$ configuration, the intervalley transfer

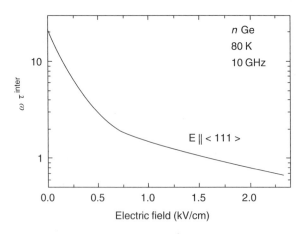

Figure 9.14. Calculated [186] dependence of $\omega \bar{\tau}^{inter}$ against electric field $\mathbf{E} \parallel \langle 111 \rangle$ for n-type Ge at 80 K lattice temperature; $f = 10$ GHz.

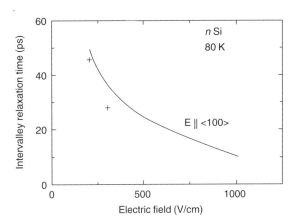

Figure 9.15. Dependence of intervalley relaxation time on electric field in $\mathbf{E} \parallel \langle 100 \rangle$ configuration for n-type silicon ($n_0 = 3 \times 10^{13}$ cm^{-3}) at 80 K lattice temperature: ++ are experimental results [75], solid curve stands for Monte Carlo simulation data [163].

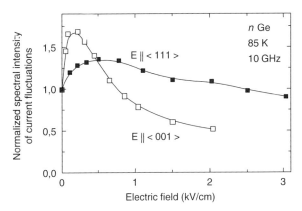

Figure 9.16. Experimental dependence on electric field of normalized spectral intensity of longitudinal current fluctuations $S_I(E)/S_I(0)$ at 10-GHz frequency for n-type germanium, $\rho(300\ \text{K}) = 10$ ohm·cm, in $\mathbf{E} \parallel \langle 111 \rangle$ and $\mathbf{E} \parallel \langle 001 \rangle$ configurations at 85 K lattice temperature [177]. Solid curves are guides to the eye.

takes place; it is responsible for (i) the intervalley noise and (ii) the reduced electron mobility (the electrons tend to leave the "hot" high-mobility valleys for the "cold" low-mobility valleys). The latter effect leads to a weaker hot-electron "thermal" noise, observed in the range of fields $\mathbf{E} < 0.5$ kV/cm, where the intervalley fluctuations cannot manifest themselves at 10-GHz frequency because the intervalley relaxation time becomes too long: $\omega \bar{\tau}^{\text{inter}}(\mathbf{E}) > 1$ (cf. Fig. 9.14).

The condition $\omega \bar{\tau}^{\text{en}} \sim 1$ is reached at moderate fields near 200 V/cm in the $\mathbf{E} \parallel \langle 001 \rangle$) configuration (for independent experimental data on the energy relaxation time, see reference 204). This implies the cutoff of the convective contribution as the electric field decreases: The spectral intensity of current fluctuations passes the maximum near $\omega\bar{\tau}^{en} \sim 1$ (for a similar field-dependent behavior in p-type germanium and p-type silicon, see Figs. 9.3 and 9.4).

A similar maximum forms in the configuration $\mathbf{E} \parallel \langle 111 \rangle$. However, this configuration supports the intervalley fluctuations as well; their cutoff condition $\omega\bar{\tau}^{inter} \sim 1$ is reached at fields $E \sim 1$ kV/cm (see Fig. 9.14). At higher fields, the intervalley contribution becomes the dominant one; it compensates for the negative convective contribution exceeding the latter in the absolute value. As a result, $S_{I\parallel}(\omega)$ for $\mathbf{E} \parallel \langle 111 \rangle$ exceeds $S_{I\parallel}(\omega)$ for $\mathbf{E} \parallel \langle 001 \rangle$) at high electric fields.

Interplay of Noise Sources in E-Spectroscopy: *n*-Type Si

In *n*-type silicon, the field applied in $\langle 111 \rangle$ direction finds all valleys oriented at equivalent angles to the field: No hot-electron intervalley noise is agitated (Fig. 9.11). In this configuration, the spectral intensity of longitudinal current fluctuations demonstrates the interplay of convective and hot-electron "thermal" noise (Fig. 9.17) appearing at a fixed 10-GHz frequency due to the dependence of the energy relaxation time on the electric field. The intervalley noise in *n*-type silicon manifests itself in the $\mathbf{E} \parallel \langle 100 \rangle$ configuration (Fig. 9.17). The condition $\omega\bar{\tau}^{inter} \sim 1$ is met at 10-GHz frequency near 1 kV/cm [211]. The condition $\omega\bar{\tau}^{en} \sim 1$ is satisfied at fields $E \sim 200$ V/cm (cf. Fig. 9.13). As a result, the intervalley noise is cutoff at $E < 1$ kV/cm, and the convective noise is cutoff at $E < 200$ V/cm. The cutoffs form two maxima of $S_{I\parallel}(\omega)$ observed in the $\mathbf{E} \parallel \langle 001 \rangle$) configuration (Fig. 9.17).

Figure 9.17 also shows that the spectral intensity of the transverse current fluctuations exceeds those in the longitudinal direction. This effect is twofold. First of all, the electron mobility is extremely high in the direction normal to the long axes of the valleys. Consequently, the electron heating by electric field $\mathbf{E} \parallel \langle 001 \rangle$

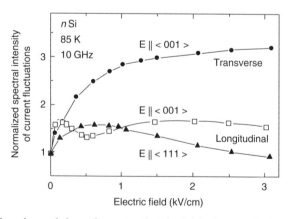

Figure 9.17. Experimental dependence on electric field of normalized spectral intensity $S_I(E)/S_I(0)$ of longitudinal (□, ▲) and transverse (●) current fluctuations at 10-GHz frequency for *n*-type silicon, $\rho(300 \text{ K}) = 30$ ohm·cm, at 85 K lattice temperature in $\mathbf{E} \parallel \langle 111 \rangle$ and $\mathbf{E} \parallel \langle 001 \rangle$ configurations [177]. Solid curves are guides to the eye.

is very efficient in the valleys $\langle 100 \rangle$, $\langle \bar{1}00 \rangle$, $\langle 0\bar{1}0 \rangle$, and $\langle 010 \rangle$. Moreover, there is no negative contribution of the convective noise in the transverse direction.

9.8 LONGITUDINAL AND TRANSVERSE DIFFUSION COEFFICIENTS

Tensor of Diffusion Coefficients

The first experimental results on diffusion coefficient tensor components for hot majority carriers were obtained using the noise technique [154]. The longitudinal and transverse hot-electron noise temperatures were measured at 9.6 GHz for n-type Ge, and the Price relation was used to obtain the tensor components of the hot-electron diffusion coefficient (Fig. 9.18). The transverse component was found to exceed its value at equilibrium. The longitudinal component decreases as the electric field increases. As mentioned earlier (see Section 9.3), the energy fluctuations contribute to the longitudinal rather than the transverse fluctuations of velocities. For nonuniform electron distribution, the electron density gradient causes a change in electron energy responsible for the corresponding contribution to the diffusion coefficient in the direction of the steady current (see reference 112). Consequently, Fig. 9.18 gives an experimental evidence of the negative contribution of the convective noise to the spectral intensity of longitudinal current fluctuations and of the corresponding negative contribution to the longitudinal diffusion coefficient.

Noise and Time-of-Flight Experiments

It would be interesting to compare the results on hot-carrier diffusion obtained by the noise technique to those available from other experiments. The time-of-flight

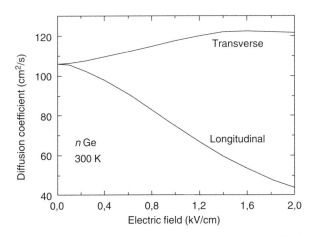

Figure 9.18. Hot-electron diffusion coefficient components in longitudinal and transverse directions to the steady current for n-type germanium ($n_0 = 2 \times 10^{14}$ cm^{-3}) at 300 K lattice temperature [154].

technique (see references 31 and 212) provides us with a direct observation of the longitudinal diffusion. In this technique, a sheet of electrons (or holes) drifts in an electric field in a semi-insulating plate placed into a charged condenser. The shape of the pulse of the discharge current contains information on the average time of flight and its dispersion, with the latter being dependent on the sheet spreading, the hot-electron diffusion is among other possible causes.

Noise and time-of-flight experiments are difficult to perform on exactly the same material because of almost incompatible requirements inherent to these techniques. Insulating or semi-insulating samples with blocking electrodes are preferable in the time-of-flight experiment, while the noise experiment must be performed on doped (though better on lightly doped) samples with ohmic electrodes. As mentioned, the latter requirements are important for matching the sample to the input circuit of the microwave radiometer and in order to avoid shot noise. Despite these difficulties, a few experiments allowed valuable comparisons. Figure 9.19 compares the longitudinal components of hot-hole diffusion coefficient available from experiments on noise (closed circles) and spreading (open circles) performed for silicon at 300 K [10]. The agreement is good in all the range of electric fields where the results are available from the both experiments.

Time-of-Flight Experiments and Monte Carlo Simulation

Figure 9.20 illustrates the crystallographic anisotropy of longitudinal diffusion of hot-electrons in germanium measured at 77 K temperature by time-of-flight technique [179, 213]. The experiments confirm that the configuration $\mathbf{E} \parallel \langle 111 \rangle$ supports the intervalley contribution to diffusion: The longitudinal diffusion coefficient in this configuration exceeds that in the configuration $\mathbf{E} \parallel \langle 100 \rangle$. The results

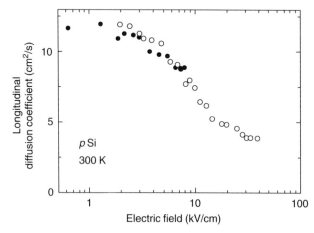

Figure 9.19. Longitudinal diffusion coefficient of hot holes in silicon at 300 K lattice temperature. The results obtained from noise experiments (●) match those available from time-of-flight experiments (○) [10].

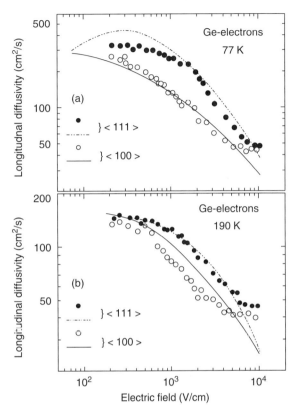

Figure 9.20. Hot-electron longitudinal diffusion coefficient for germanium at 77 K (a) and 190 K (b) lattice temperatures in $\mathbf{E} \parallel \langle 111 \rangle$ and $\mathbf{E} \parallel \langle 100 \rangle$ configurations. Time-of-flight experiments (symbols) and Monte Carlo simulation (curves) are shown [179, 213].

of Monte Carlo simulation [213, 214] give a satisfactory description of the experimental data in n-type germanium (Fig. 9.20) and other semiconductors (for the details on Monte Carlo technique see reference 31). The experimental results on the diffusion coefficient (Fig. 9.20) agree with those on the intensity of current fluctuations (Fig. 9.16) at fields exceeding 1 kV/cm where the frequency dependence of the hot-electron noise is not important in the X-band of microwave frequencies (see Section 9.5).

CHAPTER 10

HOT-ELECTRON MICROWAVE NOISE IN GaAs AND InP

Gallium arsenide and indium phosphide are the basic materials for devices operating in microwave frequency range. Their noise properties are of direct interest for device performance, and this chapter considers sources of hot-electron noise manifesting themselves in this frequency range. Hot-electron characteristics are sensitive to the zero-field mobility, and a special attention is paid to the effects of doping and lattice temperature. High-frequency noise of long samples are under discussion; their properties can serve as a reference for understanding noise in small-size and low-dimensional structures (Chapters 11, 16, 17, and 20).

10.1 INTERVALLEY NOISE IN GaAs AND InP

Intervalley Transfer and Associated Phenomena

Conduction band of direct-band-gap compound semiconductors differs essentially from that of silicon and germanium. Equilibrium electrons occupy the lowest single valley — Γ valley — where the electron mobility is high. The upper low-mobility valleys with minima at L and X points are located on $\langle 111 \rangle$ and $\langle 100 \rangle$ crystallographic axes correspondingly (see Chapter 15). The upper valleys are usually empty at equilibrium (unless the lattice temperature is high). The edge energy of L valleys with respect to that of Γ valley in GaAs was determined as $\mathcal{E}_{L\Gamma} = 310 \pm 10$ meV, and the same for X valleys is $\mathcal{E}_{X\Gamma} = 485 \pm 10$ meV at 2 K lattice temperature [215]. Naturally enough, a high electric field is needed for the electron transfer to the upper valleys at room (and a lower) temperature.

The intervalley transfer of hot electrons plays an important role in the high-field transport of compound semiconductors, and many devices are based on this phenomenon. The transfer into the upper valleys leads to a strongly sublinear current–voltage characteristic. The differential conductivity changes its sign at a certain threshold field, and the effect of the negative differential mobility takes place at high electric fields [216–218] leading to the Gunn effect [219] and the associated instabilities (see references 220–222). The intervalley transfer of hot electrons is responsible for the longitudinal fluctuations of the instantaneous (fluctuations contained) drift velocity averaged over all valleys. The intervalley noise of this type appears in GaAs [181] and InP [182] in the range of sub-threshold fields for the Gunn effect.

Intervalley Noise in Lightly Doped GaAs and InP

Figure 10.1 presents the field-dependent spectral intensity of longitudinal velocity fluctuations $S_{V_\parallel}(\omega, \mathbf{E})$ measured in a long ($L = 1$ mm) n-type GaAs sample at 10-GHz frequency [181]. An essential increase in the intensity occurs at electric fields exceeding 2 kV/cm. The phenomenon is explained in terms of the intervalley fluctuations caused by random reversible transitions of hot electrons from Γ valley into the upper L valleys and backwards as confirmed by Monte Carlo simulation assuming the three-valley (Γ–L–X) conduction band model [223].

Since L valleys are located approximately 0.3 eV above the conduction band edge, a strong electric field is required for the electrons to gain sufficient energy before undergoing the intervalley transfer. Solid curves in Fig. 10.2 support the above interpretation: the electron number in L valleys, n_L/n_0, becomes of importance at $E > 2$ kV/cm; at the same electric field the noise temperature $T_{n\parallel}$ is found to exceed considerably the electron temperature T_e in Γ valley. The contribution of X valleys becomes important at extremely high electric fields (for details see Chapter 11).

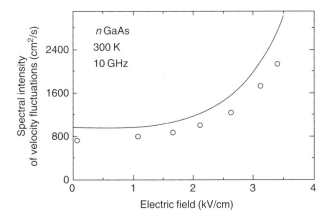

Figure 10.1. Field-dependent spectral intensity of longitudinal hot-electron velocity fluctuations S_v at 10-GHz frequency for GaAs at 300 K lattice temperature. ○ represents experimental results [181] ($n_0 = 3 \times 10^{15}$ cm^{-3}, $\mu_0 = 6000$ cm^2/(V·s)), solid line represents results of Monte Carlo simulation within three-valley Γ–L–X model [223].

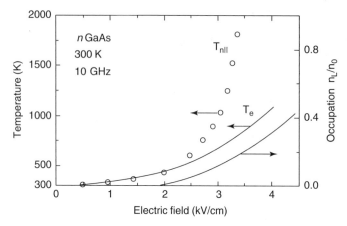

Figure 10.2. Dependence on electric field of the experimental (○) hot-electron longitudinal noise temperature at 10 GHz frequency for GaAs ($n_0 = 3 \times 10^{15}$ cm^{-3}, $\mu_0 = 6000$ cm^2/(V·s)) at 300 K lattice temperature [181] together with the calculated [223] electron temperature T_e in Γ valley and the relative electron density n_L/n_0 in L valleys (solid curves).

There is a close fit of the longitudinal noise temperature $T_{n\parallel}$ and the electron temperature T_e in the range of electric fields below the onset of the intervalley noise. This feature gives an opportunity to estimate the effective energy relaxation time in Γ valley. The estimation based on Eq. (9.2) leads to $\bar{\tau}^{en} \approx 0.7$ ps at $E < 2$ kV/cm for GaAs ($n_0 = 3 \times 10^{15}$ cm^{-3}, $\mu_0 = 6000$ cm^2/(V·s) at 300 K).

The electron mobility in InP is lower, and the intervalley separation energy is wider, than those in GaAs. Consequently, higher electric fields are required for the hot-electron intervalley transfer to appear [219]. Indeed, the threshold field for the negative differential mobility due to the intervalley transfer is around 12 kV/cm in InP at 300 K in contrast to 3.5 kV/cm in GaAs. The intervalley noise appears at the subthreshold electric fields and becomes the dominant source among hot-electron sources at fields over 2 kV/cm in GaAs [181] and over 6 kV/cm in InP [182] (Fig. 10.3). This is an illustration that even a small number of high-energy electrons (available, e.g., at $E \sim 6$ kV/cm in InP) is essential for the intervalley noise to prevail over the other sources of hot-electron noise.

The experimental results on the longitudinal noise temperature for n-type InP [182] (Fig. 10.4, symbols) in the frequency range from 100 MHz to 10 GHz can be interpreted in terms of sources of noise caused by electron number and velocity fluctuations (Fig. 10.4, solid lines). The electron number fluctuations dominate at frequencies below 1 GHz. Essentially the same behavior is observed in n-type GaAs [224]. At X-band frequencies, the longitudinal noise is caused by the kinetic processes inside the conduction band; the contribution of the generation–recombination noise is negligible. The latter fluctuations are cut off because of the long relaxation time of the generation–recombination processes. Again, 10-GHz frequency proves to be quite suitable for investigation of dependence

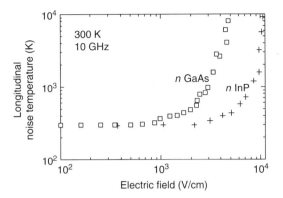

Figure 10.3. Hot-electron longitudinal noise temperature $T_{n\parallel}$ measured at 10-GHz frequency at 300 K lattice temperature [184]. □ represents GaAs, $L = 7.5\,\mu\text{m}$, $n_0 = 9 \times 10^{14}\,\text{cm}^{-3}$, $\mu_0 = 7500\,\text{cm}^2/(\text{V·s})$; + represents InP, $L = 10\,\mu\text{m}$, $n_0 = 3.2 \times 10^{15}\,\text{cm}^{-3}$, $\mu_0 = 4600\,\text{cm}^2/(\text{V·s})$.

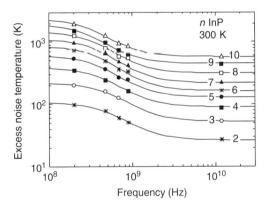

Figure 10.4. Spectra of longitudinal excess noise temperature $(T_{n\parallel} - T_0)$ at 300 K lattice temperature [182] for lightly doped n-type InP ($n_0 = 2.7 \times 10^{15}\,\text{cm}^{-3}$) at different electric field in kV/cm (numbers at the symbols). Contribution due to generation–recombination noise is important at $f < 1$ GHz (curves).

of hot-electron velocity fluctuations on electric field, on lattice temperature, on doping, and on other semiconductor parameters.

Dependence of Fluctuation Intensity on Intervalley Coupling

According to Eq. (9.6), the intervalley contribution to the spectral intensity of velocity fluctuations is proportional to the intervalley relaxation time. The latter depends on the squared intervalley coupling constant; and the inverse proportionality holds, provided that the intravalley processes are faster than the intervalley transfer. The coupling constant can be estimated by comparing the experimental

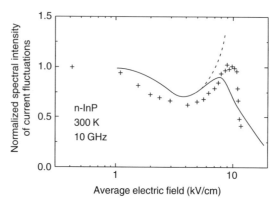

Figure 10.5. Normalized spectral intensity of longitudinal current fluctuations $S_{I\parallel}(E)/S_I(0)$ at 10-GHz frequency for n-type InP at room temperature. Experimental results: + represents $L = 10$ μm, $n_0 = 3.2 \times 10^{15}$ cm^{-3}, $\mu_0 = 4600$ cm^2/(V·s) [194]. Results of simulation [226] assuming different intravalley coupling constant: solid curve — 1×10^9 eV/cm; broken curve — 3×10^8 eV/cm.

results to those of model calculations. Simulation of hot-electron velocity fluctuations [225] predicted the intervalley-related maximum of the spectral intensity of longitudinal velocity fluctuations in n-type InP to appear at electric field strength near 8 kV/cm. The experimental results [194] ("plus" signs in Fig. 10.5) are in reasonable agreement to those of simulation [226] performed under the assumption that the coupling constant equals 1×10^9 eV/cm (Fig. 10.5, solid line). A lower value of the coupling constant leads to a longer intervalley relaxation time and to a higher intervalley contribution to the intensity of velocity fluctuations [cf. Eq. (9.6)], as illustrated by broken curve in Fig. 10.5.

The spectral intensity of the intervalley fluctuations in lightly doped GaAs [181] is essentially higher than that in InP. Correspondingly, a lower value of the intervalley coupling constant is expected. The problem was considered in the framework of a three-valley (Γ–L–X) model; a rather low Γ–L coupling constant, 1.8×10^8 eV/cm, was proposed [223] to fit the time-of-flight data on the hot-electron diffusion [161]. The model predicted a strong frequency dependence at around 10 GHz; this behavior was not confirmed by the experimental data [183], and an intermediate value of the Γ–L coupling constant, 3×10^8 eV/cm, was assumed to avoid contradictions of the three-valley model to the experimental data on the hot-electron noise (see reference 12 and references therein). The hot-electron luminescence experiments performed at 2 K for heavily acceptor-doped GaAs lead to $(6.5 \pm 1.5) \times 10^8$ eV/cm [227].

10.2 THERMALLY QUENCHED HOT-ELECTRON INTERVALLEY NOISE

In this section we shall consider the effect of electron scattering in the central Γ valley on the intervalley noise. In lightly doped semiconductors at high electric fields, the scattering is controlled by phonons, and the scattering rate can be easily changed with the lattice temperature; cooling of the lattice is accompanied by a weaker electron–phonon scattering. Stronger hot-electron effects are expected at lower lattice temperatures.

Figure 10.6 presents the effect of the lattice temperature on the spectral intensity of the velocity fluctuations for lightly doped n-type GaAs [181, 228]. According to the Nyquist theorem, valid in the absence of an electric field, the intensity of velocity fluctuations is directly proportional to the lattice temperature and the electron mobility. Since the mobility strongly increases under cooling in lightly doped semiconductors, the net effect is often decided by the mobility rather than the lattice temperature. On the other hand, a higher mobility causes stronger hot-electron effects. In particular, the deviations from Ohm's law are better pronounced at the lower temperature. A stronger dependence of the intensity of the velocity fluctuations on electric field is observed at the lower temperature (Fig. 10.6, filled circles).

The field-dependent longitudinal noise temperature measured for the lightly doped n-type GaAs at two lattice temperatures is given in Fig. 10.7 (unfilled and filled circles) [184, 192]. The increase of the noise temperature with the applied electric field is steeper at the lower temperature; the curves cross at a certain electric field value. This illustrates once more the stronger hot-electron effects at cryogenic temperatures: At high fields (where the intervalley noise dominates) the noise temperature is lower at a higher lattice temperature. In other words, the intervalley noise is thermally quenched as the lattice temperature is increased. This is in line with simple considerations based on the power consumed and dissipated by the electrons present in Γ valley. However, the dependence of the longitudinal noise temperature on the electric field shows features at intermediate fields that cannot be interpreted in terms of competition of hot-electron "thermal,"

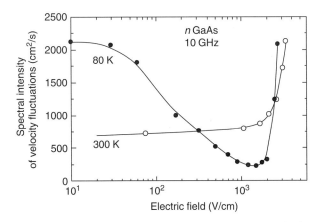

Figure 10.6. Spectral intensity of longitudinal hot-electron velocity fluctuations $S_{v_\parallel}$ at 10-GHz frequency for n-type GaAs at different lattice temperature. ○ represents 300 K, $n_0 = 3 \times 10^{15}$ cm^{-3}, $\mu_0 = 6000$ cm^2/(V·s) [181]; ● represents 80 K, $n_0 = 9 \times 10^{14}$ cm^{-3}, $\mu_0 = 77{,}300$ cm^2/(V·s) [228]. Solid curves are to guide the eye.

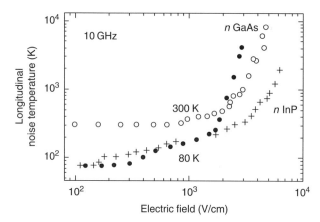

Figure 10.7. Hot-electron longitudinal noise temperature $T_{n\|}$ at 10-GHz frequency: ○ represents GaAs, 300 K, $n_0 = 9 \times 10^{14}$ cm^{-3}, $\mu_0 = 7500$ cm^2/(V·s) [184]; ● represents GaAs, 80 K, $n_0 = 9 \times 10^{14}$ cm^{-3}, $\mu_0 = 77,300$ cm^2/(V·s) [192]; + represents InP, 80 K, $n_0 = 3 \times 10^{15}$ cm^{-3}, $\mu_0 = 48,700$ cm^2/(V·s) [192].

convective, and intervalley noise in the way discussed in Chapter 9. We shall deal with this problem in more detail in the following sections.

10.3 TERMINATED RUNAWAY OF LUCKY ELECTRONS

A relatively weak noise source is resolved [184, 192] over a wide range of electric fields: 0.3 kV/cm < E < 2 kV/cm in GaAs and 0.15 kV/cm < E < 4 kV/cm in InP at 80 K lattice temperature (Fig. 10.7). The source is better expressed in InP (crosses in Fig. 10.7). No analytic model has been proposed to consider hot-electron noise of this type. According to Monte Carlo simulation, the resonant behavior due to the streaming motion is heavily damped in the range of fields $E > 200$ V/cm in GaAs at 78 K lattice temperature [229]: The electron does not emit an optical phonon immediately, having reached the energy $\varepsilon = \hbar\omega^{opt}$, and its acceleration by the field continues for some time. This penetration of electrons into the active region (cf. Fig. 7.2) leads to a strong damping of the periodic emission of optical phonons [cf. Eq. (9.4)], and the associated resonance-type fluctuations disappear. No noise due to the streaming motion (discussed in Section 9.4) takes place in InP at 80 K lattice temperature as well. Optical phonon scattering, dominating in n-type GaAs and InP in the range of fields in question, supports the source of noise of different origin.

The simulation of hot-electron fluctuations in InP shows that the noise source at the intermediate fields is sensitive to electron–acoustic-phonon coupling [230]. It is not straightforward to understand this because the acoustic phonons scattering is relatively weak at high electric fields. An explanation is as follows. Under the dominant *small-angle* optical phonon scattering, some "lucky" electrons gain more energy than can dissipate — a "polar runaway" of "lucky" electrons develops. This nearly collimated acceleration of a "lucky" electron is disturbed by rare *large-angle* scattering events (often due to acoustic phonons). The process is steady: No breakdown and no current instabilities appear in this range of fields. The number of "lucky" electrons being low, their contribution to the average drift velocity is difficult to resolve. Nevertheless, the noise technique is sensitive enough to deal with the fluctuations caused by the low number of the "lucky" electrons present at the subthreshold fields for the intervalley transfer. The runaway termination by acoustic phonons is expected [230] to be responsible for the observed shoulder of hot-electron noise at the intermediate electric fields (Fig. 10.7).

Of course, acoustic phonons have a limited possibility to terminate the electron runaway, and the runaway causes an onset of the intervalley transfer at sufficiently high electric fields (Fig. 10.7). A wider range of the fields for acoustic-phonon terminated runaway is observed in InP; it is a natural consequence of its essentially wider intervalley energy gap as compared to GaAs — more **k**-space is given for the "lucky" electrons to perform their runaway fluctuations.

It is noteworthy that the identification of the terminated-runaway noise is an excellent illustration of the fluctuation-based methods of diagnostics — the fluctuations are usually much more sensitive to subtle details of band structure and scattering than the averages, like mobility and drift velocity. Exploitation of noise as a diagnostic tool is widely known; the method is successfully used, for example, for early diagnostics of electric breakdown in devices under the pre-breakdown bias — the device damage is avoided in this way and in many other cases. Fluctuations often serve as "fingerprints" for identification of the underlying nonequilibrium kinetic processes difficult to resolve by other means.

10.4 DOPING-DEPENDENT HOT-ELECTRON NOISE

Doping-Dependent Hot-Electron Velocity Fluctuations

The experimental data presented in the previous sections of this chapter correspond to GaAs and InP containing a relatively low density of electrons (10^{15} cm^{-3}); field-effect transistors exploit channels containing an essentially higher density of electrons. Doping introduces at least two scattering mechanisms: electron scattering by impurities and interelectron collisions. The ionized impurity scattering is the main impurity scattering mechanism important for electron transport in doped semiconductors. According to Conwell–Weisskopf and Brooks–Herring models [31, 231], the ionized impurity scattering manifests itself at low lattice temperatures at low and moderate electric fields.

Effects of ionized impurity scattering and interelectron collisions on hot-electron transport and noise in GaAs were repeatedly considered by the Monte Carlo technique [232–247]. The interelectron collisions play an important role eventually restricting the applicability of the hot-electron fluctuation–diffusion relation, as emphasized in Chapter 6. In the case of standard-doped GaAs channels for field-effect transistors (where the electron density is high, exceeding

10^{17} cm^{-3}), the problem of the interelectron collisions is considered in detail [246, 247], and the recent advances are described in Chapters 13 and 14. The developed approach is of direct interest also for quantum well channels containing a very high density of mobile electrons. Hot-electron noise in selectively doped quantum well channels is considered in Chapters 16 and 17.

The spectral intensity of the hot-electron velocity fluctuations for the lightly doped and the standard-doped GaAs at 80 K temperature are compared in Fig. 10.8 [228]. According to the Nyquist theorem, the spectral intensity at zero field is proportional to the electron mobility that is quite sensitive to doping. Of course, the doping effect remains important at low and moderate fields, but it tends to diminish as the electron energy increases with electric field. This experimental result is in qualitative agreement with the simulation, taking into account ionized impurity scattering [234, 242]. For quantitative agreement, one should consider also interelectron collisions manifesting themselves at moderate electric fields, ranging from 5 V/cm to 500 V/cm in the standard-doped GaAs channels at 80 K lattice temperature [246, 247].

Now, let us demonstrate the effect of the interelectron collisions in the case of comparatively light doping ($n_0 \sim 10^{15}$ cm^{-3}). Figure 10.8 compares the experimental results for GaAs (unfilled circles) with those of Monte Carlo simulation [248], taking into account (solid curve) the interelectron collisions in the framework of the model developed for the standard-doped channels [246, 247] (cf. Chapter 13). Provided that the interelectron collisions were neglected, no agreement of the experiment and the simulation would be reached at low/moderate electric fields, as illustrated by the dotted curve in Fig. 10.8. Consequently, the

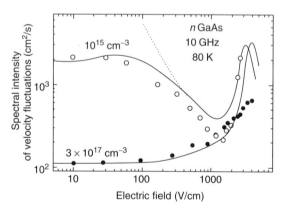

Figure 10.8. Spectral intensity of longitudinal hot-electron velocity fluctuations $S_{v_\|}$ at 10-GHz frequency at 80 K lattice temperature for lightly doped and standard-doped GaAs. Experimental data [228]: ○ represents $n_0 = 9 \times 10^{14}$ cm^{-3}, $\mu_0 = 77{,}000$ cm^2/(V·s); ● represents $n_0 = 3 \times 10^{17}$ cm^{-3}, $\mu_0 = 4000$ cm^2/(V·s). Monte Carlo simulation within Γ–L–X model assuming 5×10^8 eV/cm for Γ–L coupling constant [248]. Dotted curve indicates that interelectron collisions are neglected; solid curves indicate that interelectron collisions are taken into account ($n_0 = 1 \times 10^{15}$ cm^{-3}, $N^{\text{ion}} = 2.5 \times 10^{15}$ cm^{-3}; $n_0 = 3 \times 10^{17}$ cm^{-3}, $N^{\text{ion}} = 7.5 \times 10^{17}$ cm^{-3}).

interelectron collisions are important in GaAs even under comparatively light doping.

According to the results of Monte Carlo simulation, the effect of doping is strong at low and moderate electric fields (Fig. 10.8, solid curves). At high electric fields, the results of simulation at two doping levels tend to merge; the model predicts only a small shift, due to doping, of the maximum caused by the intervalley transfer. In contrast to a satisfactory agreement of the experimental results and those of simulation for the lightly doped GaAs in the whole range of fields and for the standard doped GaAs channels at fields below 1 kV/cm, no agreement is obtained for the standard doped GaAs at high electric fields. Indeed, a strong effect of doping on hot-electron noise is observed in the range of high electric fields where the intervalley fluctuations dominate and form the maximum of the intensity in the case of light doping (cf. symbols and solid lines in Fig. 10.8). More experimental results are given in the next subsection.

High Electric Fields

The typical effect of doping on the longitudinal noise temperature at high electric fields is illustrated by Fig. 10.9, where the room temperature experimental results are given for InP containing different density of electrons [192, 249]. The onset field for the intervalley noise slightly increases in the result of doping. The observed small shift of the onset field with doping is in agreement with the expectations based on the model for ionized impurity scattering. However, this is not the case for GaAs. Figures 10.10 and 10.11 show [11, 184, 250] that the effect of the doping on the onset field is much stronger for GaAs than for InP: The onset field increases more than twice as the electron density increases from $\sim 10^{15}$ cm^{-3} to 3×10^{17} cm^{-3} (Fig. 10.11).

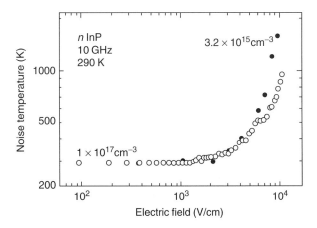

Figure 10.9. Doping-dependent hot-electron longitudinal noise temperature $T_{n\parallel}$ at 10-GHz frequency at 300 K lattice temperature for InP with different doping: ● represents $n_0 = 3.2 \times 10^{15}$ cm^{-3} [184]; ○ represents $n_0 = 1 \times 10^{17}$ cm^{-3} [249].

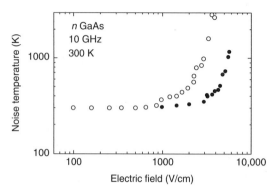

Figure 10.10. Doping-dependent hot-electron longitudinal noise temperature $T_{n\parallel}$ at 10-GHz frequency at 300 K lattice temperature for GaAs. ● represents $n_0 = 9 \times 10^{14}$ cm^{-3}, $\mu_0 = 7500$ cm^2/(V·s) [184]; ○ represents $n_0 = 3 \times 10^{17}$ cm^{-3}, $\mu_0 = 3900$ cm^2/(V·s) [11].

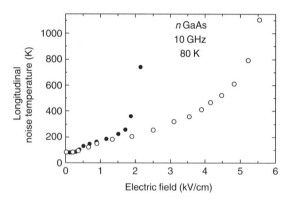

Figure 10.11. Doping-dependent hot-electron longitudinal noise temperature $T_{n\parallel}$ at 10-GHz frequency at 80 K lattice temperature for GaAs [250]: ● represents $n_0 = 3 \times 10^{15}$ cm^{-3}, $\mu_0 = 70,000$ cm^2/(V·s); ○ represents $n_0 = 3 \times 10^{17}$ cm^{-3}, $\mu_0 = 4000$ cm^2/(V·s).

The experimental results on the onset field for the intervalley noise in n-type GaAs (Figs. 10.10 and 10.11) contradict the results obtained for the models considering doping effects in terms of ionized impurity scattering and interelectron collisions. As one can see in Chapters 13 and 14, the interelectron collisions are not important at high electric fields. The simulation taking into account ionized impurity scattering, acoustic deformation potential, optical phonon, and intervalley scattering [234] predicts the increase by only ∼10% of the onset field for the intervalley diffusion of hot electrons in GaAs as the doping is increased within the range of interest. The problem of doping-dependent intervalley noise in standard-doped GaAs calls for considering *other* impurity scattering mechanisms, the mechanisms that are efficient for high-energy electrons. This is done in the following section.

10.5 RESONANT IMPURITY SCATTERING IN GaAs

The dependence on doping of the intervalley noise of hot electrons in GaAs was associated with electron scattering on resonant impurity levels [250, 251]. This scattering mechanism is important at high energies; it differs essentially from the electron scattering on ionized impurities. The resonant impurity levels located inside the conduction band were introduced by Bate in 1962 [252] in order to interpret doping-dependent low-field transport properties that did not fit into the standard scheme based on ionized impurity scattering in semiconductors. The resonant scattering was also thought [253] to contribute to the hot-electron intervalley transfer. Now, we shall briefly consider the effect of resonant impurity levels on hot-electron noise.

Quantitative interpretation of the experimental results on doping-dependent intervalley noise in doped GaAs (Figs. 10.10 and 10.11) requires information on resonance scattering by impurities. Standard techniques for impurity diagnostics fail to detect impurities inside the conduction band, but resonance energies of impurity levels in the conduction band of GaAs are available from optoelectronic modulation spectroscopy [254], high-pressure luminescence spectroscopy [255], and some other techniques. Hot-electron drift velocity, diffusion coefficient, and noise temperature as a function of electric field was simulated by Monte Carlo technique in the framework of the model taking into account ionized impurity, resonant impurity, and phonon scattering mechanisms [256–258]. The model interpreted the observed doping-dependent behavior of hot-electron fluctuations at high electric fields in silicon-doped GaAs [250]. According to the simulation, the resonant scattering became important, provided that its rate at the resonance energy exceeded the rate of the spontaneous optical phonon emission — the main scattering mechanism at the electron energies of interest. A satisfactory fit of the results of simulation to the experimental data was obtained, assuming that the maximum resonant impurity scattering rate is nearly three times higher than that of the spontaneous emission. This interpretation is supported by the length-dependent behavior of the hot-electron noise in the standard-doped n-type GaAs channels for field-effect transistors (see Sections 11.3 and 11.4).

Let us estimate the resonance scattering time in GaAs. As mentioned, the fitting of the experimental data on hot-electron noise and those of Monte Carlo simulation is achieved [257], assuming that the minimum impurity scattering time is nearly three times shorter than the spontaneous optical phonon emission time. Hot-electron luminescence experiments provide a well-established value for the spontaneous emission time of an optical phonon by a high-energy electron, $\bar{\tau}^{opt} \sim 160\text{--}190$ fs [227, 259, 260]. Consequently, for the silicon-doped GaAs at 80 K lattice temperature in the field range from 2 kV/cm to 3 kV/cm, the estimate gives the value $\tau(\varepsilon^{res}) \sim 50$ fs for the hot-electron relaxation time at the resonance energy. Because of the short relaxation time, this source of hot-electron noise is in a strong competition with the intervalley noise at high electric fields in doped GaAs.

CHAPTER 11

LENGTH-DEPENDENT HOT-ELECTRON NOISE

This chapter discusses experimental investigation of short-channel effect on hot-electron noise in n-type GaAs and n-type InP — important materials for contemporary high-speed electronics. The speed of operation of field-effect transistors and of many other devices depends on the electron drift velocity and the length of the active region in the direction of current; the speed can be increased if the length is reduced and/or a higher electric field is applied. Of course, semiconductors demonstrating a higher mobility are most promising from the same point of view. On the other hand, high electric fields together with a high mobility favor hot-electron effects, with excess noise of hot electrons being one of them.

Shrinking dimensions make analysis of the hot-electron transport and noise more complicated. Since no universal relation between drift velocity and excess noise holds under nonequilibrium conditions, the discussion of noise–speed tradeoff in a short channel device calls for length-dependent data on hot-electron fluctuations in addition to those on drift velocity. The experimental data on hot-electron noise in short channels for different doping, lattice temperature, applied electric field, and other conditions seem to be important for considering physical limitations of low-noise device performance at microwave and millimeter-wave frequencies.

As far as the theoretical basis is concerned, the equations of Chapter 3 in principle describe the fluctuations in samples of any length, if the collision duration and the scatterer dimensions remain insignificant. Important length-dependent changes are expected at frequencies $\omega \tau_m < 1$, where τ_m is the relaxation time of the mth process responsible for the corresponding noise source in the direction of the current. The mth contribution to the longitudinal noise is cutoff as the electron transit time $t^{tr} = L/\bar{v}^{dr}$ is made essentially shorter than the relaxation time τ_m (here $\bar{v}^{dr}$ is the drift velocity properly averaged over the sample length).

Thus, changing the sample length L as well as the electric field strength E, one can control the sources of hot-electron excess noise. The experimental techniques (see Chapter 8) originally developed to study hot-electron noise in long samples are applicable also to short channels. X-band microwave frequencies are well-suited for demonstration of the short-channel effects. Of course, an important difference arises: The electric field cannot be considered as uniform in a current carrying state, provided that the hot-electron drift velocity depends on the distance traveled by the electrons in the high-field region. The nonuniformity of the electric field is an inevitable feature under steady current in short channels. In experiments, one is forced to introduce the average electric field $\overline{E} = U/L$ (U is the applied voltage) as the parameter characterizing electron heating.

11.1 SHORT-CHANNEL EFFECT ON HOT-ELECTRON NOISE

Length-Dependent Intensity of Current Fluctuations

The spectral intensity of current fluctuations depends on channel length for two reasons. First, the intensity S_I depends on L explicitly according to relation (8.4); this is the only dependence, provided that the cross section Q, the electron density n_0, and the intensity of velocity fluctuations S_V do not depend on L. Next, the intensity S_V itself depends on the sample length, as one can expect from the general theory.

Figure 11.1 gives the room temperature data on the longitudinal current fluctuations $S_{I\parallel}$ for the lightly doped n-type GaAs obtained from the measured noise temperature $T_{n\parallel}$ and the differential conductivity $\sigma_\parallel$ by means of Eq. (8.2). Symbols in Fig. 11.1 are the experimental data, and the dotted lines are $1/L$

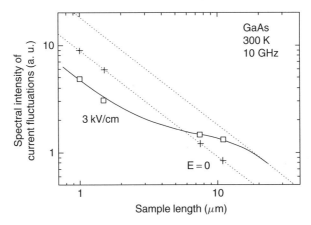

Figure 11.1. Length-dependent spectral intensity of longitudinal current fluctuations $S_{I\parallel}$ in lightly doped n-type GaAs at 300 K. Dotted lines are the $1/L$ dependencies, and the solid curve is a guide to the eye. Symbols are experimental data [183, 261]: + represents $E = 0$, □ represents $\overline{E} = 3.25$ kV/cm.

dependencies. The long sample data (cf. Fig. 10.1) are used to plot the $1/L$ dependencies at $E = 0$ and $E = 3.25$ kV/cm. The experimental results at the fixed average field $U/L = 3.25$ kV/cm for samples of different length are given by squares in Fig. 11.1 [183, 261]. (Solid curve is to guide an eye).

The increase in $S_{I\parallel}$ with the decrease of the sample length L (Fig. 11.1, squares) is less steep than expected from the simple considerations neglecting the hot-electron effects (dotted line). In the shortest channels the values of $S_{I\parallel}$ are below those at the thermal equilibrium (Fig. 11.1, cf. dotted line and squares). In order to deal with the short-channel effect, it is convenient to eliminate the explicit $1/L$ dependence [see Eq. (8.4)]. This is achieved by introducing the normalized spectral intensity of current fluctuations: $S_{I\parallel}(E, L)/S_I(0, L)$, where $S_I(0, L)$ is the length-dependent value at zero bias.

Convective Noise

At low/moderate electric field strengths, hot-electron noise in the direction of current results from two main sources: the hot-electron "thermal" noise due to velocity component fluctuations, and the convective noise due to energy fluctuations (see Sections 7.2 and 9.3). Let us analyze the length-dependent behavior in terms of the effective energy relaxation time $\bar{\tau}^{en}$ [defined by Eq. (9.3)], and the momentum relaxation time $\bar{\tau}$. In long samples, the convective noise is important at frequencies $\omega \bar{\tau}^{en} \ll 1$ under the condition of quasi-elastic scattering $\bar{\tau}^{en} \gg \tau$. The sample length being reduced, the transit time t^{tr} comes into play — eventually it becomes shorter than the energy relaxation time: $t^{tr} = L/\bar{v}^{dr} < \bar{\tau}^{en}$. Under this condition, the hot electrons leave the sample before the energy relaxation is completed — that is, before the electron energy distribution becomes close to that realized in the long sample. This has an effect on the drift velocity: In the case of a sublinear current–voltage characteristic the velocity in a short sample overshoots its steady value [262]. The longitudinal noise temperature $T_{n\parallel}$ of hot electrons in the short sample approaches the transverse noise temperature $T_{n\perp}$ of the long sample, and the contribution of the convective noise is partly suppressed [263] (cf. Section 7.2). Since the convective contribution to the spectral intensity of the longitudinal velocity fluctuations is negative, the suppression means the increase in the normalized intensity as the sample length is reduced.

Figure 11.2 illustrates the experimental results on the length-dependent behavior in n-type GaAs at 80 K lattice temperature at 10-GHz frequency [184, 264]. In order to eliminate the dependence of the zero-field resistance on the sample dimensions, the intensity of the longitudinal current fluctuations $S_{I\parallel}$ is normalized to its value at zero electric field. The experimental data on the normalized intensity of the fluctuations are given for a short sample and a long one. At moderate electric fields, a higher normalized intensity is observed in the short sample. This behavior at fields $E < 300$ V/cm is in agreement with the expected suppression of the negative convective contribution in short samples. Similar results were reported [184, 264] for n-type InP (Fig. 11.3).

Of course, one should be aware that the two-source model (taking into account the hot-electron "thermal" noise and the convective noise) is applicable in a

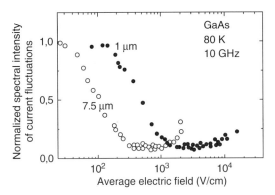

Figure 11.2. Normalized spectral intensity of longitudinal current fluctuations $S_{I\parallel}(\overline{E})/S_I(0)$ at 10-GHz frequency plotted against average electric field $\overline{E}$ for n^+-n-n^+ GaAs structures of various length [184]. ○ represents $L = 7.5$ μm, $n_0 = 0.9 \times 10^{15}$ cm^{-3}, $\mu_0 = 77{,}000$ cm^2/(V·s); ● represents $L = 1$ μm, $n_0 = 1.5 \times 10^{15}$ cm^{-3}, $\mu_0 = 47{,}000$ cm^2/(V·s).

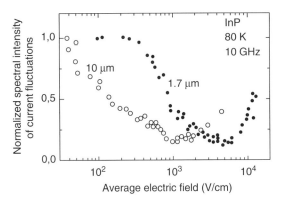

Figure 11.3. Normalized spectral intensity of longitudinal current fluctuations $S_{I\parallel}(\overline{E})/S_I(0)$ at 10-GHz frequency at 80 K plotted against average electric field $\overline{E}$ for n^+-n-n^+ InP structures of various length [184]. ○ represents $L = 10$ μm, $n_0 = 3.3 \times 10^{15}$ cm^{-3}, $\mu = 49{,}000$ cm^2/(V·s); ● represents $L = 1.7$ μm, $n_0 = 4.9 \times 10^{15}$ cm^{-3}, $\mu = 18{,}000$ cm^2/(V·s).

restricted range of low/moderate electric fields. As discussed in Section 10.3 (see Fig. 10.7), the third source—electron runaway noise—appears in the range of fields $0.3 < E < 2$ kV/cm in n-type GaAs and in the range of fields $0.15 < E < 4$ kV/cm in n-type InP at 80 K lattice temperature.

Terminated Runaway Noise

Polar runaway is an important hot-electron effect in compound semiconductors appearing because of dominant polar-optical scattering. At high electric fields, exceeding certain threshold, the average electron gains more energy than dissipates on optical phonons [265]. At the subthreshold fields, only "lucky" electrons enter into the runaway conditions, but their runaway is terminated by large-angle scattering events (see Section 10.3). An incubation precedes the runaway, and an electron drifts a long distance before the conditions to become a "lucky" one in terms of the runaway are satisfied. Therefore, the associated fluctuations need time and space to develop—the fluctuations are suppressed, provided that the required space is not given.

Figure 11.4 illustrates the suppression of the acoustic-phonon-terminated-runaway fluctuations in short samples of n-type InP [184] (see also reference 9). At the intermediate electric fields (0.15 kV/cm < E < 4 kV/cm at 80 K lattice temperature) the parallel noise temperature $T_{n\parallel}$ decreases as the sample length is reduced. Essentially the same behavior is observed at room temperature in the field range 2 kV/cm < E < 6 kV/cm [184]. The fluctuations in question are also resolved and their length-dependent suppression is observed in n-type GaAs, but the range of field is quite narrow in this semiconductor [184].

To interpret these results, Monte Carlo simulation was performed for InP within a one-valley model [230]. The results of simulation for electron scattering on acoustic and optical phonons show (Fig. 11.5) that the velocity correlation function has a long tail in the long sample, but no long-time correlation survives in the short sample. In the considered one-valley model, the long-time correlation is not associated with the intervalley transfer, but is instead associated with the fluctuations due to the considered combined action of optical-phonon and acoustic-phonon scattering mechanisms. The results of simulation can be interpreted in terms of the terminated runaway: Because the space is limited in short samples, the electrons leave for the electrode before entering/completing the runaway process.

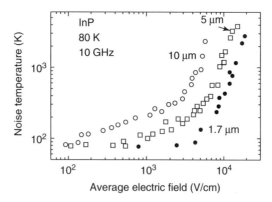

Figure 11.4. Hot-electron longitudinal noise temperature $T_{n\parallel}$ at 10-GHz frequency plotted against average electric field for n^+–n–n^+ InP structures of various length [184]. ○ represents $L = 10$ μm, $n_0 = 3.3 \times 10^{15}$ cm^{-3}, $\mu_0 = 49{,}000$ cm^2/(V·s); □ represents $L = 5$ μm, $n_0 = 2.2 \times 10^{15}$ cm^{-3}, $\mu_0 = 24{,}000$ cm^2/(V·s); ● represents $L = 1.7$ μm, $n_0 = 4.9 \times 10^{15}$ cm^{-3}, $\mu_0 = 18000$ cm^2/(V·s).

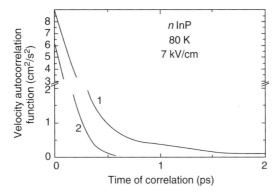

Figure 11.5. Calculated electron drift velocity autocorrelation function at 7 kV/cm electric field at 80 K lattice temperature plotted against time for a single-valley model of InP in infinitely long (1) and short (2, $L = 1$ µm) undoped channels [230].

11.2 SUPPRESSION OF INTERVALLEY NOISE

Length-Dependent Noise Temperature

Intervalley noise is a threshold type noise offering a possibility to analyze length-dependent noise suppression of a different type [184, 250, 261, 266, 267] (see also reference 9). Figure 11.6 presents the longitudinal noise temperature measured at 10 GHz for lightly doped n-type GaAs at 300 K lattice temperature. Long samples demonstrate the steep increase of noise due to intervalley transfer of hot electrons at around 2 kV/cm electric field (cf. Figs. 10.2 and 10.3). The noise temperature is about the same for 1-mm bulk samples and for 7.5-µm n^+–n–n^+ structure (filled triangles and unfilled circles in Fig. 11.6). The experimental results for the 7.5-µm sample (circles in Fig. 11.6 are in satisfactory agreement with those of Monte Carlo simulation [244, 245].

Figure 11.6 shows that the noise temperature is length-dependent: For a fixed average electric field, say $U/L = 3$ kV/cm, the noise suppression exceeds 10 dB as the sample length L is reduced from 7.5 µm (unfilled circles) to 1.5 µm (unfilled squares). In other words, the same value of noise temperature is reached at an essentially higher electric field as the channel length is reduced. Provided that we fix the noise temperature (say, $T_n = 400$ K), a higher electric field is required in a shorter sample for the intervalley noise to appear. In particular, the distance of 1 µm is too short for an onset of the intervalley noise in GaAs even at quite high fields up to 7 kV/cm. Figure 11.7 illustrates the relation between the onset field and the channel length (symbols); the relation is useful to get the feeling of the range of lengths where the noise suppression takes place. These experimental results also demonstrate that it becomes harder to eliminate the intervalley noise at a higher applied electric field — one has to operate a shorter channel in order to obtain the length-dependent suppression of the intervalley noise.

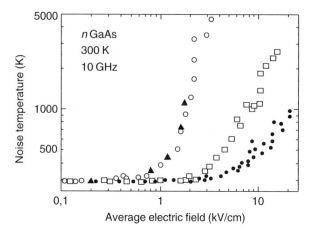

Figure 11.6. Suppression of intervalley noise at 300 K lattice temperature in short n^+–n–n^+ GaAs structures [181, 261]. ▲ represents $L = 1000$ μm, $n_0 = 3 \times 10^{15}$ cm^{-3} $\mu_0 = 6000$ cm^2/(V·s); ○ represents $L = 7.5$ μm, $n_0 = 0.9 \times 10^{15}$ cm^{-3}, $\mu_0 = 7500$ cm^2/(V·s); □ represents $L = 1.5$ μm, $n_0 = 1.3 \times 10^{15}$ cm^{-3}, $\mu_0 = 7500$ cm^2/(V·s); ● represents $L = 1$ μm, $n_0 = 1.6 \times 10^{15}$ cm^{-3}, $\mu_0 = 7600$ cm^2/(V·s).

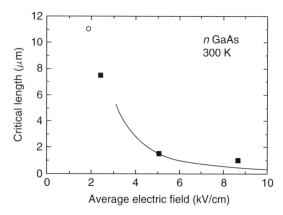

Figure 11.7. Total critical length versus average electric field for lightly doped n-type GaAs at room temperature. Experiment: ■ [261], ○ [183]. Monte Carlo simulation: solid line [271].

Let us note that the suppression of the intervalley transfer in short channels not only reduces the associated noise but simultaneously leads to an enhancement of the drift velocity [261]. The velocity enhancement in short channels in the presence of the intervalley transfer was considered through simulation by the Monte Carlo technique [268, 269] (see also reference 31). The results on the velocity overshoot and noise suppression suggest an interesting noise–speed tradeoff in short channels of lightly doped n-type GaAs.

Length-Dependent Spectral Intensity of Current Fluctuations

The experimental data on the normalized spectral intensity of longitudinal current fluctuations for the lightly doped n-type GaAs are presented by unfilled symbols in Fig. 11.8 [181, 183, 261]. As noticed earlier (see Section 10.1), the current fluctuations in long samples at fields over 2 kV/cm result from the hot-electron Γ–L intervalley transfer. This source of fluctuations is heavily suppressed in a 1-μm sample (Fig. 11.8, triangles), and the monotonously decreasing dependence on electric field (typical for one-valley semiconductors) is observed.

The hot-electron intervalley fluctuations observed in long samples of InP at fields over 6 kV/cm are also suppressed in short samples (Fig. 11.9). Indeed, the maximum of spectral intensity in 10 μm samples (Fig. 11.9, unfilled circles) diminishes and almost disappears as the sample length L is reduced down to 1.7 μm (filled squares).

Monte Carlo Simulation

A detailed interpretation of the length-dependent fluctuations is reached by comparing the experimental results [261, 267] to those obtained by Monte Carlo

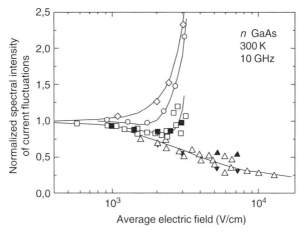

Figure 11.8. Transition from monotonously increasing dependence on electric field ($\Diamond$) to the monotonously decreasing ($\triangle$) dependence on electric field of the normalized longitudinal spectral intensity of current fluctuations $S_{I\parallel}(\overline{E})/S_I(0)$, illustrating suppression of the intervalley fluctuations in short (micrometer) samples. Experimental results correspond to lightly doped n-type GaAs (unfilled symbols and curves are guide to the eye). $\Diamond$ represents $L = 1000$ μm, $\mu_0 = 6000$ cm^2/(V·s) [181]; $\bigcirc$ represents $L = 11$ μm, $\mu_0 = 5200$ cm^2/(V·s) [183], $\square$ represents $L = 7.5$ μm, $\mu_0 = 7500$ cm^2/(V·s) [261] $\triangle$ represents $L = 1$ μm, $\mu_0 = 7600$ cm^2/(V·s) [261]. Monte Carlo simulation data (filled symbols) correspond to different values of sample length and Γ–L intervalley coupling constant [270]. ■ represents 7.5 μm, 1.8×10^8 eV/cm, ▼ represents 1 μm, 1.8×10^8 eV/cm, ▲ represents 1 μm, $1 \cdot 10^9$ eV/cm.

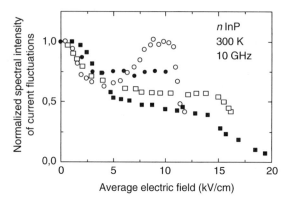

Figure 11.9. Normalized spectral intensity of longitudinal current fluctuations $S_{I_\parallel}(E)/S_I(0)$ for InP samples of different length. ○ represents $L = 10$ μm, $n_0 = 3.2 \times 10^{15}$ cm^{-3}, $\mu_0 = 4600$ cm^2/(V·s) [194], ● represents $L = 5$ μm, $n_0 = 2.7 \times 10^{15}$ cm^{-3}, $\mu_0 = 4500$ cm^2/(V·s) [182], □ represents $L = 5$ μm, $n_0 = 2.3 \times 10^{15}$ cm^{-3}, $\mu_0 = 4600$ cm^2/(V·s) [194], ■ represents $L = 1.7$ μm, $n_0 = 5.4 \times 10^{15}$ cm^{-3}, $\mu_0 = 4600$ cm^2/(V·s).

simulation in the framework of three-valley model [270]. The simulation takes into account the nonuniformity of the electric field and the space charge fluctuations. Filled symbols in Fig. 11.8 correspond to the results of Monte Carlo simulation of hot-electron fluctuations. The results of the simulation are in satisfactory agreement with the experimental data.

An attempt was made to discriminate between the models assuming different values of Γ–L coupling [270]. Once an electron accumulates enough energy, it has at least two possibilities: Either it jumps into an L valley or it continues energy accumulation. The choice depends on the coupling constant and the electric field strength. In short samples, the electron can also leave the channel for the electrode. At electric fields above 3.5 kV/cm, the electrons prefer undergoing the intervalley transfer before exiting the channel if a strong coupling is assumed — the intervalley contribution to the current fluctuations is enhanced (cf. filled symbols, up- and down-facing triangles in Fig. 11.8). This result is opposite to that discussed in Section 10.1 in relation to long samples, where a weaker coupling leads to a higher intensity of the intervalley noise. We shall discuss this problem in more detail in the following subsections of this chapter.

11.3 DOPING-DEPENDENT SHORT-CHANNEL EFFECT

Length-Dependent Noise Temperature

It is well known that electron heating by a given electric field depends on electron mobility (the consumed power is proportional to the drift velocity). According to such simple considerations, based on the balance of the consumed and the dissipated power, the hot-electron noise temperature is expected to be lower in

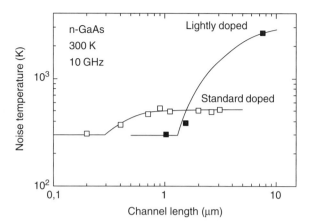

Figure 11.10. Length-dependent longitudinal noise temperature $T_{n\parallel}$ of hot electrons in GaAs at 10-GHz frequency for $\overline{E} = 4$ kV/cm at 293 K lattice temperature [251]. Experimental points: ■ represents $n_0 = 0.9 \times 10^{15}$ cm^{-3}, $\mu_0 = 7500$ cm^2/(V·s); □ represent $n_0 = 3 \times 10^{17}$ cm^{-3}, $\mu_0 = 3900$ cm^2/(V·s). Solid lines are approximations based on Eq. (11.2) [11].

a low-mobility channel as compared to that in the high-mobility channel. This point of view is supported by the experimental results on long channels (see Figs. 10.10 and 10.11). Let us discuss if the same considerations apply for the short channels.

Figure 11.10 presents the longitudinal noise temperature for n-type GaAs as a function of the channel length at the fixed electric field strength [251]. The experimental results for lightly doped (10^{15} cm^{-3}) channels are compared with those for standard-doped channels (used in field-effect transistors, 3×10^{17} cm^{-3}). As expected, the noise temperature in the long lightly doped channel is higher than that in the long standard-doped one (cf. Figs. 10.10 and 10.11). However, the same does not hold for short channels despite the fact that the low-field electron mobility is independent of the channel length in the range of lengths of interest. While the results for the long channels agree with the simple considerations based on the balance of the consumed and the dissipated power, the results for the short channels need a more detailed consideration. In the following subsection we shall give a simplified understanding of the noise suppression in terms of energy-loss-free and dissipative critical lengths.

Energy-Loss-Free and Dissipative Critical Lengths

Under steady current, hot electrons are constantly leaving the channel for the anode, and the same amount of equilibrium electrons are entering at the cathode. This "exchange" opens an additional (external) energy loss mechanism for the electron ensemble present in the channel. The external losses are negligible in long channels, but their relative weight increases if the sample length L is reduced. Evidently, at certain critical length the external losses become of primary importance. This suggests that the critical length is shorter if the internal losses are heavier.

The minimum distance L_1 needed to accumulate the threshold energy for the intervalley transfer in the electric field $\overline{E}$ is given by

$$L_1 = \frac{\mathcal{E}^{\text{thr}} - \varepsilon^{\text{in}}}{e\overline{E}}, \tag{11.1}$$

where $\mathcal{E}^{\text{thr}}$ is the threshold energy, and ε^{in} is the input energy.

It is important for the following that relation (11.1) can be applied in the presence of impurity scattering. Of course, elastic collisions on impurities reduce the consumed power, and the energy accumulation on the given distance takes more time. Nevertheless, the energy gained on the given distance in the energy-loss-free process is independent of the elastic collisions. Let us call L_1 the energy-loss-free critical length. Monte Carlo simulation within the one-valley model [271] shows that some electrons avoid energy loss on phonons before they reach the intervalley energy in GaAs and InP at electric fields of importance for the intervalley noise.

Most electrons drift a longer distance until they accumulate the same energy. For the given channel length L, the number of the electrons reaching the threshold while drifting in the channel from one electrode to another can be approximated by [271]:

$$\begin{aligned} N(L) &= 0 & \text{for } L < L_1 \\ N(L) &= N_0(1 - e^{-(L-L_1)/L_2}), & \text{for } L > L_1, \end{aligned} \tag{11.2}$$

where the dissipative critical length L_2 is introduced. According to the simulation, $L_2 > L_1$ at $E < 7$ kV/cm and at $E < 20$ kV/cm in lightly doped n-type GaAs and n-type InP, respectively [271]. The dissipative critical length is found insensitive, within 5%, to nonuniformity of the electric field, unless the local deviations exceed $\pm$ 50% from the average electric field. Let us call $L^{\text{tot}} = L_1 + L_2$ the total critical length; after Eq. (11.2) the major part, $1 - e^{-1}$, of all electrons gain enough energy for the intervalley transfer having drifted the distance equal to the total critical length.

Interpretation of Experimental Data

The total critical length L^{tot} available from the simulation [271] is given as a function of the electric field in Fig. 11.7 (solid curve). The results of simulation within the considered simple model are in good agreement with the results deduced from the experimental data [261] on noise suppression in lightly doped n-type GaAs (symbols in Fig. 11.7). Consequently, the energy limitation is very important for suppression of the intervalley noise in short channels.

After the discussion on the critical length in the lightly doped n-type GaAs, let us return back to Fig. 11.10. Assuming that the excess noise temperature is proportional to the number of the electrons that reach the threshold energy [see Eq. (11.2)], the experimental data (symbols) can be approximated by the curves (Fig. 11.10) containing two fitting parameters: the energy-loss-free length L_1 and the dissipative length L_2. The fitting parameters for $\overline{E} = 4$ kV/cm field are chosen as follows: $L_1 = 1.3$ µm, $L_2 = 3$ µm for the lightly doped GaAs and $L_1 = 0.3$ µm, $L_2 = 0.2$ µm for the standard-doped GaAs. It is noteworthy that the critical lengths are shorter in the standard-doped GaAs channels. By inserting the fitting parameter L_1 into Eq. (11.1), one obtains the threshold energy of the dominant source of noise subject to suppression in the short channels. The threshold energies appear to be different in lightly doped and standard-doped GaAs. This means that the origin of the noise sources differ as well.

It has been demonstrated [251] that the noise due to L valleys is difficult to resolve in short samples of lightly doped GaAs; the associated excess noise temperature hardly exceeds 20 K in 1-µm samples. Indeed, because of weak coupling, most electrons avoid the Γ–L intervalley transfer though their energy is sufficient for the transfer. Being accelerated further on, the electrons reach the X-valley energies and can take part in the Γ–X transfer. The X-valley edge is approximately 0.48 eV above the conduction band edge, and Eq. (11.1) yields $L_1 = 1.2$ µm, which is comparable with the value chosen for the fitting parameter for lightly doped GaAs in Fig. 11.10.

The same argumentation fails in the case of the standard-doped GaAs. It is discussed in Section 10.5 that impurity resonant scattering of hot electrons (probably on DX levels located inside the conduction band [258]) is important in silicon-doped GaAs. The resonance energy is estimated to be between 0.14 eV [254] and 0.24 eV [251] at room temperature; these values and Eq. (11.2) yield the range 0.3 µm $< L_1 <$ 0.5 µm for the silicon-doped GaAs. It is noteworthy that the fitting parameter $L_1 = 0.3$ µm satisfies the experimental results for the standard-doped channels (Fig. 11.10, unfilled squares). Consequently, the results on the short-channel effect in doped GaAs channels provide us with an independent evidence for the existence of the resonant impurity scattering confirming the results of Section 10.5 on long channels.

11.4 DOPING-STIMULATED INTERVALLEY NOISE

Impurity scattering hinders electron heating, and a lower hot-electron noise temperature is often observed in the low-mobility channels. This point of view is supported by the experimental results on long channels (see Sections 10.4 and 10.5 and Fig. 11.10, $L > 2$ µm). However, the same considerations fail in the submicrometer channels: A higher noise temperature (Fig. 11.10, unfilled squares) is observed in the standard-doped n-type GaAs channels as compared to the lightly doped ones of similar length. This phenomenon has been called the hot-electron noise stimulated with doping [11].

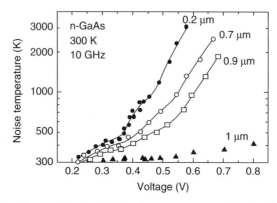

Figure 11.11. Hot-electron longitudinal noise temperature $T_{n\parallel}$ in short channels of GaAs at room temperature against the applied voltage. Standard-doped channels [250] ($n_0 = 3 \times 10^{17}$ cm^{-3}). ● represent $L = 0.2$ μm, ○ represent $L = 0.7$ μm, □ $L = 0.9$ μm. Lightly doped sample [230] ▲ represents $L = 1$ μm, $n_0 = 9 \times 10^{14}$ cm^{-3}.

Let us plot the noise temperature as a function of the applied voltage (Fig. 11.11). Three shoulders of excess noise are resolved [250] in the standard-doped channels of submicrometer length (circles and squares). The two shoulders at $U > 0.3$ V and $U > 0.5$ V are attributed to the Γ–L and Γ–X transfer of hot electrons. The third shoulder appears at $U \approx 0.2$ V [250]. As the sample length is increased, the structure of the noise sources tends to fade away but remains observable even in the 0.9-μm sample (Fig. 11.11, squares). This structure has not been observed in lightly doped samples of the same length [251]: the shoulder at 0.3 V is extremely weak, and the 0.2-V shoulder has not been observed at all (see filled triangles in Fig. 11.11 and filled circles in Fig. 11.6).

The noise sources associated with the resonant impurities and the upper valleys is better resolved in the plot of the voltage-dependent normalized intensity of longitudinal current fluctuations $S_{I\parallel}(U)/S_{I\parallel}(0)$, as reported for hydrogenated GaAs channels at 80 K lattice temperature (Fig. 11.12) [250, 272].

The ionized impurity scattering fails to explain these results. According to the results of Monte Carlo simulation, this scattering is not important at high fields [256]. Impurities can assist an intervalley transfer but this process is known to be weak as compared to the phonon-assisted one. However, one is free to consider resonant impurity scattering as suggested in [250]. Let us discuss the possible effects caused by almost isotropic elastic resonant scattering of hot electrons on impurity levels located inside the conduction band.

It is noteworthy that hot-electron noise is quite sensitive to the large-angle scattering, as discussed in relation to the runaway noise (see Section 11.1). In the absence of any impurity scattering, the optical phonon scattering supports a well-collimated acceleration of hot electrons in the Γ valley (the runaway effect). At high electric fields, many electrons reach the energies sufficient for the Γ–L transfer. Supposing that Γ–L coupling is not strong enough, some of the electrons avoid Γ–L transfer: Either they reach X valleys and take part in Γ–X transfer

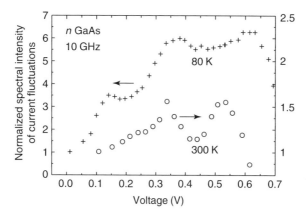

Figure 11.12. Normalized spectral intensity of longitudinal current fluctuations $S_{I\|}(E)/S_I(0)$ at 10-GHz plotted against applied voltage for standard-doped ($n_0 = 3 \times 10^{17}$ cm^{-3}) n^+–n–n^+ GaAs channels: ○ represents $L = 0.2$-μm channels at room temperature [250]; + represents hydrogenated channels $L = 0.4$ μm at 80 K lattice temperature [272].

or they leave the channel for the anode. The latter possibility is quite probable in short channels.

The scattering by the resonant impurity levels introduces a strong momentum dissipation mechanism at the resonance energies $\mathcal{E}^{res}$ (the scattering time is estimated to be shorter than 50 fs [257]; see Section 10.5). This leads to at least two consequences. First, the excess noise source appears at $U \sim \mathcal{E}^{res}/e$. Next, a high-energy electron, scattered at $\varepsilon = \mathcal{E}^{res}$, moves at a large angle to the electric field and gains the energy slowly ($e\mathbf{E} \cdot \mathbf{v}^{dr} = 0$, if $\mathbf{v}^{dr} \perp \mathbf{E}$). The electron spends more time at energies $\mathcal{E}_{L\Gamma} < \varepsilon < \mathcal{E}_{X\Gamma}$ in the energy gap between the bottom edges of L and X valleys, and consequently has better chances for the Γ–L transfer, as compared to the typical electron performing the well-collimated runaway in an undoped GaAs. Consequently, the intervalley Γ–L transfer and the associated intervalley noise can be stimulated by the resonant impurity scattering. Experiments show that the Γ–X transfer noise is stimulated as well unless the field is extremely high ($E > 100$ kV/cm). For the discussion of hot-electron noise at extremely high electric fields, see the following section.

11.5 Γ–X AND Γ–L INTERVALLEY TRANSFER TIMES FOR GaAs

In this section the experimental results on microwave noise temperature measured [201] for n-type GaAs at electric fields up to 300 kV/cm are presented and discussed. The results at room temperature demonstrate that the intervalley noise temperature tends to saturate at fields exceeding 50 kV/cm until impact ionization noise becomes the dominant source of noise at fields over 200 kV/cm. The role of L valleys decreases as the field strength is increased, and the noise experiments provide information on the Γ–X transfer. The experimental results yield an independent determination of the Γ–X transfer time.

Different optical techniques were used [227, 273, 274] to measure the intervalley transfer time. The obtained experimental results were impressive but partially controversial [275], so that investigations by complementary techniques were called for. Noise spectroscopy, being a powerful tool to study hot-electron kinetic processes, is suited for estimation of the intervalley scattering rates. To investigate the time of intervalley transfer by high-energy electrons from Γ valley into X valleys in GaAs, the nanosecond pulsed radiometric noise technique [201] applicable to doped semiconductors subjected to extremely high electric fields was developed (see Fig. 8.11 in Section 8.2).

As known, hot-electron fluctuations at microwave frequencies in GaAs at high electric fields are dominated by intervalley transfer of hot electrons. The intervalley Γ–L fluctuations appear at fields over 2 kV/cm and remain the most important source of microwave noise in long samples [9] until the hot-electron velocity fluctuations are obscured by the Gunn-effect-type instabilities. The latter can be avoided in short channels, and measurements of the hot-electron noise can be performed at higher fields. Three distinct sources of noise are resolved [11, 250] in the field range up to 25 kV/cm in short channels of GaAs (see Fig. 11.11 and Fig. 11.12).

Separation of Γ–L and Γ–X Noise Sources

Microwave noise was investigated in recessed silicon-doped n-type GaAs channels of submicrometer length supplied with two ohmic electrodes of Au–Ge evaporated onto the n^+-doped cap layer. The electrons were heated up by pulsed voltage to reduce the channel overheat. The principle of measurements is discussed in Section 8.2. The dependence of equivalent noise temperature on average electric field up to 300 kV/cm is given by Fig. 11.13 [276]. The exploitation of the 100-ns voltage pulses opened an essentially wider range of fields and noise temperatures (Fig. 11.13, unfilled squares) as compared to the microsecond time-domain data (Fig. 11.13, filled circles in the left-hand bottom corner). Since the electron drift velocity in short samples depends on coordinate, and the field is not uniform, the noise temperature is plotted against the average field defined as $\overline{E} = U/L$. An essential increase of the hot-electron noise takes place at $\overline{E} \sim 25$ kV/cm average field (Fig. 11.14, unfilled squares). This increase is followed by the noise temperature saturation at fields around 100 kV/cm. Another, strong source of noise dominates at fields exceeding 200 kV/cm (Fig. 11.14).

Figure 11.14 shows the excess noise temperature $\Delta T_n = T_n - T_0$, plotted as a function of the applied voltage U at $T_0 = 293$ K ambient temperature [201, 276]. A decomposition of the $\Delta T_n(U)$ dependence into four sources of noise is illustrated: thin lines in Fig. 11.14 represent possible contributions of each source. The contributions sum up into the solid curve, which fits the experimental results pretty well (Fig. 11.14).

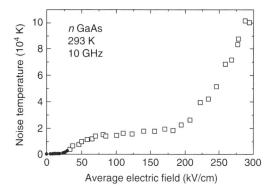

Figure 11.13. Field-dependent hot-electron longitudinal noise temperature $T_{n\|}$ at 10-GHz frequency at room temperature for short standard-doped GaAs channels ($L = 0.2$ μm, $n_0 = 3 \times 10^{17}$ cm^{-3}) subjected to electric field pulses of different duration: ● represents 2 μs, □ represents 100 ns [201].

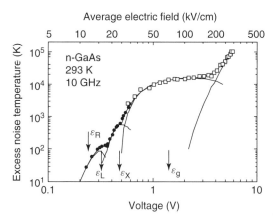

Figure 11.14. Voltage-dependent excess noise temperature $T_{n\|} - T_0$ of hot electrons at 10-GHz frequency at room temperature for short standard-doped GaAs channels ($L = 0.2$ μm, $n_0 = 3 \times 10^{17}$ cm^{-3}) subjected to electric field pulses of different duration: ● represents 2 μs, □ represents 100 ns [201]. Possible contributions of different sources of noise (thin lines) sum up into the total noise power (solid line).

The lowest threshold appears near 0.2 V. It has been associated [250, 272] with the resonant scattering of hot electrons by the impurity levels located inside the conduction band (see Section 11.3). The next two thresholds near 0.3 V and 0.5 V are caused by the intervalley transfer of hot electrons (the L and X valley threshold energies are close to 0.3 eV and 0.5 eV, respectively). The experimental results of Fig. 11.14 show that the maximum contribution to the noise temperature due to Γ–L transfer is essentially lower than that due to Γ–X transfer; this supports the discussed model, assuming a strong Γ–X coupling and the intermediate Γ–L coupling (see Section 10.1).

The extrapolated $\Delta T_n(U)$ dependence at fields $\overline{E} \sim 200$ kV/cm yields the threshold energy for the fourth source; the energy exceeds the forbidden gap (Fig. 11.14). This source of noise, accompanied by the steep increase of the current and the consumed power, can be ascribed to impact ionization.

Estimation of Γ–X Intervalley Transfer Time

The saturation of hot-electron noise temperature at fields over 100 kV/cm turns into the field-controlled dependencies at the lower fields and in the field range over 200 kV/cm (cf. Fig. 11.13). This very specific behavior suggests two independent ways to estimate the time constant of the intervalley scattering experienced by high-energy electrons present in the Γ valley at the energy sufficient for the Γ–X transfer ($\varepsilon > \mathcal{E}_{X\Gamma} \approx 0.5$ eV). The strong dependence of the intervalley noise on the field at $E < 100$ kV/cm means that the intervalley noise is bottlenecked by the electron acceleration rather than by Γ–X transfer of the already accelerated electrons.

Indeed, to have electrons with energies larger than $\mathcal{E}_{X\Gamma}$, we need electric field in any case not less than that capable to accelerate a "lucky" electron up to the energy $\mathcal{E}_{X\Gamma}$ during the time comparable with the electron relaxation time inside the Γ valley, τ_Γ (or the electron transit time from one electrode to another; this time is longer in our case). The electron acceleration time up to the energy $\mathcal{E}_{X\Gamma}$ is $t_\Gamma(0, \mathcal{E}_{X\Gamma}) \approx \sqrt{2m_\Gamma \mathcal{E}_{X\Gamma}}/eE$. The experimental results show (Figs. 11.13) that the Γ–X transfer becomes effective at the field about $E \sim 50$ kV/cm; at this field we obtain $t_\Gamma(0, \mathcal{E}_{X\Gamma}) \approx 120$ fs.

We argue that the intravalley relaxation time τ_Γ is of the same order of magnitude as $t_\Gamma(0, \mathcal{E}_{X\Gamma})$. This does not contradict the available data on the main intravalley scattering process: The time for spontaneous emission of optical phonons is estimated as 190 fs [260] or 160 fs [227] at low temperatures of the lattice. Of course, the contribution of the Γ–L transfer, resonant impurity scattering, and other scattering mechanisms make the time constant τ_Γ somehow shorter. On the other hand, in the field range where the intervalley Γ–X transfer is important, the intervalley transfer time $\tau_{\Gamma X}$ is much shorter than the intravalley relaxation time τ_Γ (otherwise the transfer into X valleys would not be effective). This leads to

$$\tau_{\Gamma X} < 120 \text{ fs}. \tag{11.3}$$

The noise temperature is saturated in the range 100–200 kV/cm. In this range of fields, the competition of the intervalley Γ–X transfer and the impact ionization takes place. At $\overline{E} \sim 150$ kV/cm, most electrons (present in Γ valley at energy $\varepsilon > \mathcal{E}_{X\Gamma}$) prefer jumping into an X valley rather than acceleration to the impact ionization energy $\mathcal{E}^{ii}$. Consequently, the time $t_\Gamma(\mathcal{E}_X, \mathcal{E}^{ii})$ needed for the acceleration of the Γ-valley electron from the energy $\mathcal{E}_{X\Gamma}$ to the impact ionization energy $\mathcal{E}^{ii}$ exceeds the intervalley Γ–X transfer time $\tau_{\Gamma X}$:

$$t_\Gamma(\mathcal{E}_X, \mathcal{E}^{ii}) > \tau_{\Gamma X} \quad \text{at} \quad E \sim 150 \text{ kV/cm}. \tag{11.4}$$

Supposing that large $\tau_{\Gamma X}$ were assumed, the kink at $\overline{E} \sim 100$ kV/cm would not be observed — only a small number of electrons would jump into X valleys before gaining the impact ionization energy $\mathcal{E}^{ii}$.

The band structure data [277] can be used to estimate $t_\Gamma(\mathcal{E}_X, \mathcal{E}^{ii})$; at $E = 150$ kV/cm, one obtains $t_\Gamma(\mathcal{E}_X, \mathcal{E}^{ii}) \approx 60$ fs. The condition (11.4) appears to be more rigid than (11.3), and it has been concluded [201] that

$$\tau_{\Gamma X} < 60 \text{ fs.} \qquad (11.5)$$

At $\overline{E} = 150$ kV/cm, only a small number of secondary electrons are created by the impact ionization to contribute to the dc current, consumed power, and noise [201]. Sufficiently higher fields are needed for many electrons to avoid scattering into the X valleys and to reach impact ionization energy. This takes place as the time $t_\Gamma(\mathcal{E}_X, \mathcal{E}^{ii})$ becomes less than $\tau_{\Gamma X}$. One can see from Figs. 11.13 and 11.14 that the impact ionization noise prevails over the intervalley noise at $\overline{E} \sim 300$ kV/cm, and we conclude that

$$t_\Gamma(\mathcal{E}_X, \mathcal{E}^{ii}) < \tau_{\Gamma X} \quad \text{at} \quad \overline{E} \sim 300 \text{ kV/cm}, \qquad (11.6)$$

Having estimated that at this field $t_\Gamma(\mathcal{E}_X, \mathcal{E}^{ii}) \sim 30$ fs, one has

$$\tau_{\Gamma X} > 30 \text{ fs.} \qquad (11.7)$$

Consequently, the very structure of the field dependence of the noise temperature dictates the conclusion that the time of transfer from Γ valley into X valleys of electrons with energy $\mathcal{E}_{X\Gamma} < \varepsilon < \mathcal{E}^{ii}$ falls into the interval [201]

$$30 \text{ fs} < \tau_{\Gamma X} < 60 \text{ fs}, \qquad (11.8)$$

or, in other words, $\tau_{\Gamma X} = (45 \pm 15)$ fs at $E \sim 200$ kV/cm. This estimation, based on the hot-electron noise data, does not contradict data obtained from the cw spectroscopy data [227], providing with its independent confirmation. On the other hand, the experiment illustrates that electron acceleration by extremely high electric field serves as an efficient internal clock operating in the femtosecond time scale.

Estimation of Γ–L Intervalley Transfer Time

It was demonstrated in the foregoing section that a source of noise caused by a faster relaxation process tends to dominate over the slower one provided a high electric field is applied to a short channel. This suggests a possibility for the following rough estimate of the relaxation time for the intervalley Γ–L transfer noise — the parameter important for the competition of the electron acceleration inside Γ valley and their transfer into the satellite L valleys. Figure 11.14 shows that the maximum noise temperature due to Γ–L transfer is at least ten times lower than the saturated value caused by Γ–X transfer. The latter noise source overshadows the Γ–L transfer noise at electric fields exceeding 25 kV/cm. The competition of the electron acceleration and the intervalley transfer (see the foregoing section) means that the relaxation time for the Γ–L transfer at 20 kV/cm is at least ten times longer than that for the Γ–X transfer estimated as $\tau_{\Gamma X}$ (45 ± 15) fs. Thus, the relaxation time for the Γ–L transfer is $\tau_{\Gamma L} \sim 0.4$ ps at $E \sim 20$ kV/cm. Another independent experimental argument is as follows: For the intervalley Γ–L transfer to take place in a short channel the electron transit time should be comparable with (longer than) the Γ–L transfer time. The transit time is estimated to be ~0.5 ps in the 0.2-μm channel at 20-kV/cm field.

Concluding this section, let us summarize the advantages of noise spectroscopy of short channels. As the channel length is reduced, a higher electric field is needed to give rise to the threshold processes such as intervalley transitions to upper valleys or impurity resonant scattering. If the sample is so short that even at highest available field the electron does not gain the threshold energy during its transit time, the noise associated with the corresponding process does not reveal itself. In other words, one has the possibility to suppress the dominant sources of noise — to "clean" a large enough "window" of the electric fields for other noise sources to manifest themselves. The richness of the noise characteristics as compared to the standard transport ones demonstrates advantages inherent to the hot-electron microwave noise spectroscopy of submicrometer samples used as a diagnostic tool of ultrafast kinetic processes in semiconductors. The noise spectroscopy has good perspectives (and is already used) as a tool while seeking the optimization of parameters of high-speed devices. The noise–speed tradeoff can be controlled by choosing the material, the doping level, and the degree of electron heating as well as the active-region length. The investigations performed for short channels demonstrate the possibility of the increased speed of operation and the reduced noise level; this supports the favorable noise–speed tradeoff in the channels for microwave devices.

CHAPTER 12

HOT-ELECTRON NOISE IN DOPED SEMICONDUCTORS: THEORY

In Chapters 4 to 6, we presented the general theory of fluctuations taking into account electron–electron collisions as well. It was found that in a nonequilibrium electron gas with pair collisions the latter lead to an additional correlation between the electrons, as a result of which the fluctuation–diffusion relation is violated. It was shown that, in general, the contribution of the additional correlation into the current noise under nonequilibrium conditions is not parametrically small. It remains to present explicit results illustrating the general theory — that is, to decode the formal expressions (4.12), (6.6), and (6.9) for the spectral intensities of the current fluctuations and for the diffusion and correlation tensors in few simple but more or less realistic cases, opening the door for interpretation of the available and future experimental results on noise in semiconductors and semiconductor structures with moderate (moderately high) carrier density.

12.1 EFFECTIVE ELECTRON TEMPERATURE AND ITS FLUCTUATIONS

A detailed study of the case where collisions between electrons govern the electron distribution in energy but not in quasi-momentum is possible:

$$\tau \ll \tau^{ee} \ll \tau^{en}. \tag{12.1}$$

Here τ and τ^{en} are the characteristic electron quasi-momentum and energy relaxation times due to electron collisions with the lattice (i.e., with impurities, phonons, etc; cf. Section 7.2), and τ^{ee} is the characteristic time for energy and quasi-momentum transfer *within* the system of electrons given by the order of magnitude of the linearized electron–electron collision operator (2.20) while acting on an arbitrary function of the quasi-momentum. While the rates of quasi-momentum and energy transfer to the thermal bath can differ essentially, those for quasi-momentum and energy exchange within the electron system are known not to differ drastically from each other. Hence, in some region of electron concentrations, the rate at which the electron–electron collisions redistribute energy within the electron system can happen to be larger than the rate at which the electron system transfers the energy to the thermal bath:

$$\tau^{ee} \ll \tau^{en}, \qquad (12.2)$$

while the fast relaxation, due to scattering on the thermal bath, of the *odd part* of the electron distribution function can remain unaffected by electron–electron collisions, in this sense still comparatively rare:

$$\tau \ll \tau^{ee}. \qquad (12.3)$$

This particular region of carrier densities (the typical values for bulk semiconductors are $n_0 \sim 10^{15}$–10^{17} cm^{-3}) is the subject of investigation in this section. In the case, the analytic expressions are available both for the spectral intensities of current fluctuations [131, 132] and for the diffusion tensor [278] (see also references 70 and 105). A comparison of these expressions demonstrates the degree of violation of the fluctuation–diffusion relation for the particular case.

Effective Electron Temperature, Its Relaxation

The theoretical investigation of effects of carrier–carrier scattering on high-electric-field properties of semiconductors goes back to Compton (1923) [279], Davydov (1937) [111], Fröhlich and Paranjape (1956) [265], and Stratton (1957–1958) [280, 281]. Interelectron collisions, though conserving energy and quasi-momentum in the electron system, have an indirect effect on high-field transport. When the rate at which the electron–electron collisions redistribute energy within the electron system is larger than the rate at which the electron system transfers the energy to the lattice, the energy distribution, irrespective of the initial distribution, after a time interval of the order of τ^{ee}, turns out to be nearly Maxwellian:

$$F(\varepsilon_\mathbf{p}) \propto \exp(-\varepsilon_\mathbf{p}/T_e). \qquad (12.4)$$

Correspondingly, the situation where inequalities (12.1) hold is referred to as the *case of effective electron temperature*. The *effective electron temperature* T_e differs, in general, from the temperature T_0 of the thermal bath and changes comparatively slowly in time due to the gain of energy by the electron system from the external field and the transfer of energy to the lattice:

$$\frac{3n_0}{2} \frac{\partial T_e}{\partial t} = \mathbf{j} \cdot \mathbf{E} - P(T_e, T_0), \qquad (12.5)$$

where
$$j_\alpha = \tilde{\sigma}_{\alpha\beta}(T_e) E_\beta \qquad (12.6)$$

is the current density and $P(T_e, T_0)$ is the rate of energy loss by the electron system. The explicit form of the functions $\tilde{\sigma}_{\alpha\beta}(T_e)$ and $P(T_e, T_0)$ — the "chord" conductivity and the rate of energy loss by the electron system — depends on the details of electron interaction with the thermal bath. The *steady-state value of the effective electron temperature* T_e is determined from the condition that the rate at which the electron system gains energy from the field, $\tilde{\sigma}_{\alpha\beta}(T_e) E_\alpha E_\beta$, equals the rate $P(T_e, T_0)$ at which the electron system loses energy through scattering by the thermal bath:

$$\tilde{\sigma}_{\alpha\beta}(T_e) E_\alpha E_\beta = P(T_e, T_0). \qquad (12.7)$$

For small deviations from the steady state,

$$\Delta T_e(t) \equiv T_e(t) - T_e \ll T_e, \qquad (12.8)$$

Eq. (12.5) takes the form

$$\partial(\Delta T_e)/\partial t = -(\Delta T_e/\tau_T), \qquad (12.9)$$

with the *differential time for the electron temperature relaxation* τ_T being given by the expression [282]

$$\tau_T = \frac{3n_0/2}{\partial P(T_e, T_0)/\partial T_e - E_\alpha E_\beta \partial \tilde{\sigma}_{\alpha\beta}(T_e)/\partial T_e}. \qquad (12.10)$$

The values of derivatives in Eq. (12.10) should be taken at the stationary value of the electron temperature found from Eq. (12.7).

Small-Signal Conductivity

From what was said, it follows that at not too high frequencies, namely, at those less than the characteristic frequency of interelectron collisions,

$$\omega \tau^{ee} \ll 1, \qquad (12.11)$$

the frequency dependence of the kinetic coefficients is conditioned by time evolution of the electron temperature. The tensor of AC small-signal conductivity defined by Eq. (2.32) at frequencies complying with inequality (12.11) is given by the expression (see, e.g., reference 70)

$$\sigma_{\alpha\beta}(\omega) = \tilde{\sigma}_{\alpha\beta} + \frac{2E_\gamma E_\delta}{3n_0} \frac{\partial \tilde{\sigma}_{\alpha\gamma}}{\partial T_e} \frac{\tilde{\sigma}_{\beta\delta} + \tilde{\sigma}_{\delta\beta}}{-i\omega + \tau_T^{-1}}. \qquad (12.12)$$

In the case of an isotropic medium, where

$$\mathbf{j} = \tilde{\sigma}(T_e) \cdot \mathbf{E}, \tag{12.13}$$

the longitudinal and transverse (with respect to the direction of the applied electric field) AC small-signal conductivities are given by the expressions [283]

$$\sigma_\perp(\omega) = \tilde{\sigma}(T_e), \tag{12.14}$$

$$\sigma_\parallel(\omega) = \tilde{\sigma}\left(1 + \frac{4E^2\, \partial\tilde{\sigma}/\partial T_e}{3n_0(-i\omega + \tau_T^{-1})}\right). \tag{12.15}$$

In particular, the *static* longitudinal differential conductivity $\sigma_\parallel \equiv dj/dE$ is given by the expression [see Eq. (12.13)]

$$\sigma_\parallel \equiv \frac{d}{dE}(\tilde{\sigma}(T_e) \cdot E) = \tilde{\sigma}(T_e) + 2\frac{\partial\tilde{\sigma}}{\partial T_e} \cdot \frac{dT_e}{dE^2}E^2 \tag{12.16}$$

or, in the equivalent form following from expression (12.15),

$$\sigma_\parallel \equiv \sigma_\parallel(\omega)|_{\omega\tau_T \ll 1} = \tilde{\sigma}\frac{\partial P/\partial T_e + E^2\, \partial\tilde{\sigma}/\partial T_e}{\partial P/\partial T_e - E^2\, \partial\tilde{\sigma}/\partial T_e}. \tag{12.17}$$

Later we shall make use of these expressions in the limiting case of comparatively low electric fields (the case of *weakly heated electron gas*, also referred to as *warm electrons*). This is the case where the effective electron temperature T_e only slightly exceeds the lattice temperature T_0:

$$T_e - T_0 \ll T_0, \tag{12.18}$$

where the difference $T_e - T_0$ is proportional to the field strength squared:

$$T_e - T_0 \propto E^2. \tag{12.19}$$

Neglecting terms of the order higher than E^2 in the expression (12.16), we have

$$\sigma_\parallel = \tilde{\sigma}\left(1 + \frac{2(T_e - T_0)}{T_0}\left(\frac{\partial \ln \tilde{\sigma}}{\partial \ln T_e}\right)_{T_e=T_0}\right). \tag{12.20}$$

The dimensionless quantity $d \ln \tilde{\sigma}/d \ln T_e$ can be referred to as *coefficient of electric sensitivity to electron heating of the system* in question.

Expressions for Diffusion Tensor

In the effective electron temperature case, side by side with the expressions for the small-signal conductivity, the expression for the tensor of electron diffusion coefficients ("tensor of diffusivities") defined by Eq. (6.6) is available [70]:

$$D_{\alpha\beta} = \frac{T_e}{e^2 n_0} \left(\tilde{\sigma}_{\alpha\beta} + \frac{2\tau_T}{3n_0} E_\gamma E_\delta \frac{\partial \tilde{\sigma}_{\alpha\gamma}}{\partial T_e} \left(\tilde{\sigma}_{\beta\delta} + \tilde{\sigma}_{\delta\beta} + T_e \frac{\partial \tilde{\sigma}_{\beta\delta}}{\partial T_e} \right) \right). \quad (12.21)$$

In the case of an isotropic medium, the expressions for the longitudinal and transverse (with respect to the direction of the external electric field) electron diffusion coefficients were obtained in [278] (see also reference 105):

$$D_\perp = T_e \tilde{\sigma}/e^2 n_0, \quad (12.22)$$

$$D_\| = D_\perp \left[1 + \frac{4\tau_T}{3n_0} E^2 \frac{\partial \tilde{\sigma}}{\partial T_e} \left(1 + \frac{1}{2} \frac{\partial \ln \tilde{\sigma}}{\partial \ln T_e} \right) \right]. \quad (12.23)$$

By comparing Eqs. (12.15) and (12.23), we rewrite the latter as follows:

$$D_\| = D_\perp \left[1 + \left(\frac{\sigma_\|}{\tilde{\sigma}} - 1 \right) \left(1 + \frac{1}{2} \frac{\partial \ln \tilde{\sigma}}{\partial \ln T_e} \right) \right]. \quad (12.24)$$

The magnitude of the coefficient of electric sensitivity to electron heating, $d \ln \tilde{\sigma}/d \ln T_e$, determines to what extent the anisotropy of the electron diffusivity differs from that of differential conductivity.

Expressions for Additional-Correlation Tensor

We have noticed in Chapter 4 that the source of the extra correlation the electron–electron collisions create, the term $I_{pp_1}^{ee}\{F, F\}$, vanishes after substitution of the Maxwellian distribution [see Eq. (4.7)]. In the case of comparatively frequent electron–electron collisions we are now investigating, the *main part* of the distribution function is Maxwellian. May we conclude that the extra correlation is unimportant in the case where the inequalities (12.1) hold, and that the spectral intensities of the current fluctuations at low frequencies in this case are proportional to the diffusion coefficients given by Eqs. (12.22) and (12.23)?

It would be careless to jump to such a conclusion without a detailed investigation of the problem. Electron–electron collisions play a double role, shaping the distribution function and creating the additional correlation. The more frequent the interelectron collisions are, the more considerable the additional correlation should be. On the other hand, in the case of more frequent interelectron collisions, the energy distribution is Maxwellian with greater accuracy, and so the extra correlation should more effectively vanish. As was shown in reference 132, these opposing tendencies counterbalance each other, and the extra correlation is, generally speaking, as important in the formation of the longitudinal current fluctuations in a heating electric field as the electron temperature fluctuations are.

To obtain a comparatively simple explicit expression for the additional-correlation tensor $\Delta(\omega)$ defined by Eq. (6.9), one should assume, side by side with inequalities (12.1), a *weak inelasticity* of the scattering of electrons by the

152 HOT-ELECTRON NOISE IN DOPED SEMICONDUCTORS: THEORY

lattice ($\Delta\varepsilon \ll \bar{\varepsilon}$, where $\Delta\varepsilon$ is the characteristic change of the electron energy upon collision; cf. Section 7.2). The following expression was derived [70] for $\omega\tau^{ee} \ll 1$:

$$\Delta_{\alpha\beta}(\omega) = \frac{4T_e^2\tau_T E_\gamma E_\delta}{3n_0^2 e^2(1+\omega^2\tau_T^2)} \frac{\partial\tilde{\sigma}_{\alpha\gamma}}{\partial T_e} \frac{\partial\tilde{\sigma}_{\beta\delta}}{\partial T_e}\left(1 - \frac{2\tau_T}{3n_0}\frac{P}{T_e - T_0}\right), \qquad (12.25)$$

from which, in the case of an isotropic medium, we obtain the expressions obtained in reference 132:

$$\Delta_\perp(\omega) = 0, \qquad (12.26)$$

$$\Delta_\parallel(\omega) = -\frac{D_\perp}{1+\omega^2\tau_T^2}\left(\frac{\sigma_\parallel}{\tilde{\sigma}} - 1\right)\left[\frac{T_e}{2(T_e - T_0)}\left(\frac{\sigma_\parallel}{\tilde{\sigma}} - 1\right) - \frac{\partial \ln \tilde{\sigma}}{\partial \ln T_e}\right]; \quad (12.27)$$

The extra correlation contributes only to the longitudinal current fluctuations. The contribution is of Lorentz form shaped by the electron temperature relaxation time; as far as the inequalities (12.1) and (12.11) hold, only the extra correlation of electron energies play a role.

It is quite easy to see that, thanks to Eq. (12.20)), valid up to linear in $(T_e - T_0)$ terms, the right-hand side of Eq. (12.27) *vanishes in the linear [with respect to $(T_e - T_0)$] approximation*, with the additional correlation term $\Delta(\omega)$ appearing only as a correction of the order of $(T_e - T_0)^2 \propto E^4$ (cf. references 9, 70 and 136). In other words, for warm electrons in the effective electron temperature approximation the additional correlation contribution vanishes. On the contrary, when $(T_e - T_0)$ is comparable to T_0 or even larger, the longitudinal additional-correlation term $\Delta_\parallel(\omega)$ in general is of the same order of magnitude as the diffusion coefficients are [cf. Eqs. (12.27) and (12.22), (12.23)].

Expressions for Spectral Intensities of Current Fluctuations

For the tensor of spectral intensities of current fluctuations as given by Eq. (6.8), one has the following from Eqs. (12.21) and (12.25):

$$(\delta j_\alpha \delta j_\beta)_\omega = \frac{T_e}{V_0}\left\{\tilde{\sigma}_{\alpha\beta} + \tilde{\sigma}_{\beta\alpha} + \frac{2E_\gamma E_\delta}{3n_0}\left[\frac{\partial\tilde{\sigma}_{\alpha\gamma}}{\partial T_e} \cdot \frac{\tilde{\sigma}_{\beta\delta} + \tilde{\sigma}_{\delta\beta}}{-i\omega + \tau_T^{-1}}\right.\right.$$
$$\left.\left. + \frac{\partial\tilde{\sigma}_{\beta\gamma}}{\partial T_e} \cdot \frac{\tilde{\sigma}_{\alpha\delta} + \tilde{\sigma}_{\delta\alpha}}{i\omega + \tau_T^{-1}} + \frac{\partial\tilde{\sigma}_{\alpha\gamma}}{\partial T_e}\frac{\partial\tilde{\sigma}_{\beta\delta}}{\partial T_e} \cdot \frac{4\tau_T^2/3n_0}{1+\omega^2\tau_T^2} \cdot \frac{T_e P}{T_e - T_0}\right]\right\} (12.28)$$

(see reference 70). In the case of an isotropic medium, this expression reduces to those obtained in reference 131 (see also reference 132):

$$(\delta j_\perp^2)_\omega = 2T_e\tilde{\sigma}/V_0, \qquad (12.29)$$

$$(\delta j_\parallel^2)_\omega = \frac{2T_e\tilde{\sigma}}{V_0}\left\{1 + \frac{\sigma_\parallel/\tilde{\sigma} - 1}{1+\omega^2\tau_T^2}\left[1 + \frac{T_e}{4(T_e - T_0)}\left(\frac{\sigma_\parallel}{\tilde{\sigma}} - 1\right)\right]\right\}. \quad (12.30)$$

The frequency dependence of the spectral intensity of the longitudinal current fluctuations reflects the way in which fluctuations of the electron temperature die down, with the second term in the curly brackets in Eq. (12.30) having the Lorentz shape. In the low-frequency limit $\omega \tau_T \ll 1$ we obtain

$$(\delta j_\parallel^2)_{\omega \tau_T \ll 1} = \frac{2T_e \tilde{\sigma}}{\mathcal{V}_0} \left\{ 1 + \left(\frac{\sigma_\parallel}{\tilde{\sigma}} - 1 \right) \left[1 + \frac{T_e}{4(T_e - T_0)} \left(\frac{\sigma_\parallel}{\tilde{\sigma}} - 1 \right) \right] \right\}. \quad (12.31)$$

In the effective electron temperature case the spectral intensities of current fluctuations in the isotropic medium in the low-frequency limit $\omega \tau_T \ll 1$ are expressible in terms of the longitudinal differential conductivity $\sigma_\parallel$, the chord conductivity $\tilde{\sigma}$, the electron temperature T_e, and the ambient temperature T_0 only [this is not true for the diffusivity D and the additional correlation Δ taken separately, see Eqs. (12.24) and (12.27)].

Expressions for Noise Temperature

Now we are in position to calculate the noise temperature, defined by Eq. (7.20), in the effective electron temperature case. It follows from Eqs. (7.20) and (12.29) that, in the case of an isotropic medium, the transverse noise temperature *is equal* to the electron temperature:

$$T_{n\perp}(\omega) = T_e, \quad (12.32)$$

while the frequency dependence of the longitudinal noise temperature is given by the expression [70]

$$T_{n\parallel}(\omega) = T_e \left[1 + \frac{T_e}{4(T_e - T_0)} \left(\frac{\sigma_\parallel}{\tilde{\sigma}} - 1 \right)^2 \frac{1}{(\sigma_\parallel/\tilde{\sigma}) + \omega^2 \tau_T^2} \right] \quad (12.33)$$

so that

$$T_{n\parallel}(\omega) \geq T_{n\perp}(\omega) = T_e. \quad (12.34)$$

The longitudinal noise temperature exceeds (or, in special cases, is equal to) the transverse one; the latter, in turn, is equal to the electron temperature. At low frequencies, where $\omega \tau_T \ll 1$, we obtain

$$T_{n\perp} = T_e \quad (12.35)$$

$$T_{n\parallel} = \frac{T_e}{T_e - T_0} \left\{ \frac{T_e}{4} \left[\left(\frac{\sigma_\parallel}{\tilde{\sigma}} \right)^{1/2} + \left(\frac{\tilde{\sigma}}{\sigma_\parallel} \right)^{1/2} \right]^2 - T_0 \right\}. \quad (12.36)$$

It follows from Eqs. (12.35) and (12.36) that the degree of anisotropy of the low-frequency noise temperature is the function only of the anisotropy of the differential conductivity,

$$\frac{\sigma_\parallel}{\sigma_\perp} = \frac{\sigma_\parallel}{\tilde{\sigma}}, \quad (12.37)$$

and of the degree of heating of the electron system characterized by the ratio T_e/T_0:

$$\frac{T_{n\|}}{T_{n\perp}} = \frac{T_{n\|}}{T_e} = 1 + \frac{1}{4(1 - T_0/T_e)} \frac{\tilde{\sigma}}{\sigma_\|} \left(\frac{\sigma_\|}{\tilde{\sigma}} - 1\right)^2. \qquad (12.38)$$

Relations (12.38) enable one to verify, to some extent, the validity of the approximations leading to those relations. If the quantities $\sigma_\perp$, $\sigma_\|$ and $T_{n\perp}$, $T_{n\|}$ are measured and/or computed through a simulation procedure, the validity of relation (12.38) can be checked up. Remarkable deviations would definitely mean that the electron energy distribution is rather far from the Maxwellian one.

In the case of weakly heated electron gas, where the effective electron temperature T_e only slightly exceeds the lattice temperature T_0 [i.e., $T_e - T_0 \ll T_0$, see Eq. (12.18)], we have

$$\frac{T_{n\|}}{T_{n\perp}} = 1 + \frac{T_e - T_0}{T_0} \left(\frac{\partial \ln \tilde{\sigma}}{\partial \ln T_e}\right)^2_{T_e=T_0}; \qquad (12.39)$$

that is, in the warm-electron case the sensitivity coefficient *squared* enters the expression for the anisotropy of the noise temperature.

The theoretical results presented in this section will be used below while seeking an interpretation of the available experimental and Monte Carlo simulation results on noise in moderately doped semiconductor.

12.2 FLUCTUATIONS IN DRIFTED MAXWELLIAN APPROXIMATION

Analytic expressions are also available in the case where interelectron collisions are so frequent that they control the electron distribution both in energy and in quasi-momentum:

$$\tau^{ee} \ll \{\tau, \tau^{en}\}, \qquad (12.40)$$

where τ and τ^{en} are the characteristic electron quasi-momentum and energy relaxation times due to electron collisions with the lattice, and τ^{ee} is the characteristic time for energy and quasi-momentum transfer within the system of electrons.

It is known that in the semiconductors of some classes, especially in lead chalcogenides, the screening (owing to lattice polarization) of the static Coulomb potential of ionized impurities is more effective than the same type of screening of the electron–electron Coulomb interaction [284, 285]. Therefore, in uncompensated lead-chalcogenide samples at high carrier densities and low temperatures, one can expect inequality (12.40) to be realized; that is, one can expect the electron–electron interaction to be effective enough to control the electron distribution both in energy and in quasi-momentum. Under these conditions, the nonequilibrium noise was investigated in reference 135.

When inequality (12.40) is fulfilled, electron–electron collisions shape the distribution function, making it close to the *drifted Maxwellian distribution*

$$F_{\mathbf{p}}^M \propto \exp[-(\varepsilon_{\mathbf{p}} - \mathbf{p} \cdot \mathbf{V})/T_e]. \qquad (12.41)$$

The stationary values of the temperature T_e and of the drift velocity **V** of the electron gas are obtainable from the energy and momentum balance equations that follow from the Boltzmann equation. The diffusion-coefficient tensor in the case of distribution (12.41) can be easily shown to be proportional to the differential-conductivity tensor. This means that, under conditions (12.40), the spectral intensity of the current fluctuations is not proportional to the AC small-signal conductivity (and thus the noise temperature is anisotropic) *only*, provided that the correlation tensor $\Delta_{\alpha\beta}$ does not vanish. In fact, the influence of the additional correlation created by electron–electron collisions on the current fluctuations in the case (12.40) turned out to be substantial [135]. Under the condition $\tau \sim \tau^{en}$ the additional correlation affects not only the longitudinal but also the transverse current fluctuations. The influence of the additional correlation as well as the anisotropy of the noise temperature takes place in a wide range of frequencies up to $\omega \sim (\tau^{ee})^{-1}$.

If the energy of the electron system relaxes more slowly than the momentum,

$$\tau^{ee} \ll \tau \ll \tau^{en}, \tag{12.42}$$

then the additional correlation affects only the longitudinal current fluctuations, and only at frequencies $\omega \lesssim (\tau^{en})^{-1}$. In the case of nearly elastic scattering by the lattice ($\Delta\varepsilon \ll \bar{\varepsilon}$), the longitudinal spectral intensity of the current fluctuations and the longitudinal noise temperature can be expressed in terms of the electron and lattice temperatures, the static differential conductivities, and the electron temperature relaxation time just as in the case of effective electron temperature (12.1); that is, Eqs. (12.30) and (12.33) also hold in the case (12.42). However, the contribution of the additional correlation between the occupation numbers of the single-electron states due to the interelectron collisions is different in cases (12.1) and (12.42). In the latter case,

$$\Delta_\|(\omega) = \frac{T_e^2(\sigma - \tilde{\sigma})^2}{2e^2 n_0 \tilde{\sigma}(T_e - T_0)(1 + \omega^2 \tau_T^2)} \tag{12.43}$$

while in case (12.1) Eq. (12.27) holds. One can easily see from Eq. (12.43) that for "warm" electrons the correlation contribution in the drifted Maxwelian case does not vanish, contrary to the electron temperature case (12.1) (see Section 12.1).

In the drifted Maxwellian case the relative contribution of the additional correlation, that is, the ratio

$$\frac{\Delta_\|}{2D_\|} = -\frac{1}{4(1 - T_0/T_e)} \left[\left(\frac{\sigma_\|}{\tilde{\sigma}}\right)^{1/2} - \left(\frac{\tilde{\sigma}}{\sigma_\|}\right)^{1/2} \right]^2, \tag{12.44}$$

depends *only* on the degree of heating of the electron system and on the nonlinearity of the current–voltage characteristic [in case (12.1) it depended, through $\partial\sigma/\partial T_e$, also on the details of scattering mechanisms — see Eq. (12.27)].

Contrary to Eq. (12.27), the sign of $\Delta_\|(\omega)$ as given by Eq. (12.43) is fixed: Independently of the details of the scattering mechanisms, we obtain

$$\Delta_\|(\omega) \leq 0, \qquad (12.45)$$

provided that the scattering on the thermal bath is nearly elastic. We conclude that in the case of frequent electron–electron collisions [Eq. (12.10)] and quasi-elastic scattering by the lattice, the additional correlation does not reveal itself in the transverse current fluctuations and enhances the longitudinal current fluctuations. The inequality

$$T_{n\|}(\omega) \geq T_{n\perp}(\omega) = T_e \qquad (12.46)$$

holds, being ensured in this case exclusively by the additional correlation created by electron–electron collisions.

To conclude the section we note that, in the case of the drifted Maxwellian distribution (12.41) ensured by the inequalities $\tau^{ee} \ll \{\tau, \tau^{en}\}$, the spectral intensity of the current fluctuations $(\delta j_\alpha \delta j_\beta)_\omega$ contains two characteristic terms: a term proportional to AC small-signal conductivity $\text{Re}\{\sigma_{\alpha\beta}(\omega)\}$ (in the low-frequency limit, also to the symmetrized diffusion tensor $D_{\alpha\beta} + D_{\beta\alpha}$) and a term proportional to the correlation tensor $\Delta_{\alpha\beta}(\omega)$. Contrary to this, in the electron temperature case (12.1) the spectral intensity of the current fluctuations $(\delta j_\alpha \delta j_\beta)_{\omega\tau_\varepsilon \ll 1}$ consisted of three characteristic terms: that proportional to $\sigma_{\alpha\beta}$, that "transforming" $\sigma_{\alpha\beta}$ into $D_{\alpha\beta}$, and the correlation tensor $\Delta_{\alpha\beta}$.

Therefore the experimental separation of an interesting physical effect — the contribution made to the current fluctuations by the additional correlation due to collisions between the electrons — may turn out to be relatively simple in the case of frequent collisions between the electrons. If it is established independently that the case $\tau^{ee} \ll \{\tau, \tau^{en}\}$ is realized in experiment, then to investigate the additional correlation it is sufficient to compare $(\delta j^2)_\omega$ with $\text{Re}\,\sigma(\omega)$ (i.e., it is not mandatory, as in other cases, to have independently measured $(\delta j^2)_\omega$ and D). This makes an experimental investigation of the current fluctuations in the case $\tau^{ee} \ll \{\tau, \tau^{en}\}$ quite enticing.

12.3 FLUCTUATIONS IN WEAKLY HEATED ELECTRON GAS OF INTERMEDIATE DENSITY

It follows from what was said that the problem of electronic noise in semiconductor is solvable analytically, under some simplifying assumptions, in three limiting cases. The first is that of low electron densities, when effects of interelectron collisions can be neglected (see Section 7.2). The analytic approach, as was shown in Sections 12.1 and 12.2, is also possible in two other cases: those of frequent interelectron collisions shaping either (a) the distribution of electron energy or (b) the distribution of energy as well as quasi-momentum.

Of course, it would also be interesting to investigate *intermediate* cases, for example, to follow how, with increasing electron density, the additional correlation becomes important. In the case of "warm" electrons this program was accomplished in reference 136. Detailed calculations of the diffusion tensor $D_{\alpha\beta}$ and the correlation tensor $\Delta_{\alpha\beta}$ were performed for intermediate electron densities.

It follows from the kinetic theory of fluctuations in a nonequilibrium state that in the general case the correlation tensor $\Delta_{\alpha\beta}$ consists of two different parts [see Eq. (6.9)]:

$$\Delta_{\alpha\beta} = A_{\alpha\beta} + B_{\alpha\beta}. \qquad (12.47)$$

One of them, $A_{\alpha\beta}$, is caused by the equal-time electronic-state-occupancy cross-correlation generated by the interelectron collisions — that is, by the term $I^{ee}_{pp_1}\{\overline{F},\overline{F}\}$. Up to this moment, we have paid principal attention only to this part of the tensor $\Delta_{\alpha\beta}$, since in the electron temperature approximation [see inequality (12.1)] and the drifted Maxwellian approximation [see inequality (12.42)] the distribution function $\overline{F}$ is a linear function of concentration. Then $N\,\partial \overline{F}_p/\partial N = \overline{F}_p$, and the second part, $B_{\alpha\beta}$, vanishes. But in the general case the second part, $B_{\alpha\beta}$, can be essential, also being conditioned by interelectron collisions, but not through the correlation created by them. In fact, interelectron collisions violate the Price relation, not only creating the correlation but also changing the dependence of the form of the stationary distribution function $\overline{F}_p$ on the electron density. Vanishing in the electron temperature case as well as in the drifted Maxwellian case, the term $B_{\alpha\beta}$ could play an essential role for intermediate carrier densities. Indeed, as was shown in reference 136, in the case when electron–electron scattering and the energy relaxation via electron–phonon scattering is of the same efficiency — that is, the conditions

$$\tau \ll \tau^{ee} \sim \tau^{en} \qquad (12.48)$$

are satisfied — the term $B_{\alpha\beta}$ in the warm-electron region is even more essential than the pure extra correlation term $A_{\alpha\beta}$.

In accordance with the main idea of the warm-electron theory, the longitudinal and transverse components of the diffusion and correlation tensors were expanded into a Taylor series with respect to the strength of the electric field up to the quadratic terms:

$$D_{\perp,\|} = D^{eq}[1 + \gamma^D_{\perp,\|}(E/E_0)^2], \qquad (12.49)$$

$$\Delta_{\perp,\|} = D^{eq}\gamma^\Delta_{\perp,\|}(E/E_0)^2, \qquad (12.50)$$

where D^{eq} is the equilibrium isotropic diffusion coefficient, and the dimensionless coefficients γ^D and γ^Δ characterize the nonequilibrium correction to the diffusion and correlation tensors, correspondingly.

It was found that the tensor $\Delta_{\alpha\beta}$ in the weak-heating region in the case (12.48) receives an *isotropic* contribution proportional to E^2 (i.e., $\gamma^\Delta_\| = \gamma^\Delta_\perp = \gamma^\Delta$). It was shown that in the weak-heating region in the case of quasi-elastic scattering, it is necessary to retain only the term $B_{\alpha\beta}$ in the tensor $\Delta_{\alpha\beta}$, with the term $A_{\alpha\beta}$ being parametrically small. In the case of effective electron temperature where the inequality (12.1) holds, the additional correlation term for warm electrons was shown to vanish (see Section 12.1). Consequently, under weak-heating conditions, violation of the Price relation in the case of quasi-elastic scattering is possible only due to the dependence of the *shape* of the stationary distribution function on the electron density at intermediate carrier densities, where the inter-electron relaxation time τ^{ee} and the energy relaxation on the lattice time τ^{en} have the same order of magnitude.

For an actual estimate of the extent to which the Price relation can be violated in the weak-heating region, we cite some numerical-calculation results of reference 136. The coefficients $\gamma^D_{\|,\perp}$ and $\gamma^\Delta = \gamma^\Delta_B$ were calculated for quasi-elastic interactions, when the equations for the symmetric parts of the distribution and correlation functions contain the electron–electron collision operator and the Davydov-type operator of electron collisions with the heat bath (cf. Section 7.2). The asymmetric parts of the collision terms were written in the momentum-relaxation time approximation $\tau(\varepsilon) \propto \varepsilon^{-1/2}$. The Davydov-type operator is a differential operator. The electron–electron collision operator was used in the so-called Landau form, which takes suitable account of the scattering with small momentum change, which is most significant in the case of weak heating. In case (12.48), only the symmetric part of this operator is essential, with its integral part being a Volterra operator. Consequently, the equations can be reduced to a system of differential equations, and these were solved numerically by the Runge–Kutta method.

The dependencies of the coefficients $\gamma^D_\|$, $\gamma^D_\perp$, and γ^Δ_B, on the only parameter W, calculated for this model, are shown in Fig. 12.1. The parameter W characterizes the relative intensity of the electron–electron scattering compared with the heat-bath mechanisms of energy relaxation. In the limit $W \to 0$, the coefficients $\gamma^D_\|$ and $\gamma^D_\perp$ go over into the values obtained earlier by neglecting electron–electron collisions [110] (see Section 7.2), while as $W \to \infty$ the values approach those

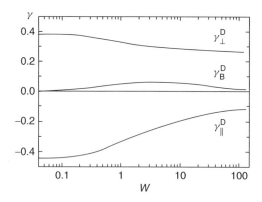

Figure 12.1. Plots of the coefficients $\gamma^D_\perp$, $\gamma^D_\|$, and γ^Δ_B against the parameter W that determines the relative intensity of the electron–electron scattering compared with the (quasi-elastic) scattering by the lattice [136].

obtained in the electron-temperature approximation. The coefficient γ_B^Δ, as it should be, vanishes in both limiting cases. It has a maximum when the rate at which electron–electron collisions redistribute energy within the electron system is of the same order as the rate at which the electron system transfers energy to the thermal bath. Indeed, the time τ^{ee} is equal to τ^{en} for $W \approx 6$. The maximum of γ_B^Δ corresponds approximately to the same value of W (Fig. 12.1). It can be seen from Fig. 12.1 that in the case of quasi-elastic scattering the warm-electron coefficients (the plots of $\gamma_\perp^D$ and $\gamma_\parallel^D$) are sensitive to the presence of electron–electron collisions. Violation of the Price relation (curve γ_B^Δ) also occurs, but is not strongly pronounced.

It is known that in the case of scattering of electrons at low lattice temperatures by optical phonons the electron–electron collisions provide a new energy-relaxation channel ("composite" scattering sets in). As a result, electron–electron collisions effectively influence both the energy relaxation itself and the form of the stationary distribution function. In this situation, the coefficients $\gamma_{\parallel,\perp}^D$ and γ_B^Δ should be more sensitive to the presence of electron–electron collisions, and the coefficient γ_A^Δ should not vanish because the collision operator is not quasi-elastic in the presence of scattering by optical phonons.

The computation of these coefficients was performed [136] for a compensated semiconductor with parameters corresponding to the heavy-hole band in p-Ge at $T_0 = 80$ K. It was demonstrated that hole–hole collisions at intermediate densities ($p \sim 10^{14} - 10^{16}$ cm^{-3}) influence the diffusion coefficients substantially. The coefficients γ_A^Δ and γ_B^Δ are plotted in Fig. 12.2. In the region of intermediate densities the coefficient γ_A^Δ is still rather small compared with γ_B^Δ, apparently an indirect indication that the inequality $\tau^{ee} \gg \tau$ (but not $\bar{\varepsilon} \gg \Delta\varepsilon$)

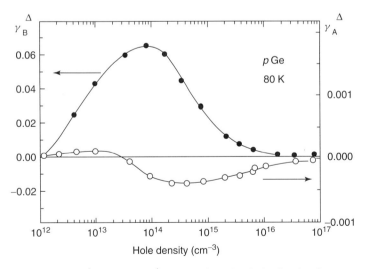

Figure 12.2. Plots of γ_A^Δ (○), and γ_B^Δ (●) against the hole density for p-type Ge at $T_0 = 80$ K [9]. Solid curves are guide to the eye.

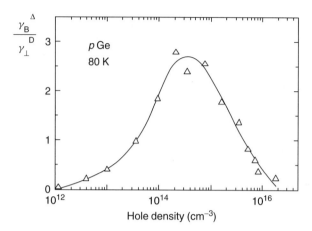

Figure 12.3. Plot of the ratio $\gamma_B^\Delta/\gamma_\perp^D$ against the hole density for p-type Ge at $T_0 = 80$ K [9]. Solid curve is the guide to the eye.

still holds. Figure 12.3 shows the ratio of the coefficient γ_B^Δ and the nonequilibrium correction to the diffusion coefficient $\gamma_\perp^D$. It is seen to be substantial at $p \sim 10^{14} - 10^{15}$ cm^{-3}. This leads to the hope that there is a possibility of observing experimentally violations of the Price relation in semiconductors at intermediate densities under conditions of weak as well as intermediate heating.

CHAPTER 13

ELECTRONIC NOISE IN STANDARD-DOPED n-TYPE GaAs

During recent years, progress has been achieved in experimental investigation, simulation, and theoretical interpretation of noise properties of doped semiconductors at moderate electric fields. Noise characteristics of silicon-doped n-type gallium arsenide (free electron concentration $n = 3 \times 10^{17}$ cm^{-3} at 80 K lattice temperature) have been obtained experimentally [11, 228] (see also reference 247). The performed Monte Carlo calculations [246] have showed that the obtained experimental results at moderate fields of few hundreds volts per centimeter cannot be even qualitatively interpreted within a framework of the model neglecting interelectron collisions. Thus, microscopic simulation at moderate fields for a realistic model of an electron gas in a doped semiconductor with necessary electron–lattice and interelectron scattering mechanisms taken into account was called for in order to fit the available experimental data. On these lines, effects of interelectron collisions on observables have been resolved [246]. Moreover, interpretation in terms of the effective electron temperature and its fluctuations — in terms of the theory presented in Section 12.1 — was achieved [247]. In this chapter, these and some other achievements related to experimental investigation and interpretation of noise in doped semiconductors will be described in detail.

13.1 MONTE CARLO SIMULATION

In this section, the procedure of simulation of current fluctuations for a realistic model of an electron gas in a doped semiconductor [246] is described, with the results of computation to be compared to the experimental data [11, 228, 247] in the subsequent section.

Let us introduce the time-dependent drift velocity of N free electrons weakly interacting among themselves and with an unperturbed thermal bath—that is, the velocity of the mass center of the electron system:

$$V(t) = \frac{1}{N} \sum_i v_i(t), \qquad (13.1)$$

where $v_i(t)$ is the instantaneous velocity of the ith electron. Under a steady state (at equilibrium as well), the drift velocity of the electron system fluctuates around its averaged (over time, or over an ensemble of the systems) value

$$V = \overline{V(t)}. \qquad (13.2)$$

The total energy and quasi-momentum of the electrons being conserved during an interelectron collision, the fluctuations of the drift velocity,

$$\delta V(t) = V(t) - V, \qquad (13.3)$$

are caused only by electron interaction with the thermal bath (phonons and impurities in the case of a semiconductor). On the other hand, the instantaneous velocity of the ith electron $v_i(t)$ with respect to its long-time average V,

$$\delta v_i(t) = v_i(t) - V, \qquad (13.4)$$

is influenced by all scattering mechanisms in action. Here, as far as electron velocity fluctuations are considered along a chosen direction (that in which a constant electric field is applied), the vector indices are omitted. The time-displaced drift-velocity to drift-velocity correlation function is

$$\Phi(t) = N\overline{\delta V(t_1 + t)\delta V(t_1)}, \qquad (13.5)$$

where the average is taken over t_1 with the time interval between two observations, t, being kept fixed. The function $\Phi(t)$ can be presented as

$$\Phi(t) = \Phi_{\text{auto}}(t) + \Phi_{\text{cross}}(t), \qquad (13.6)$$

where

$$\Phi_{\text{auto}}(t) = \frac{1}{N} \sum_i \overline{\delta v_i(t_1 + t)\delta v_i(t_1)} \qquad (13.7)$$

and

$$\Phi_{\text{cross}}(t) = \frac{1}{N} \sum_{i \neq j} \overline{\delta v_i(t_1 + t)\delta v_j(t_1)} \qquad (13.8)$$

here will be referred to as autocorrelation and cross-correlation function, respectively. The main features of time-displaced correlation of the electron velocities in the presence of interelectron collisions can be illustrated [286] by the (rather artificial) case when the interaction of electrons with the thermal bath is weak as compared to that between themselves — that is when the interelectron relaxation time τ^{ee} is shorter than the electron quasi-momentum relaxation time τ conditioned by an interaction with the thermal bath: $\tau^{ee} \ll \tau$. In this case the autocorrelation function, starting from its equal-time ($t = 0$) value,

$$\Phi_{\text{auto}}(0) = \overline{v_i^2}, \qquad (13.9)$$

decreases with t mainly due to interelectron collisions: Any collision causes a loss of one-electron velocity autocorrelation; and the shortest time constant, τ^{ee}, dominates the rate of decay of the autocorrelation function $\Phi_{\text{auto}}(t)$ in a short time scale.

In equilibrium there is no equal-time cross-correlation: $\Phi_{\text{cross}}(0) = 0$. On the other hand, any interelectron collision, conserving energy and quasi-momentum, causes the correlation of velocities of the two electrons involved, and, for small t, the cross-correlation function $\Phi_{\text{cross}}(t)$ grows proportionally to t. The opposite tendencies in the evolution of $\Phi_{\text{auto}}(t)$ and $\Phi_{\text{cross}}(t)$ counterbalance each other, and the resultant total (drift velocity to drift velocity) correlation function $\Phi(t)$ changes slowly, with its decay being caused only by the interaction of the electrons with the thermal bath. So, frequent interelectron collisions tend to redistribute the correlation between the diagonal and off-diagonal terms in favor of the cross-correlation, with the total correlation function for $t \ll \tau$ being kept approximately constant.

At $t \gg \tau^{ee}$ the autocorrelation function becomes small enough, and the cross-correlation function $\Phi_{\text{cross}}(t)$ follows closely the total correlation function $\Phi(t)$, both decaying with the time constant τ. So, the cross-correlation function passes over the maximum, with its maximum value being under $k_B T_0/m$ at equilibrium [286].

Let us simulate motion in a uniform electric field of N electrons undergoing different types of scattering events. All electrons move without scattering for the time between two successive "events in the electron system". By the event we mean either a scattering of an electron by the thermal bath or a mutual collision between two electrons. The time between two successive events in the electron system will be referred to as the "time of free flight of the system."

For independent scattering events, the time of free flight of the system is defined by the combined scattering rate [246]:

$$\lambda_{\text{comb}}(\mathbf{v}_1, \mathbf{v}_2, \ldots, \mathbf{v}_N) = \sum_{i=1}^{N} \lambda_i(\mathbf{v}_i) + \frac{1}{N-1} \sum_{i=1}^{N-1} \sum_{j=i+1}^{N} \lambda_{ij}^{ee}(\mathbf{v}_i, \mathbf{v}_j), \qquad (13.10)$$

where λ_i and λ_{ij}^{ee} are the integral rates of scattering of the ith electron by the thermal bath and by the jth electron, respectively. The factor $(N-1)^{-1}$ normalizes the interelectron scattering rate, so that each electron under simulation is weighted by n_0/N, where n_0 is the electron density. Equation (13.10) reduces to that written down in reference 238 for $N = 2$.

The combined scattering rate λ_{comb} depends on the electron velocity distribution. The Monte Carlo procedure deals with the instantaneous velocity distribution rather than its time-average function. Thus the fluctuations of electron distribution are not ignored. In order to cope with the time-dependent combined scattering rate, the Rees self-scattering procedure (see reference 31) is applied: A fictitious "scattering rate" is added to the instantaneous combined scattering rate to make the resultant total scattering rate independent of time.

Now, the simulation of synchronous motion of N electrons is straightforward. The procedure starts from a chosen electron velocity distribution. A random number is generated to simulate a realization of the "time of free flight of the system" determined by the total scattering rate. Since all electrons move without any scattering during the time of free flight, the velocities of all electrons before the scattering event are available, and this is sufficient to calculate the integral scattering rates of each possible scattering event, to make up the combined scattering rate, and to determine the self-scattering rate. Now, another random number is generated to select a type of the scattering event in proportion to its integral scattering rate. Provided that the choice falls on the self-scattering, nothing happens in the system, and another free flight is simulated. Provided that the choice falls on the ith electron to be scattered by one of the lattice-related mechanisms (phonon, impurity), the consequences of the collision are simulated in the standard way [31]. If the pair collision of the ith electron with the jth electron is selected, then a random number is generated to choose the scattering angle in the $(\mathbf{v}_i, \mathbf{v}_j)$ plane according to the differential interelectron scattering rate, and the final velocities of the two electrons involved are determined with respect to the energy and quasi-momentum conservation.

The final velocity of the electron or the velocities of the electrons of the pair are used to renew the set of initial velocities for the next free flight. So, the velocities of all electrons are known at any time, and the simulation continues as long as required.

The simulated realization of the events in the electron system contains information on fluctuations around the steady state. The velocity correlation functions $\Phi(t)$, $\Phi_{auto}(t)$, and $\Phi_{cross}(t)$ [Eqs. (13.6)–(13.8)] are obtained as averages over the simulation time t_1 for any fixed time difference t.

The proposed "combined scattering rate" technique avoids the short-time-step procedure inherent to conventional Ensemble Monte Carlo methods [31, 235, 236, 237, 240]. Since the time step in the Ensemble Monte Carlo technique should be chosen essentially shorter than the mean time of free flight of the system, while each step is accompanied by a selection of a type of scattering event (the latter procedure is the same in both techniques), the combined scattering rate technique seems to be beneficial.

Evidently, Eq. (13.10) can be modified to consider interparticle collisions of different quasi-particles: electrons and holes, light and heavy holes, Γ and X electrons, and so on.

13.2 EXPERIMENTAL DATA

The electronic noise was measured at 10-GHz frequency [11, 228, 247] in typical silicon-doped n-type GaAs channels with essentially the same free electron density (around $n_0 = 3 \times 10^{17}$ cm^{-3}), at 80 K lattice temperature. The electron mobility in the channels was essentially lower than that predicted for uncompensated n-type GaAs (see reference 287), indicating acceptor contribution. Compensation of donors in silicon-doped n-type GaAs can be important (silicon is an amphoteric impurity in GaAs, and silicon atoms in Ga sites act as donors while those occupying As sites act as acceptors). The technique to evaluate the degree of compensation in GaAs and other compound semiconductors is based on the low-field mobility measurements (see reference 288 and references therein, as well as reference 289). The degree of compensation of the doped GaAs samples for which the experimental data on noise were available [11, 228, 247] was determined [246, 247] from comparison of the measured dependence of current on electric field with those calculated at different densities of ionized impurities, assuming the same experimentally determined free electron density, $n_0 = 3 \times 10^{17}$ cm^{-3}. The results of this procedure are illustrated by Fig. 13.1 for two samples, with their zero-field mobility at 80 K being 4080 cm^2/(V·s) and 2540 cm^2/(V·s), respectively. A reasonable fit to the experimental data on current–voltage characteristics was obtained assuming the ionized-impurity density $N^{\text{ion}} = 7.5 \times 10^{17}$ cm^{-3} for sample 1 and $N^{\text{ion}} = 12 \times 10^{17}$ cm^{-3} for sample 2.

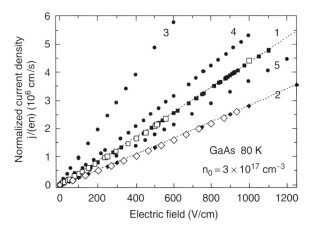

Figure 13.1. Measured current–voltage characteristic of GaAs samples ($\square$ represents sample 1, $\diamond$ represents sample 2) compared with those obtained from Monte Carlo simulation (solid symbols) taking into account interelectron collisions, $n_0 = 3 \times 10^{17}$ cm^{-3}, for different ionized impurity density N^{ion} [247]. 1 represents 7.5×10^{17} cm^{-3}, 2 represents 12×10^{17} cm^{-3}, 3 represents 3×10^{17} cm^{-3}, 4 represents 6×10^{17} cm^{-3}, 5 represents 9×10^{17} cm^{-3}. Dotted lines present current–voltage characteristics calculated for samples 1 and 2, neglecting interelectron collisions.

It remained to perform the above-described Monte Carlo simulation of fluctuations in the framework of a realistic model of an electron gas in a doped semiconductor in the case where experimental data on microwave noise were available, namely, for silicon-doped n-type GaAs, $n_0 = 3 \times 10^{17}$ cm^{-3} at 80 K lattice temperature. Since interelectron collisions were expected to be important at not-too-high electric fields (as is well known, the Coulomb scattering mechanisms gradually switch off with increase of electron energies), calculations were performed within a framework of a parabolic one-valley (Γ-valley) model, thus ignoring intervalley transitions. Nonelastic acoustic and optical scattering by phonons (acoustic deformation potential, polar optical) was considered, with the phonons being supposed to remain in thermal equilibrium. The ionized impurity scattering and interelectron pair collisions were taken into account in the screened Coulomb approach. The effect of electron heating on the screening was neglected. The results on the spectral intensity of drift-velocity fluctuations for sample 1 are presented in Fig. 13.2. The spectral intensity of current fluctuations was measured in references 11 and 228 at 10-GHz frequency. The frequency was high enough to avoid $1/f$ and generation–recombination noise, but it was low in comparison to the inverse time constants of the kinetic processes related to electron scattering in the conduction band. The experimental data on the spectral intensity of drift-velocity fluctuations (Fig. 13.2, unfilled squares) were obtained from the current fluctuation data through normalization at zero field by using mobility data and the Nyquist formula. The closed circles in Fig. 13.2 give the electric field dependence of the spectral intensity resulting from the calculated total correlation function. For comparison, the results of the simulation neglecting interelectron collisions are shown (Fig. 13.2, diamonds).

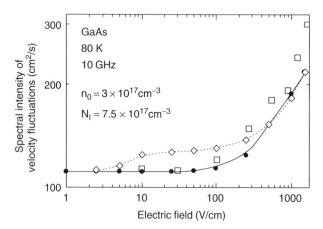

Figure 13.2. Dependence of the spectral intensity of electron drift velocity fluctuations at 10 GHz for Si-doped n-type GaAs at 80 K ($n_0 = 3 \times 10^{17}$ cm^{-3}, $N_{\text{imp}} = 7.5 \times 10^{17}$ cm^{-3}). Monte Carlo simulation [246]: ● — with phonon, impurity, and interelectron scattering taken into account, ◊ — neglecting interelectron scattering. Experimental data: □ — from reference 228. Solid and dotted lines are guides to the eye.

One can see that the interelectron collisions have little influence on the fluctuation spectra at very low electric fields and at high electric fields (cf. diamonds and filled circles in Fig. 13.2). The most pronounced effect is obtained at intermediate fields ranging from 5 V/cm to 500 V/cm. The interelectron collisions cause an essential increase (up to two decades) in the field strength required for the excess hot-electron fluctuations to manifest themselves. For example, one can see from Fig. 13.2 that the field value for 5% enhancement in the spectral intensity of current fluctuations shifts from about 10 V/cm predicted neglecting interelectron collisions to about 200 V/cm obtained from measurements as well as from Monte Carlo calculation taking into account interelectron collisions.

With the interelectron collisions being taken into account, the calculated dependence (Fig. 13.2, filled circles) fits the experimental one (Fig. 13.2, unfilled squares), indicating an important role of the interelectron Coulomb scattering mechanisms in the formation of the spectral intensity of current fluctuations at the intermediate electric fields.

The spectral intensity of velocity fluctuations remains nearly constant at fields up to 200 V/cm (Fig. 13.2, filled circles and unfilled squares). This behavior can be explained by enhanced energy loss by electrons on optical phonons in the presence of interelectron scattering [290]: The interelectron-collision-dependent energy losses include spontaneous optical phonon emission by the electrons having acquired enough energy as a result of collisions with other electrons. The role of interelectron scattering diminishes at higher fields: An increase of electron energy causes the interelectron scattering rate to decrease and causes the spontaneous optical phonon emission to become the dominant scattering mechanism.

Thus, the proposed Monte Carlo procedure was demonstrated to be an efficient tool for studying hot-electron noise in the presence of carrier–carrier scattering. The fluctuation properties of nonequilibrium electron gas in a semiconductor were shown to be sensitive to the presence of interelectron collisions. Taking them into account is crucial for an explanation of experimental data on microwave noise in doped gallium arsenide at 80 K at moderate electric fields (5–500 V/cm).

13.3 ANALYTIC APPROACH

In this section, experimental results on microwave noise in n-type GaAs ($n_0 = 3 \times 10^{17}$ cm^{-3}, 80 K) at electric fields up to a few hundred volts per centimeter will be interpreted combining analytic approach and Monte Carlo simulation. The analytic approach is based on applicability (due to high frequency of interelectron collisions) of the electron temperature approximation described in Section 12.1. Monte Carlo simulation shows (see Fig. 13.1) a rather low sensitivity of the conductance to the electron heating, accompanied by small deviations from Ohm's law. These circumstances were shown to favor a nearly isotropic noise temperature and — as will be demonstrated in the next chapter — an approximate validity of the Price fluctuation–diffusion relation.

In the preceding section, it was demonstrated that noise properties of a doped semiconductor can be remarkably influenced by interelectron collisions. Using a model that takes into account interelectron collisions was shown to be indispensable while seeking to fit the results of Monte Carlo calculations to the experimental data on microwave noise in typical silicon-doped n-type GaAs channels (donor density exceeding 10^{17} cm^{-3}) in the field range below 400 V/cm.

The obtained agreement of the experimental results and those of Monte Carlo simulation stimulated an analytic investigation of the situation [247] (see also references 291 and 292). Indeed, the situation in the doped n-GaAs channels is suitable for an analytic treatment. The energy distribution at low and intermediate electric fields was proved to be quite close to a Maxwellian one. This allowed, for the first time, an analytic treatment of the experimental results on noise in the nonequilibrium in the situation where interelectron collisions make a difference. On the lines described in Section 12.1, the role of theoretically predicted interelectron-collision-caused correlation between occupancies of electronic states in nonequilibrium was estimated quantitatively for typical doped GaAs channels (see Chapter 14).

Main Features of Transport and Noise Characteristics of Doped n-Type GaAs

To guarantee the experimental basis for the analysis, the measurements of the noise as well as of the current–voltage characteristic in n-type GaAs channels containing different density of ionized impurities and essentially the same density of free electrons (around 3×10^{17} cm^{-3}) were performed [247]. The density of electrons was high enough for a well-pronounced effect of interelectron collisions. On the other hand, the density was low enough for degeneracy effects in the electron gas not to influence the results remarkably (insignificant degeneracy effects on noise in the case was demonstrated by direct Monte Carlo simulation).

Two features of the typical channel current–voltage characteristic will prove to be quite important for what follows. The experimental results and those of Monte Carlo simulation, presented in Fig. 13.1, demonstrate a low sensitivity of conductance to electric field strength; the deviations from the Ohm's law are rather small in the considered range of electric fields. Another important circumstance is a weak dependence of the current–voltage characteristic on the frequency of interelectron collisions demonstrated by Fig. 13.1 (cf. dotted lines for samples 1 and 2 with solid squares and solid diamonds, respectively). Figure 13.3 presents (unfilled squares) the field dependence, measured at 80 K ambient temperature, of the longitudinal noise temperature, $T_{n\|}$, of the sample 1, with the "longitudinal" referring to the quantity determined in the direction of the steady current caused by the applied electric field. The same figure presents the longitudinal noise temperature and the transverse one (determined in the direction perpendicular to the steady current) as simulated by the Monte Carlo technique, neglecting and taking into account interelectron collisions. The most important features of the noise characteristics of Fig. 13.3 are as follows:

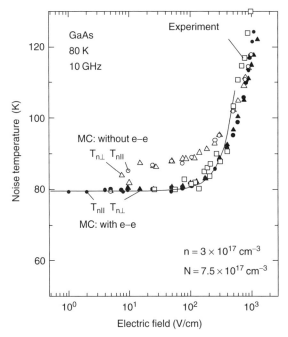

Figure 13.3. Comparison of measured and calculated noise temperatures [292]. Measured longitudinal noise temperature $T_{n\|}$ ($\square$, [247]) compared with that calculated [247] taking into account ($\bullet$) and neglecting ($\circ$) interelectron (e–e) collisions. Transverse noise temperature $T_{n\perp}$ calculated taking into account ($\blacktriangle$) and neglecting ($\triangle$) interelectron collisions is also presented. Measurements were performed under DC bias (up to 260 V/cm) and under pulsed bias (over 100 V/cm). Solid line is a parabolic approximation of the experimental data.

(i) The experimental results on the noise temperature (contrary to those on the current–voltage characteristic) cannot be explained reasonably without taking into account interelectron collisions.

(ii) In the entire field range where the interelectron collisions are essential, the electron heating remains small enough for the experimental dependence $T_{n\|}(E)$ to be almost parabolic — that is, for the hot electron problem to be treatable in the framework of the "warm electron" approach (this is not the case when fluctuations are simulated, neglecting interelectron collisions; see Fig. 13.3).

(iii) The simulated transverse noise temperature almost coincides with the longitudinal one; in other words, the noise temperature is found to be almost *isotropic* (the *same* in the longitudinal and the transverse directions).

The obtained near-isotropy of the noise temperature is in contrast to the pronounced anisotropy of the noise in nonequilibrium electron gas observed/calculated in the majority of cases investigated earlier (see references

9, 70, 113, 135 and 136). Below we use the analytic approach presented in Section 12.1 to reveal how features (i)–(iii) are interrelated. In particular, it will be shown that the near-isotropy of noise is related to the observed relatively weak dependence of the conductivity on electric field at moderate fields in GaAs.

Electron Energy Distribution in Doped n-Type GaAs. Validity of Effective Electron Temperature Approximation. Relation to Noise Temperature

As already mentioned, an understanding, on the microscopic level, of the role of interelectron collisions was achieved in terms of the electron temperature and its fluctuations as described in Section 12.1. At high electron densities, the shape of electron energy distribution at low/moderate fields is controlled by interelectron collisions (except, maybe, the high-energy tail). Free-electron density $n_0 = 3 \times 10^{17}$ cm^{-3} is high enough to expect, at not-too-high electric fields, the electron energy distribution to be close to Maxwellian. The direct Monte Carlo simulation [246, 247] in the framework of the model described in the previous section confirmed this conjecture for fields up to 400–500 V/cm for sample 1 and up to 700 V/cm for sample 2. For sample 1 this is illustrated by Fig. 13.4. Indeed, interelectron collisions set a good one-temperature distribution at 100 V/cm, a rather good one-temperature distribution is established at fields up to 400–500 V/cm, while a kink in the energy distribution gradually develops at the optical phonon energy (0.036 eV), being quite well detectable at 1000 V/cm. So, for sample 1, the effective electron temperature approximation is good enough at fields up to 400–500 V/cm, with the dependence of the electron temperature on the electric field strength being found from the slope of curves of Fig. 13.4.

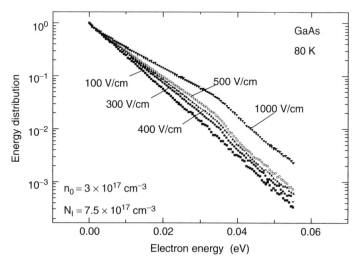

Figure 13.4. Electron energy distribution function at different applied electric fields, calculated by the Monte Carlo procedure [292].

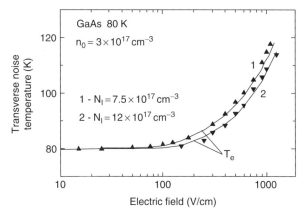

Figure 13.5. Field-dependent transverse noise temperature, for samples 1 and 2, determined from direct Monte Carlo simulation of fluctuations taking into account interelectron collisions (symbols), compared to that (solid lines) obtained from the right-hand side of Eq. (13.11), with the electron temperature T_e being deduced from the slope of the energy distribution [247].

As one may conclude from Fig. 13.5, the effective electron temperature T_e *de facto* coincides fairly well with the transverse noise temperature $T_{n\perp}$ obtained directly simulating the fluctuation process (Fig. 13.5, symbols and solid lines). This is not unexpected: According to Eq. (12.32), in the electron temperature approximation the transverse noise temperature of an isotropic semiconductor at sufficiently low frequencies (of microwave range) is equal to the effective electron temperature

$$T_{n\perp} = T_e \qquad (13.11)$$

while the longitudinal noise temperature is expressible [Eq. (12.36)] in terms of the conductivity tensor components, the electron temperature, and the lattice temperature. Here we rewrite Eq. (12.36) in an equivalent form [see Eq. (12.38)]:

$$T_{n\|} = T_e \left[1 + \frac{T_e}{4(T_e - T_0)} \frac{\tilde{\sigma}}{\sigma_\|} \left(\frac{\sigma_\|}{\tilde{\sigma}} - 1 \right)^2 \right]. \qquad (13.12)$$

According to Eqs. (13.11) and (13.12), the longitudinal and the transverse noise temperatures are interrelated through the conductivity components $\tilde{\sigma}$ and $\sigma_\|$. As noted in Section 12.1, this enables one to verify independently the validity of the electron temperature approximation in the situation of interest. Supposing that the quantities $T_{n\perp}$, $T_{n\|}$, and $\tilde{\sigma}$, $\sigma_\|$ are measured and/or computed through the simulation procedures, the validity of the relation between quantities $T_{n\perp}$ and $T_{n\|}$ can be checked up. Remarkable deviations would definitely mean that the energy distribution is rather far from a Maxwellian one.

The obvious confirmation of validity of the electron temperature approximation is presented by the already mentioned fulfillment in practice of relation (13.11)

(Fig. 13.5). Further confirmation would be offered by coincidence of data on the longitudinal noise temperature $T_{n\|}$ obtained in two different ways: (i) from the right-hand side of Eq. (13.12) as calculated using the already available data on electron temperature T_e (Fig. 13.5), on $\tilde{\sigma}$, and on $\sigma_\|$ (Fig. 13.1) and (ii) from the direct Monte Carlo simulation of the fluctuation process. One can see that these data (Fig. 13.6, solid lines and symbols) are pretty close to each other reaffirming the self-consistence of the electron temperature approach at the moderate electric fields where the interelectron collisions manifest themselves.

Near-Isotropy of Noise Temperature

The most specific property to be explained by the theory is the rather unexpected nearly isotropic behavior of the noise temperature, evidenced by Fig. 13.3 throughout the field range in question. To find a reason, another already mentioned feature of the noise–voltage and the current–voltage characteristics was exploited. At fields up to 400–500 V/cm, the electron temperature and the noise temperature of sample 1 increases by less than 20 K (i.e., within 25%), while the conductivity changes even less. As already noticed, the electron energy loss, more efficient than that in absence of the interelectron collisions, is known to result from the "combined" scattering [290]: the interelectron collisions open an additional electron energy relaxation channel through emission of optical phonons (one of two colliding electrons can gain enough energy to emit an optical phonon).

The field dependencies of the measured noise temperatures, as well as those calculated taking into account the interelectron collisions, are almost parabolic for the investigated samples at the low and moderate fields. This enables one to

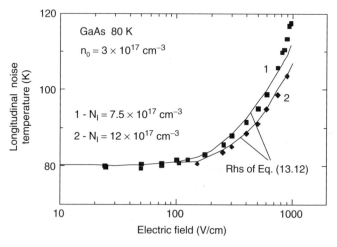

Figure 13.6. Field-dependent longitudinal noise temperature, for samples 1 and 2, determined from direct Monte Carlo simulation of fluctuations taking into account interelectron collisions (symbols), compared to that (solid lines) obtained from the right-hand side of Eq. (13.12) [247].

expand expression (13.12) for $T_{n\parallel}$ and the corresponding expression for $\sigma_\parallel$ in powers of electric field strength E to the second order, neglecting terms of the order higher than E^2. In other words, the entire region where the interelectron collisions make a difference can be treated in terms of "warm electrons." Unlike this, if the interelectron collisions were neglected, the warm-electron region would shrink dramatically—down to about 10 V/cm (see Fig. 13.3).

The expansion of $\sigma_\parallel(T_e)$ in powers of the increase in electron temperature ($\Delta T_e \equiv T_e - T_0$, the latter being proportional to E^2) in the linear approximation in ΔT_e led to the expression (12.20). As mentioned above, the electric sensitivity to electron heating is rather low at low and moderate fields (see Fig. 13.1). The coefficient of sensitivity $d \ln \tilde{\sigma}/(2d \ln T_e)$ does not exceed 0.1 for sample 1 (see Fig. 13.7, upper curve) and 0.2 for sample 2. Thus, according to Eq. (12.20), the ratio of the differential conductivity components contains the small term resulting in small values of the differential conductivity's anisotropy at the moderate electric fields:

$$\frac{\sigma_\parallel}{\tilde{\sigma}} - 1 \approx 0.4 \frac{\Delta T_e}{T_0} < 0.1$$

for sample 1 (see Fig. 13.7, lower curve). With this in mind, let us compare the longitudinal and the transverse noise temperatures of the warm electrons. According to Eq. (12.39), in the warm electron region the sensitivity coefficient *squared* decides the noise temperature anisotropy. As a result, the noise temperature anisotropy is extremely low, lower than that of the differential conductivity. Expression (12.39) tells that the noise temperature anisotropy for sample 1 should not exceed 0.01, while for sample 2 it can reach 0.04. Figure 13.3 demonstrates that this important prediction is fulfilled quite well.

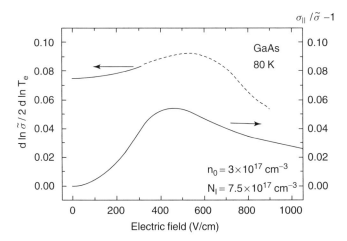

Figure 13.7. Sensitivity of conductance to electron heating, $d \ln \tilde{\sigma}/(2d \ln T_e)$ (left), and anisotropy of differential conductivity $\sigma_\parallel/\tilde{\sigma} - 1$ (right), as functions of electric field, fitting the Monte Carlo simulation data, for sample 1 (see reference 247).

Synopsis

Interelectron collisions can indirectly but significantly influence the energy losses of the electron system in GaAs channels subjected to electric field at liquid-nitrogen temperature. The resultant enhancement of the electron temperature is relatively small in the range of fields where the interelectron collisions are of importance. This leads to the changes in fluctuation and transport characteristics which are, with sufficient accuracy, quadratic in the field strength at the fields of interest ("warm electrons"). Furthermore, the results of Monte Carlo simulation compared with those of noise measurements show that the electron energy distribution at fields in question is Maxwellian with sufficient accuracy. These circumstances, being typical for the partially compensated n-type GaAs channels containing a high density of electrons, ensure applicability of the developed combined analytic and Monte Carlo approach to fluctuation phenomena leading to the following conclusion: A characteristic feature of the GaAs channels at moderate fields being the rather weak sensitivity of conductance on electron heating, the anisotropy of the noise temperature, proportional to the sensitivity coefficient squared, is extremely small at low and moderate electric fields.

CHAPTER 14

ELECTRON DIFFUSION IN STANDARD-DOPED n-TYPE GaAs

The influence of interelectron collisions on the electron diffusivity is of special interest. As described in Chapter 6, the fluctuation–diffusion relation has been proposed by Price [41] to connect the tensor of spectral intensities of current fluctuations in a uniform and stationary electron gas with that of the electron diffusion coefficients — that is, with the coefficients entering the expressions for the electric current induced by a small electron density gradient. This relation has proved to be very useful in providing information on hot-electron diffusivity from noise measurements performed in spatially homogeneous electron gas in weakly doped semiconductors (see references 9–11). However, the kinetic theory of fluctuations predicts that Price's relation in a nonequilibrium electron gas should be violated unless interelectron collisions are negligible [37, 70]. On the other hand, direct measurements of diffusion coefficients from spreading of a cloud of carriers in doped semiconductors have never been performed: The application of this technique is hindered at a high background density of electrons since the spreading is controlled by dielectric relaxation rather than diffusion. Efficient techniques of Monte Carlo simulation of diffusion process in the case of a concentration-dependent distribution function are absent [239, 293]. So, a determination of the field-dependent coefficient entering Fick's law for the diffusion current remained for a long time an unsolved problem, and quantitative data on hot-electron diffusion coefficients in doped semiconductors at high densities of electrons were lacking. Only recently it was demonstrated [247] (see also references 291 and 292) that in some cases it is possible to determine electron diffusion coefficients from noise measurements even when interelectron collisions are essential. The aim of this chapter is to present these results.

14.1 ELECTRON DIFFUSION COEFFICIENTS IN DOPED GaAs

In the electron temperature case (see Chapter 12), side by side with the expressions for the differential conductivity and the noise temperature, the expressions are available for the electron diffusion coefficients $D_{\alpha\beta}$ entering Fick's law, that is, the expression for the diffusion current $j_\alpha = -eD_{\alpha\beta}\partial n/\partial x_\beta$. The corresponding expressions have been presented in Section 12.1 [see Eqs. (12.22)–(12.24)]. These expressions enable one to calculate the field dependence of the electron diffusion coefficients from those of the conductivity, provided that the electron temperature is known from the transverse noise temperature measurement or from Monte Carlo simulation. In the absence of experimental data on the transverse noise temperature, Eq. (13.12) can be used to extract the electron temperature from experimental data on the longitudinal noise temperature and the current–voltage characteristics.

In particular, we notice that the quantity $d\ln\tilde{\sigma}/2d\ln T_e$ determines to what extent the anisotropy of the electron diffusivity is not described by the degree

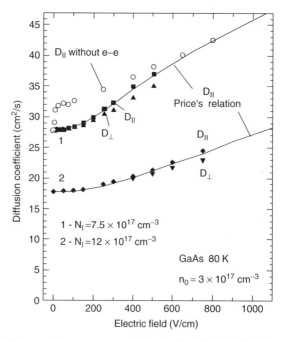

Figure 14.1. Field-dependent transverse and longitudinal Fick's diffusion coefficients, $D_\perp$ and $D_\parallel$, given by Eqs. (12.22) and (12.24) (solid symbols), for samples 1 and 2 with the zero-field mobility 4080 and 2540 cm^2/(V·s), respectively [247]. Solid lines present $D_\parallel$ as predicted by the Price relation, $D_\parallel = (\delta j_\parallel^2)_{\omega\tau\ll 1}V_0/2e^2n_0$, where the current fluctuations are directly simulated with interelectron collisions included. For comparison, $D_\parallel$ obtained from simulation of fluctuations neglecting interelectron (e–e) collisions is presented for the higher-mobility sample (○).

of anisotropy of the differential conductivity — in other words, to what extent the so-called "Robson conjecture" (see reference 113) is not valid. For the typical sample, as already noticed, the coefficient of electric sensitivity to electron heating $d\ln\tilde{\sigma}/2d\ln T_e$ does not exceed 0.1 (see the upper curve of Fig. 13.7). Such is the predicted degree of violation of Robson's conjecture in this particular case. Figure 14.1 presents the field dependence of the transverse and longitudinal diffusion coefficients obtained from Eqs. (12.22) and (12.24) using data of Figs. 13.1, 13.5, and 13.7. For comparison, the longitudinal diffusion coefficient calculated without taking into account the interelectron collisions is also presented in Fig. 14.1 for sample 1 (unfilled circles).

14.2 DEGREE OF VIOLATION OF PRICE'S RELATION

We know that interelectron collisions in the nonequilibrium state create correlation between occupancies of electronic states (Section 4.1). This leads to violation of the fluctuation–diffusion relation, known as Price's relation, between the current fluctuations and the carrier diffusivity (Section 6.2). The spectrum of the current-density fluctuations in the low-frequency limit, according to Eq. (6.8), is expressible in terms of the electron diffusion coefficients $D_{\alpha\beta}$ and the above-mentioned contribution of the additional (interelectron-collision-caused) correlation, $\Delta_{\alpha\beta}$. In the electron temperature case, the transverse contribution due to the interelectron-collision-caused correlation vanishes (see Eq. (12.26)):

$$\Delta_\perp = 0, \qquad (14.1)$$

while the longitudinal contribution is expressible [see Eq. (12.27)] in terms of the current–voltage characteristic and the field dependence of the electron temperature:

$$\frac{\Delta_\|}{2D_\|} = -\left(\frac{\sigma_\|}{\tilde{\sigma}} - 1\right) \frac{\frac{T_e}{2(T_e - T_0)}\left(\frac{\sigma_\|}{\tilde{\sigma}} - 1\right) - \frac{\partial\ln\tilde{\sigma}}{\partial\ln T_e}}{2\frac{\sigma_\|}{\tilde{\sigma}} + \left(\frac{\sigma_\|}{\tilde{\sigma}} - 1\right)\frac{\partial\ln\tilde{\sigma}}{\partial\ln T_e}} \qquad (14.2)$$

Expressions (14.1) and (14.2) determine to what extent Price's relation — the relation connecting, in the absence of interelectron collisions ($\Delta = 0$), the spectral intensities of current fluctuations and the electron diffusion coefficients — is violated by frequent interelectron collisions in the electron temperature case. The Price relation survives in the directions perpendicular to the steady current [Eq. (14.1)]. The degree of violation of the relation between the longitudinal diffusion coefficient and the longitudinal current fluctuation spectral intensity in the case under consideration can be estimated using Eq. (14.2).

As already noticed, the warm electron region in the case under consideration almost coincides with the range of fields where the interelectron collisions make a difference (see Fig. 13.3). Correspondingly, in the entire field range in question we can expand Eq. (14.2) in powers of $\Delta T_e = T_e - T_0$. It is easy to verify that, thanks to Eq. (12.20), valid up to linear in ΔT_e terms, the right-hand side of Eq. (14.2) *vanishes in the linear (with respect to ΔT_e) approximation,* with the interelectron-collision-caused correlation term $\Delta_\parallel$ as given by Eq. (14.2) appearing only as a correction of the order of ΔT_e^2 (i.e., E^4) (see Section 12.1). This is illustrated for sample 1 by Fig. 14.2, where the solid line presents the E^4 curve. At higher fields, the dimensionless contribution given by Eq. (14.2) changes its sign, with its absolute value remaining below 0.15% in the entire range of fields of interest (Fig. 14.2, dotted line). The deviation from Ohm's law and the electric sensitivity (see Section 12.1) are better pronounced in the investigated GaAs channels with a lower mobility of electrons (sample 2). Nevertheless, the contribution $\Delta_\parallel/2D_\parallel$ does not exceed 1% for sample 2. So, the typical partially compensated GaAs channels with electron mobility of about 4000 cm^2/(V·s) at 80 K demonstrate the exceptionally weak contribution of the interelectron-collision-caused correlation into noise.

The performed analysis suggests that the considered contribution is relatively small in typical GaAs channels for electron density and compensation ratio of practical interest. In other words, Price's relation between the spectral intensities of current fluctuations and Fick's diffusion coefficients is found to hold quite well in the given situation, despite the fact that the interelectron collisions appear to be important in shaping the electron energy distribution function. At high fields, where the interelectron collisions become ineffective, Price's relation works by definition. We come to an important conclusion that, with a sufficiently high accuracy, in the given situation Price's relation is valid in the entire range of fields. Consequently, the electric field dependence of Fick's diffusion coefficients of the electron system in the considered GaAs channels in a wide range of fields is attainable from noise calculations/measurements.

The validity of Price's relation with high accuracy even in the range of fields where in other respects the interelectron collisions make a difference is the

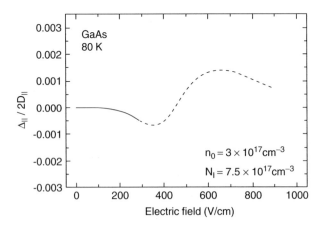

Figure 14.2. Relative contribution of the correlation due to interelectron collisions as predicted by Eq. (14.2), for sample 1 [247].

specific feature of the GaAs channels containing a high density of electrons. The result is interesting because it regenerates the possibility to obtain information about hot-electron diffusion coefficients from noise measurements even in cases when the interelectron collisions influence them. This possibility is illustrated by the solid lines (Fig. 14.1) obtained from Price's relation, because the lines are in agreement with the independently calculated longitudinal diffusion coefficient data. In particular, the line for sample 1 (Fig. 14.1) fits the simulation data in the corresponding ranges of electric fields (filled squares at the low and moderate fields and unfilled circles at the high fields). The possibility to exploit Price's relation can prove to be important because a direct measurement or calculation of Fick's diffusion coefficients when interelectron collisions influence them, as already mentioned, is by no means straightforward (cf. references 239 and 293).

The extremely small interelectron-collision-caused correlation is specific to warm electrons in the electron temperature case. Outside the warm-electron region, the higher terms in the expansion become of importance, and, in general, the ratio $\Delta_\parallel/2D_\parallel$ may not remain small. This does not happen in our case: The ratio $\sigma_\parallel/\tilde{\sigma}$ entering Eq. (14.2) remains close to unity until the interelectron collisions cease to influence the noise, and the ratio $\Delta_\parallel/2D_\parallel$ remains small in the entire field range. Supposing that the nonlinearity of the current–voltage characteristic is well-pronounced, the contribution of the interelectron-collision-caused correlation would not be necessarily small in the entire field range where the noise is affected by the interelectron collisions.

On the other hand, as was shown in Section 12.3, the exceptional smallness of the additional correlation contribution even for warm electrons cannot be guaranteed beyond the limits of validity of the electron temperature approximation $\tau \ll \tau^{ee} \ll \tau^{en}$. In Section 12.3, the results [136] were presented showing that in the intermediate cases (e.g., $\tau^{ee} \sim \tau^{en}$) a substantial violation of Price's relation can be conditioned by a non-linear dependence on the electron density of the electron quasi-momentum and/or energy distribution. This possibility of violation of Price's relation is excluded from the very beginning in the electron temperature case in which the shape of the distribution function does not depend on the electron density.

In conclusion, let us recall that Chapters 13 and 14 dealt with interelectron collision effects on noise and diffusion in standard-doped GaAs at moderate electric fields. It was demonstrated that interelectron collisions can indirectly but significantly influence the energy losses of the electron system in GaAs channels subjected to electric field at liquid-nitrogen temperature. The resultant enhancement of the effective electron temperature is relatively small in the range of fields where the interelectron collisions are of importance. This leads to the changes in fluctuation and transport characteristics which are, with sufficient accuracy, quadratic in the field strength at the fields of interest. These conditions, being typical for the partially compensated n-type GaAs channels containing a high density of electrons, ensure applicability of the developed combined analytic/Monte Carlo approach to fluctuation, noise, and diffusion phenomena leading to the following conclusions.

A characteristic feature of the channels is the rather weak sensitivity of conductance on electron heating. As a direct result, the anisotropy of the noise temperature, being proportional to the sensitivity coefficient squared, is extremely small at the low and moderate electric fields.

Price's relation between the spectral intensities of current fluctuations and Fick's diffusion coefficients is found to hold quite well, despite the fact that the interelectron collisions are important in shaping the electron energy distribution and opening the effective channel for the energy relaxation.

Price's fluctuation–diffusion relation is valid with a rather good accuracy because the contribution of the additional correlation created by the interelectron collisions in the warm electron region is proportional to the enhancement of the electron temperature squared. The degree of violation of Price's relation inside and outside the warm-electron region is estimated not to exceed 0.01 for the investigated GaAs channels, so the electric-field dependence of Fick's diffusion coefficients is attainable from noise calculations and/or measurements over a wide range of fields.

CHAPTER 15

ELECTRONIC SUBBANDS IN QUANTUM WELLS

The advances in semiconductor technology have shifted investigations of electronic properties from *bulk* crystals to structures of submicrometer and nanometer size. The essential feature of the nanometer-size structures is *quantum confinement* of mobile electrons. A structure containing *two-dimensional electron gas* (2DEG) confined in a quantum well (QW) is an interesting object of modern physics [50, 51, 56, 294–300] of great importance for applications in high-speed electronics [14, 29, 30, 301–305].

The electron confinement in a QW implies quantization of electron energy and introduces strong anisotropy of transport and other properties. Heterojunctions, together with electrostatic barriers, confine mobile electrons into a thin layer, where the electrons are free to perform planar motion, but their transverse freedom is limited. The latter depends on the barrier height, but can also be controlled by changing electron kinetic energy—for example, by applying electric field in the plane of electron localization. The field-dependent interplay of planar and transverse processes is responsible for new sources of hot-electron fluctuations observed in structures containing two-dimensional electron gas. Heterostructure composition, electric field strength, interelectrode distance, doping, electron sheet density, and other parameters can be changed in order to excite and resolve different sources of fluctuations.

Up to now, we needed no deeper knowledge on the electronic energy band structure of semiconductors; it was enough to mention single-valley or many-valley structure of that. This cannot necessarily be enough when describing, in the following chapters, noise properties of heterojunctions. In this chapter we present, in some detail, the well-known features of the band structure of (a) several semiconductors we dealt with in the previous chapters and (b) heterojunctions.

15.1 BAND STRUCTURE OF COMPOUND SEMICONDUCTORS

The solution of the electron wave function in a semiconductor lattice leads to the energy contour in quasi-momentum space (or wave-vector space) that is often referred to as band structure. Let us describe in some detail the band structure of GaAs — the semiconductor that was mentioned most frequently in the previous chapters. GaAs has the cubic sphalerite (zincblende) lattice, this having face-centered cubic (fcc) translation symmetry (see reference 288), with the basis of one GaAs molecule. Figure 15.1 shows the first Brillouin zone for GaAs. The zone comprises a truncated octahedron, lying within a cube (shown with dotted lines) with side length 2.223×10^8 cm^{-1} of wave-vector space (**k**-space). The most important paths through the Brillouin zone are those from the zone center Γ to the high symmetry points X, L, and K on the zone boundary.

Figure 15.2 shows for GaAs the general features of electron energy versus wave-vector dependencies along high symmetry directions in the zone, for the

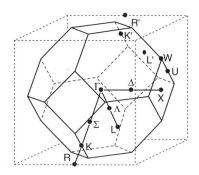

Figure 15.1. The first Brillouin zone for GaAs and other solids with the diamond and zincblende face-centered cubic (fcc) structure.

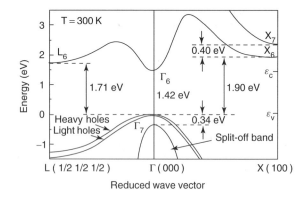

Figure 15.2. Electron energy versus reduced wave-vector curves along high-symmetry directions in the Brillouin zone for GaAs. Minima exist at Γ, L and X points referred to as valleys.

TABLE 15.1. Band Gap and Lattice Constant Data for Selected Compound Semiconductors, and Si and Ge, at Room Temperature

Material	Lattice Constant (Å)	$\mathcal{E}_g(\Gamma)$ (eV)	$\mathcal{E}_g(L)$ (eV)	$\mathcal{E}_g(X)$ (eV)
AlP	5.46	3.6	3.5	2.45
AlAs	5.66	3.0	2.36	2.15
AlSb	6.135	2.3	2.21	1.61
GaP	5.45	2.78	2.6	2.27
GaAs	5.65	1.42	1.71	1.9
GaSb	6.1	0.72	0.81	1.03
InP	5.87	1.35	1.95	2.15
InAs	6.06	0.35	1.43	2.0
InSb	6.48	0.17	1.0	1.7
Si	5.43	4.1	2.0	1.12
Ge	5.66	0.805	0.67	0.89

lowest sets of conduction band minima and for the uppermost part of the valence band system (after reference 277; see also reference 288). The lowest conduction minimum for GaAs is a single one at the zone center (in contrast to Si and Ge both being indirect-gap solids). The conduction band order is $\Gamma-L-X$, the right sequence of the secondary conduction minima being definitely established only in 1976 [306, 307].

From hot-electron photoluminescence experiments at liquid helium temperatures (see reference 227) the bottom edge energy of the L valleys over the bottom of the central Γ valley at 2 K was determined as 310 ± 10 meV, while the position of the bottom of the side X valley over the bottom of Γ valley was found to be 485 ± 10 meV. The lowest side valley in InP was attributed to the L point of the Brillouin zone, and its edge energy was determined as 860 ± 20 meV.

The three upper valence bands have maxima at the zone center, in the Γ point: The two uppermost bands (heavy-hole and light-hole) have a degenerate maximum, and the third band is separated by the spin-orbit splitting energy 341 meV in GaAs (see reference 288).

The first Brillouin zone of the face-centered cubic lattice given in Fig. 15.1 is the same for most of the III–V compound semiconductors: GaAs, AlAs, InP, AlGaAs, and InGaAs, to mention some of interest for the following chapters. The forbidden energy gap and other data are given in Table 15.1.

15.2 HETEROJUNCTIONS

During epitaxial growth of a single-crystal consisting of layers of heterogeneous semiconductors, either with lattice matching or with relatively near lattice matching, the forbidden energy gaps of each of the materials involved align themselves

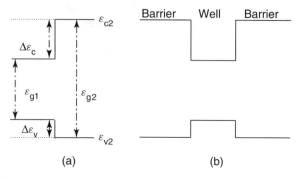

Figure 15.3. Flat bands at a heterojunction (a) and in a double-heterojunction structure (b). The structure containing a layer of a narrow-gap semiconductor together with two wide-gap semiconductors (barriers) form a rectangular potential well for electrons and holes.

TABLE 15.2. Conduction Band and Valence Band Offsets for Elementary Semiconductors, III–V Group Compound Semiconductors, and Mixed Crystals

Lattice Constant of Group (Å)	Material Systems	$\Delta\mathcal{E}_g$ (eV)	$\Delta\mathcal{E}_c$ (eV)	$\Delta\mathcal{E}_v$ (eV)
5.44	GaP/Si	1.15	0.8	0.35
5.66	GaAs/Ge	0.76	0.27	0.49
	AlAs/Ge	1.48	0.55	0.93
	$Ga_{0.7}Al_{0.3}As/GaAs$	0.39	0.26	0.13
	AlAs/GaAs	0.73	0.29	0.44
	$In_{0.49}Ga_{0.51}P/GaAs$	0.46	0.22	0.24
5.87	$In_{0.52}Al_{0.48}As/InP$	0.11	0.30	−0.19
	$InP/In_{0.53}Ga_{0.47}As$	0.60	0.20	0.40
	$In_{0.52}Al_{0.48}As/In_{0.53}Ga_{0.47}As$	0.71	0.50	0.21
6.10	AlSb/GaSb	0.89	0.49	0.40
	GaSb/InAs	0.37	0.88	−0.51
	AlSb/InAs	1.26	1.37	−0.11

so that a conduction band offset $\Delta\mathcal{E}_c$ and a valence band offset $\Delta\mathcal{E}_v$ apply (see Fig. 15.3). The band offset values are now reliably determined for most important heterojunctions as shown in Table 15.2. A thin layer of a narrow-gap semiconductor together with two layers of wide-gap semiconductors form a double-heterojunction structure (Fig. 15.3b). The potential wells for electrons and holes form in the double-heterojunction structure. The heterostructure imprisons mobile electrons and holes: The carriers can move freely inside the well, but their motion in the transverse direction is restricted by the barriers. A thin

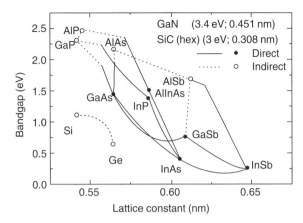

Figure 15.4. The bandgap versus lattice constant for III–V group compound semiconductors, their mixed crystals, and Si, Ge, and their mixed crystals.

narrow-gap layer confines the electrons (holes) into a two-dimensional electron gas (2DEG).

Gallium arsenide and aluminum arsenide single crystals have similar lattice constants (Table 15.1) and form almost an ideal lattice-matched heterojunction [302]. The mixed crystals $Al_xGa_{1-x}As$ of any composition are easily grown onto GaAs substrates. Most important are heterostructures of GaAs and direct-gap $Al_xGa_{1-x}As$ mixed crystals containing $x < 0.3$ of AlAs (Table 15.2). These materials, together with lattice-matched mixed crystals of $In_{0.49}Ga_{0.51}P$ and Ge, form the group with lattice constant near 0.566 nm (Table 15.2 and Fig. 15.4). Another group of great importance for high-speed electronics is based on InP (the lattice constant is near 0.59 nm). Its main counterparts for lattice-matched heterojunctions are $In_{0.52}Al_{0.48}As$ and $In_{0.53}Ga_{0.47}As$ (Table 15.2 and Fig. 15.4). Thin layers of mixed crystals of lattice-mismatched composition can be deposited and survive in a strained form; these layers are called pseudomorphic ones. In particular, InP-based pseudomorphic $In_{0.7}Ga_{0.3}As$ together with the lattice-matched InAlAs on InP form heterostructures with the conduction band offsets exceeding that in the lattice-matched heterostructures.

15.3 EMPTY QUANTUM WELLS

Electron confinement in a quantum well (QW) has a strong effect on allowed electron energies: The conduction band splits into the dimensionally quantized subbands (Fig. 15.5). Since the translation symmetry in the plane of electron confinement is conserved in perfect quantum wells, the electron wave functions in the plane remain similar to those of the original bulk semiconductor, and the dependence of the electron energy on the in-plane wave-vector components survives within each subband (Fig. 15.5).

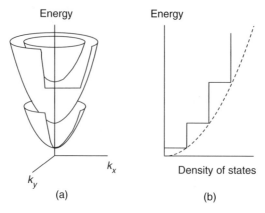

Figure 15.5. Three lowest subbands in a quantum well: (a) Dependence of electron energy on the in-plane electron wave-vector and (b) dependence of density of states on electron energy.

Rectangular Wells

An easy understanding of the subband structure can be obtained assuming a simple profile of the well potential. A rectangular quantum well is a good approximation for a double-heterojunction structure (Fig. 15.3b) containing no electrons. An illustration of the subband formation can be given within the effective mass and the envelope function approximations [294, 308]. The one-electron wave-function for the jth subband can be written as

$$\psi_j(x, y, z) = \xi_j(z) e^{i(\mathbf{k}\mathbf{r})}, \qquad (15.1)$$

where $\xi_j(z)$ is the envelope function for the jth subband, z is the coordinate transverse to the plane of confinement, and $\mathbf{k}$ and $\mathbf{r}$ are the two-dimensional in-plane wave vector and space vector, respectively.

The Schrödinger equation for the envelope function is

$$-\frac{\hbar^2}{2m}\frac{d^2}{dz^2}\xi_j(z) + [V(z) - \mathcal{E}_j]\xi_j(z) = 0, \qquad (15.2)$$

where m is the effective mass and $V(z)$ is the well potential energy. The solution of Eq. (15.2) for the envelope function in an infinitely deep rectangular well (Fig. 15.6a) is

$$\xi_j(z) = \sin\frac{j\pi z}{a}, \qquad j = 1, 2, 3, \ldots, \qquad (15.3)$$

where a is the quantum well width. The corresponding energy eigenvalues are

$$\mathcal{E}_j = \frac{\pi^2}{2}\frac{\hbar^2}{ma^2}j^2 = \frac{\pi^2}{2}\Delta\mathcal{E}j^2, \qquad (15.4)$$

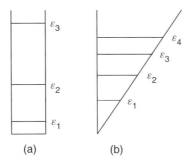

Figure 15.6. Subband edge energies in infinitely deep quantum wells (a) Rectangular well and (b) asymmetric triangular well.

where $\Delta\mathcal{E} = \hbar^2/ma^2$ is the characteristic energy. The jth eigenvalue is often called jth subband energy; in fact, $\mathcal{E}_j$ defines the energy of the bottom edge of the jth subband (see Fig. 15.5):

$$\mathcal{E}_j(\mathbf{k}) = \mathcal{E}_j + \frac{\hbar^2(k_x + k_y)^2}{2m^*}. \tag{15.5}$$

There is a remarkable difference between two-dimensional (2D) and three-dimensional (3D) systems. In a 2D system, the density of states is finite even at the bottom of the subband (Fig. 15.5b, solid line), whereas it tends toward zero in a 3D case ($\sim\sqrt{\mathcal{E}}$, see dashed line in Fig. 15.5b). This has fundamental consequences on the properties of the 2D systems, because it means that all kinetic processes remain finite at low kinetic energies and low temperatures.

An infinitely deep quantum well confines an unlimited number of subbands. The number of the subbands in a well of finite depth (Fig. 15.3b) is limited by the confining potential energy. At least one bound state (one subband) is always present in a symmetric well, even in a shallow one.

A Triangular Well

A triangular quantum well formed by a uniform electric field applied normal to the plane of an infinitely high barrier of an abrupt heterojunction is a model of special interest in electronics.

Let the confining potential energy of the triangular well be (Fig. 15.6b)

$$V(z) = \begin{cases} eEz, & z > 0, \\ \infty, & z < 0. \end{cases} \tag{15.6}$$

The envelope functions $\xi_j(z)$ in the triangular well and the corresponding subband energies $\mathcal{E}_j$ are expressible through the Airy functions $Ai(\zeta)$ and their roots α_j [303]:

$$\xi_j(z) = Ai\left(\frac{z}{a} - \frac{\mathcal{E}_j}{eEa}\right), \tag{15.7}$$

$$\mathcal{E}_j = \tfrac{1}{2}\Delta\mathcal{E}\alpha_j, \tag{15.8}$$

where the characteristic energy $\Delta\mathcal{E} = \hbar^2/(ma^2)$ is determined by the effective width of the triangular well $a = (\hbar^2/(2meE))^{1/3}$.

The first three roots α_j are $\alpha_1 \approx 2.338$, $\alpha_2 \approx 4.088$, and $\alpha_3 \approx 5.527$; the approximate expression

$$\alpha_j \approx \left(\frac{3\pi}{2}\left(j - \frac{1}{4}\right)\right)^{2/3} \tag{15.9}$$

holds as a fair estimate.

15.4 OCCUPIED QUANTUM WELLS

A quantum-mechanical treatment of occupied wells should consider contribution to the potential energy resulting from Coulomb interaction of the confined electrons among themselves and with the other charges. Detailed numerical calculations can be performed in Hartree [309], Thomas–Fermi [310], and local density [311] approximations. In the latter, the potential energy due to the space charge of the electrons and the dopants $V(z)$ is assumed to obey the solution of the Poisson equation

$$\frac{d^2}{dz^2}V(z) = \frac{4\pi e}{\varepsilon\varepsilon_0}\rho(z), \tag{15.10}$$

where $\varepsilon\varepsilon_0$ stands for the dielectric constant. The charge distribution, $\rho(z)$, is given by the sum the three-dimensional densities: the positively charged donors, $N_D^+(z)$, the negatively charged residual acceptors, $N_A^-(z)$, and the mobile electrons $n_{3D}(z)$. That is,

$$\rho(z) = e[N_D^+(z) - N_A^-(z) - n_{3D}(z)]. \tag{15.11}$$

The system of Schrödinger and Poisson equations, (15.2) and (15.10), can be solved numerically for a quantum well of interest taking into account specific charge distribution and boundary conditions.

Planar-Doped GaAs

A simple way to form a two-dimensional electron gas is to insert a very thin doping layer into a uniform undoped semiconductor (see Fig. 15.7a). This process is called planar doping or δ-doping. Both the ionized donors and the mobile electrons are confined in the vicinity of the same plane. The electron confinement

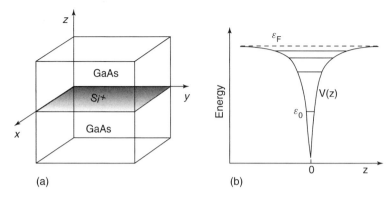

Figure 15.7. A δ-doped GaAs. (a) General view. (b) Effective potential, $V(z)$, formed by the plane of the charges of the ionized donors and of the electron gas confined near the plane.

results from electrostatic Coulomb interaction between the positive charge of the ionized donors and the negative charge of the quasi-two-dimensional gas of the mobile electrons. Negatively charged surface states and residual acceptors can be of some importance as well. At a sufficiently high sheet density of the 2DEG, the quantum well is narrow enough to induce electronic subbands (Fig. 15.7b). This means that the electrons can move freely in the plane (x–y plane, Fig. 15.7a) remaining confined in the direction (z direction) perpendicular to the doping layer.

The first observation of a 2DEG in a δ-doped structure was made by Zrenner et al. [312] in Shubnikov–de Haas magnetoresistance experiments. The structures can be grown by molecular beam epitaxy (MBE). The growth procedure is as follows. First, an undoped GaAs layer is grown, the growth is interrupted by closing the Ga shutter, and a limited number of silicon atoms (acting as n-type dopants) are introduced before the Ga source is opened again for the subsequent overgrowth of the undoped GaAs. In this way, a single monolayer of the silicon atoms can be buried into the high-quality GaAs layer. Typically, the thickness of the doping layer is less than 10 nm, and the donor sheet density exceeds 10^{12} cm^2.

For a planar-doped QW the donors are assumed ionized, with their positive charge being uniformly distributed in the $z = 0$ plane: $N_D^+(z) = N_D \delta(0)$ (see Fig. 15.7). The three-dimensional electron density $n_{3D}(z)$ is assumed to consist of the electron densities in the populated subbands

$$n_{3D}(z) = \sum_j n_j \xi_j^*(z) \xi_j(z), \qquad (15.12)$$

where the envelope functions are normalized to unity:

$$\int \xi_j^*(z) \xi_j(z) \, dz = 1. \qquad (15.13)$$

Let the population of the jth subband be defined by the electron temperature T_e [296]:

$$n_j = \frac{mk_B T_e}{\pi \hbar^2} \ln\left(1 + \exp\left(\frac{\mathcal{E}_F - \mathcal{E}_j}{k_B T_e}\right)\right). \quad (15.14)$$

The factor $m/\pi \hbar^2$ is the density of states taking the spin factor 2 into account (see Fig. 15.5).

Figure 15.8 presents the eigenvalues $\mathcal{E}_j$ and the envelope functions determined by the solution of the coupled Eqs. (15.2) and (15.10) for a δ-doped GaAs containing 2DEG with $n_0 = 1.5 \times 10^{12}$ cm^{-2} sheet density, taking into account residual acceptors and neglecting surface charge [313]. Solid lines stand for the well potential energy and the subband edge energies. Dashed lines give the normalized squared envelope functions, $\xi_j^2(z)$; they assume zero values far away from the well. Dashed–dotted line at $\mathcal{E} = 0$ is the Fermi energy. The results of Fig. 15.8 illustrate strong effects of the electron temperature on the well depth and on the subband energies in respect to the Fermi energy. At 77 K temperature, over 75% of all electrons reside in the lowest subband; the third and the fourth subbands are almost empty. The higher electron temperature (Fig. 15.8b) leads to depopulation of the lowest subband, the number of electrons in the upper subbands increases, the well becomes deeper, and the subband separation increases.

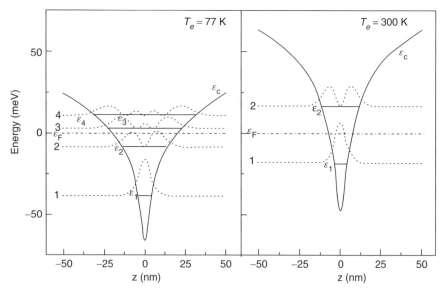

Figure 15.8. Subband structure of a δ-doped GaAs with $n_0 = 1.5 \times 10^{12}$ cm^{-2} and $N_A = 1 \times 10^{15}$ cm^{-3}, assuming different electron temperature [313]. (a) $T_e = 77$ K, (b) $T_e = 300$ K. Solid lines are the bottom edges of the subbands, and solid curves are the conduction band edges. Dotted lines are the normalized squared envelope functions vanishing far away from the well. Dashed–dotted line stands for the Fermi energy.

TABLE 15.3. Planar-Doped GaAs QW Containing $n_0 = 5.13 \times 10^{12}$ cm^{-2} Electrons[a]

	Hartree	Thomas–Fermi	Local Density
n_1/n	0.614	0.612	0.617
$\mathcal{E}_1 - \mathcal{E}_F$, meV	−112.5	−111.9	−112.4
n_2/n	0.236	0.236	0.237
$\mathcal{E}_2 - \mathcal{E}_F$, meV	−43.1	−43.4	−43.2
n_3/n	0.104	0.108	0.107
$\mathcal{E}_3 - \mathcal{E}_F$, meV	−19.3	−19.6	−19.5
n_4/n	0.036	0.038	0.038
$\mathcal{E}_4 - \mathcal{E}_F$, meV	−6.53	−6.9	−6.9

[a]Subband occupation and the edge energies are estimated using the Hartree model, the Thomas–Fermi model [310], and the local density approximation based on Eqs. (15.2 and 15.10) [313].

Table 15.3 compares the results of numerical calculation of the subband energies and the electron fraction in the individual subbands using the Hartree model, the Thomas–Fermi model [310], and the local density approximation based on coupled Eqs. (15.2) and (15.10). In the latter case, the confinement potential, the subband energies, the electron population, and the envelope functions were obtained in a self-consistent way. The exchange-correlation potential of the two-dimensional electrons was not included in the calculation (the exchange is known to have a weak influence on the subband energy and on the subband population [296]).

Further on we shall rely very often on the results obtained by this approach based on the self-consistent solution of one-dimensional Schrödinger and Poisson equations. The results (see Table 15.3) obtained in the local density approximation are similar to those obtained within more sophisticated approaches used by Ioriatti [310].

AlGaAs/GaAs Heterostructures

Heterostructure growth technology provides us with lattice-matched and pseudomorphic AlGaAs/GaAs, InAlAs/InGaAs, InP/InGaAs heterostructures containing two-dimensional electron gas (2DEG) channels for field-effect transistors (HEMTs, PHEMTs) [304, 314, 315]. High mobility of confined electrons and other advantages have been successfully exploited for ultrafast operation of 2DEG channels. The first demonstration of a 2DEG formed at an AlGaAs/GaAs interface was reported by Störmer et al. in 1979 [316]: the Shubnikov–de Haas oscillations due to the magnetic field applied perpendicular to the interface were observed, but the oscillations disappeared after rotating the field into the interface plane. The AlGaAs/GaAs system represents an ideal system for studies of unique properties of a 2DEG because the heterojunction is free from interface states. *Selective doping* ensures spatial separation of the mobile carriers from their parent impurities, leading to extremely high mobility of the carriers in parallel transport [295].

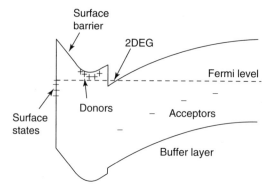

Figure 15.9. Schematic view of a band diagram of a selectively doped single-heterojunction AlGaAs/GaAs structure. The AlGaAs layer contains an undoped spacer and a uniformly doped layer. The 2DEG forms in GaAs at the interface. Low background doping with acceptors is assumed.

Let us consider a selectively doped heterostructure consisting of a GaAs layer covered by a donor-doped layer of $Al_xGa_{1-x}As$ (see Fig. 15.9). Under equilibrium the donors supply the electrons to the surface states and to the GaAs layer; and the two-dimensional electron gas forms in GaAs, provided that doping with donors is sufficient. The 2DEG is pressed to the heterojunction by the electric field of the positively charged stripped donors. High mobility of electrons is ensured by the perfect heterojunction together with the undoped spacer inserted between the 2DEG and the donor layer. The GaAs layer usually contains traces of acceptors.

The band edge discontinuities, $\Delta\mathcal{E}_c$ and $\Delta\mathcal{E}_v$, and the bandgap difference, $\Delta\mathcal{E}_g$, are among the most important parameters in determining the electrical properties of the 2DEG. The most widely used method for calculating conduction- and valence-band discontinuities is Anderson's electron affinity rule (see Common Anion Rule in reference 317). The conduction-band discontinuity, or offset, is given by

$$\Delta\mathcal{E}_c = \chi_1 - \chi_2, \qquad (15.15)$$

where χ_1 and χ_2 are the electron affinities, or energies required to promote an electron from the bottom of the conduction band into vacuum. The conduction-band discontinuity determines the confining barrier of the quantum well for the electrons.

The valence-band offset is given by

$$\Delta\mathcal{E}_v = \Delta\mathcal{E}_g - \Delta\mathcal{E}_c. \qquad (15.16)$$

Provided that all electrons resided at the donors, the potential would be step-like, and flat bands would be conserved (Fig. 15.3). The quasi-triangular quantum well is created after the transfer of N electrons into the GaAs layer. Now, the electric field at the interface is

$$E = \frac{Ne}{Q\varepsilon\varepsilon_0} = \frac{ne}{\varepsilon\varepsilon_0}. \tag{15.17}$$

where Q is the interface area, and n is the sheet density of the mobile electrons.

In the presence of a 2DEG of sufficient density, the coupled Eqs. (15.2) and (15.10) should consider the electron charge distribution in a self-consistent way, taking into account a nonzero electron temperature, effective mass and dielectric constant discontinuities, finite-barrier effects, surface charges, and other features. The detailed procedure for constructing an energy-band diagram is described elsewhere [296, 304].

The model for $Al_xGa_{1-x}As/GaAs$ heterojunctions considered by Stern and Das Sarma [318] takes into account the effective potential given by

$$V(z) = -e\varphi(z) + V_h(z) + V_{xc}(z) + V_{im}(z), \tag{15.18}$$

where $\varphi(z)$ is the electrostatic potential, $V_h(z)$ is the potential energy associated with the heterojunction discontinuity, $V_{xc}(z)$ is the local exchange-correlation potential energy, and $V_{im}(z)$ is the image potential energy. The electrostatic potential $\varphi(z)$ obeys the Poisson equation (15.10). Its derivative is taken to be zero in the GaAs layer far away from the heterojunction.

A mathematically abrupt potential barrier $\Delta\mathcal{E}_c$ is smoothed by taking a grading function that interpolates linearly between the values in the materials bounding the heterojunction. The grading of the interface barrier is taken to be $V_h(z) = [1 - G(z)]\Delta\mathcal{E}_c$, where function $G(z)$ interpolates between the value in $Al_xGa_{1-x}As$ and the value in GaAs [319]. An empirical dependence of the conduction-band discontinuity $\Delta\mathcal{E}_c$ on the AlAs mole fraction, x, in AlGaAs/GaAs heterojunction is available at 300 K [320]:

$$\begin{aligned}\Delta\mathcal{E}_c(x)[\text{eV}] &= 1.1x & \text{for} \quad 0 \leq x \leq 0.45, \\ \Delta\mathcal{E}_c(x)[\text{eV}] &= 0.43 + 0.14x & \text{for} \quad 0.45 \leq x \leq 1.0.\end{aligned} \tag{15.19}$$

The image potential energy can be written as [318]

$$V_{im}(z) = \begin{cases} [(\varepsilon_2 - \varepsilon_1)e^2]/[4\varepsilon_0\varepsilon_2(\varepsilon_2 + \varepsilon_1)z], & z > 0, \\ [(\varepsilon_2 - \varepsilon_1)e^2]/[4\varepsilon_0\varepsilon_1(\varepsilon_2 + \varepsilon_1)z], & z < 0, \end{cases} \tag{15.20}$$

where subscript 2 denotes the channel (GaAs) side of the heterojunction, and subscript 1 denotes the barrier ($Al_xGa_{1-x}As/GaAs$) side. The static dielectric constant for $Al_xGa_{1-x}As/GaAs$ is [320]

$$\varepsilon_1(x) = 13.18 - 3.12x. \tag{15.21}$$

Assuming a typical value $x = 0.3$ for the mole ratio, the dielectric constants are $\varepsilon_1 = 12.24$ for $Al_{0.3}Ga_{0.7}As$ and $\varepsilon_2 = 13.18$ for GaAs. The change is less than

10%, and consequently the image interaction is much smaller than the direct Coulomb interaction.

The simple analytic expression for the exchange-correlation potential energy V_{xc} to account for the effects of electron–electron interaction was suggested [318]:

$$V_{xc}(z) = -\frac{2}{p\alpha r_s}\left[1 + 0.7734\frac{r_s}{21} \ln\left(1 + \frac{21}{r_s}\right)\right] Ry^*, \qquad (15.22)$$

where

$$\alpha = (4/9\pi)^{1/3}, \qquad r_s(z) = \left[\frac{4}{3}\pi\alpha^{*3} n(z)\right]^{-1/3}, \qquad \alpha^* = \frac{\varepsilon\hbar^2}{me^2}.$$

Degani [311] has shown that the exchange and the correlation effects have a small influence on the energy band structure in the high-density limit.

15.5 EFFECTIVE BARRIER HEIGHT

The maximum sheet density of 2DEG confined by the high-mobility quantum well is mainly determined by the doping level, the electron temperature, and the confining barrier separating the high-mobility well from the donor-doped well located in the barrier layer. Figure 15.10 illustrates doping-dependent transformation of the wells and the confining barrier in a selectively δ-doped double-heterojunction structure. The undoped heterostructure has flat bands (Fig. 15.10a) and does not contain any two-dimensional electron gas (2DEG). The effective barrier height

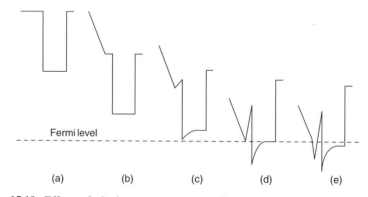

Figure 15.10. Effect of doping on quantum-well shape in a double-heterojunction structure containing a δ-doped barrier layer and deep surface states: (a) An undoped heterostructure. (b) An insufficiently doped heterostructure (no 2DEG forms). (c) A subcritically doped heterostructure containing a high-mobility 2DEG in the rectangular undoped well. (d) Critical doping. (e) A supercritically doped heterostructure demonstrating two-channel conduction.

is equal to the conduction band offset minus the energy of the lowest subband [see Eqs. (15.15) and (15.4)].

Let the donor plane be located in the left-hand barrier between the well and the surface. Under light doping, occupation of the empty surface states consumes all electrons liberated from the donors. This charge transfer is accompanied by the buildup of the electric field under the surface; as a result the rectangular well moves down, and the surface barrier forms (Fig. 15.10b). No 2DEG appears until certain doping level is reached and exceeded.

Formation of a high-mobility 2DEG in the rectangular well starts after the occupation of the deep surface levels. The space charge of the 2DEG in the rectangular quantum well modifies its shape and builds up the electric field between the donor plane and the 2DEG [see Eq. (15.17)]: A δ-doped quantum well forms in the vicinity of the donor plane. At a moderate sheet density of the donors, the δ-doped well is empty, and it is separated from the high-mobility well by the asymmetric triangular barrier (Fig. 15.10c). In consistence with Eq. (15.17), a stronger electric field builds up as the 2DEG sheet density increases. The empty δ-doped well moves down. Eventually, at certain critical donor sheet density, the bottom of the δ-doped well approaches the Fermi level close enough for some electrons to appear in it (Fig. 15.10d). Figure 15.10e illustrates the situation in supercritically doped heterostructures where the two-channel conduction takes place: The high-mobility 2DEG channel is present in the quasi rectangular well, and the low-mobility 2DEG channel is present in the quasi-triangular δ-doped well.

The dependence of the confining barrier on the doping can be considered by a self-consistent solution of the Schrödinger–Poisson equations. The most important changes take place in the vicinity of the critical sheet density of 2DEG required for the transition to two-channel conduction. Figure 15.11 illustrates the results

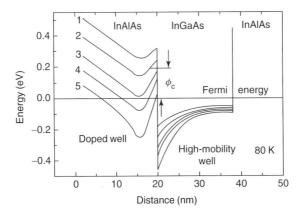

Figure 15.11. Quantum well shape in InP-based InAlAs/InGaAs/InAlAs structures containing a thin donor-doped layer in the InAlAs barrier layer [321]. The donor density: 1 represents 3.1×10^{12} cm^{-2}; 2 represents 4.1×10^{12} cm^{-2}; 3 represents 5×10^{12} cm^{-2}; 4 represents 6.6×10^{12} cm^{-2}; 5 represents 11.9×10^{12} cm^{-2}. Density of occupied surface states is 1.5×10^{12} cm^{-2}. $T_0 = 80$ K.

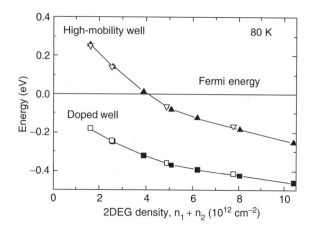

Figure 15.12. Dependence of the bottom energy on the total 2DEG density in two channels, $n_1 + n_2$ (see Fig. 15.11). The density of the occupied surface states is 0.75×10^{12} cm^{-2} (unfilled symbols) and 1.5×10^{12} cm^{-2} (filled symbols) [321].

obtained for lattice-matched InAlAs/InGaAs/InAlAs heterostructures [321]. The confining barrier tends to decrease as the 2DEG sheet density increases.

The effective barrier height ϕ_c at the critical 2DEG sheet density n_c can be estimated using a simple expression:

$$\phi_c = en_c d_s/(\varepsilon\varepsilon_0), \qquad (15.23)$$

where d_s is the spacer thickness. The critical 2DEG sheet density is found to be $\sim 4 \times 10^{12}$ cm^{-2} in the typical lattice-matched InAlAs/InGaAs/InAlAs δ-doped double-heterojunction structure containing one donor plane. The critical donor sheet density exceeds the electron sheet density by that of the surface states. However, the latter has a small (if any) effect on the critical 2DEG sheet density (Fig. 15.12).

More electrons can be confined, without the onset of two-channel conduction, provided that a higher mole ratio of InAs in the InGaAs channel is used. Now, the critical 2DEG sheet density exceeds 5×10^{12} cm^{-2} in the typical δ-doped InAlAs/InGaAs/InAlAs structures containing InAs-rich pseudomorphic channel and one donor plane. The critical electron sheet density can be doubled in two-sided δ-doped heterostructures.

CHAPTER 16

HOT-ELECTRON NOISE IN AlGaAs/GaAs 2DEG CHANNELS

This chapter deals with longitudinal noise temperature and longitudinal spectral intensity of hot-electron velocity fluctuations in selectively doped AlGaAs/GaAs heterostructure channels subjected to the electric field applied in the plane of electron confinement. The experimental results are analyzed in terms of fast and ultrafast kinetic processes associated with electron confinement in a quantum well. Of course, the processes of interest act together with those investigated in bulk semiconductors (Chapter 9 and 10) and epitaxial homostructures of micrometric/submicrometer dimensions (Chapters 11 and 13), but main efforts are taken to resolve and identify the sources of noise associated with the electron confinement in quantum well channels. As usual, X-band frequency is used; this frequency is high enough to neglect $1/f$ fluctuations and electron number fluctuations due to generation–recombination processes. Contribution of shot noise is negligible in ungated channels with perfect electrodes. Pulsed short-time-domain microwave radiometric technique seems to be appropriate in order to avoid lattice heating and thermal walkout effects (see Chapter 8).

Modern epitaxy provides with a great variety of quantum well (QW) structures containing a two-dimensional electron gas (2DEG) where kinetic processes absent in a three-dimensional electron gas take place [50, 51, 56, 294–300, 322]. Correspondingly, new sources of hot-electron noise appear. Experimental investigation of hot-electron velocity fluctuations at microwave frequencies is a way to study fast and ultrafast kinetic processes inside the conduction band — the microscopic origin of electronic noise in semiconductor structures subjected to a high electric field. The obtained information on the origin of electronic noise is important for considering how to reduce excess noise and improve reliability of data processing in high-speed electronics.

The sources of hot-electron noise in quantum-well channels containing 2DEG have been repeatedly discussed [12, 256, 323–325]. Electron heating by the electric field applied in the plane of electron confinement introduces excess microwave noise in the direction of the current [326, 327]. At sufficiently high electric fields, some of the confined electrons gain enough energy to leave the quantum well; this electron deconfinement has been named real-space transfer [298]. As expected from Monte Carlo simulation data [328], the reversible real-space transfer of the hot electrons is the microscopic origin for the real-space transfer noise resolved in AlGaAs/GaAs channels at a microwave frequency [197, 329], this origin being confirmed by the dependence of the onset field on the heterobarrier height [228, 330]. Reversible tunneling of hot electrons across the barrier of AlAs inserted between the 2DEG channel and the donor-doped layer causes the real-space transfer noise supported by transverse tunneling [331]. The electron confinement in a quantum well is known to enhance energy loss by the hot electrons on acoustic phonons. This has an effect on the excess intrasubband noise in a narrow quantum well at cryogenic temperatures at relatively low electric fields [330]. The intersubband noise is found to dominate in a δ-doped GaAs channel [329], where the upper subbands support a higher electron mobility as compared to more confined electronic states of the lower subbands [299].

16.1 2DEG CHANNELS

Figure 16.1 presents a schematic view of typical MOCVD and MBE grown structures with a high-density electron gas confined in a quantum well. In a selectively doped single-heterojunction structure (Fig. 16.1a), the mobile electrons are located in the undoped narrow-gap semiconductor (see Fig. 15.9).

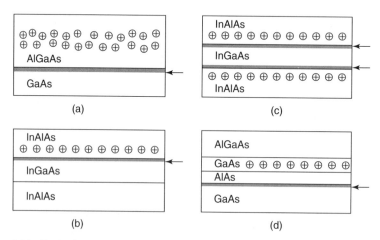

Figure 16.1. Typical structures with 2DEG channels (a) Selectively doped single heterojunction, (b) δ-doped double heterojunction, (c) two-sided δ-doped double heterojunction, (d) δ-doped triple heterojunction. Arrows indicate the 2DEG location at equilibrium.

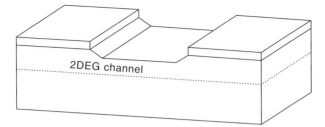

Figure 16.2. Ungated 2DEG channel supplied with two ohmic electrodes for noise temperature measurements.

The quasi-triangular quantum well is separated from the donor-doped layer by the undoped spacer. Quasi-rectangular quantum wells are typical for double-heterojunction channels (Fig. 16.1b,c). The well is almost rectangular in the case of two-sided doping (Fig. 16.1c). A triple-heterojunction channel with an undoped wide-band-gap barrier inserted between the donor-doped layer and the 2DEG is designed to prevent the real-space transfer due to jumps over the barrier at not-too-high electric fields (Fig. 16.1d). The samples for noise measurements are ungated channels supplied with two ohmic electrodes (Fig. 16.2). The electron heating by the electric field applied along the channel is a prerequisite for the real-space transfer in the subcritically doped channels (cf. curves 1 and 2 in Fig. 15.11). On the other hand, the effective barrier is very low at the high 2DEG density (curve 5 in Fig. 15.11), and the electron exchange between the high-mobility channel and the low-mobility channel can take place without electron heating.

16.2 HOT-ELECTRON REAL-SPACE TRANSFER NOISE

Dependence on Heterobarrier Height

The experimental data on hot-electron noise temperature for AlGaAs/GaAs subcritically doped channels are given in Fig. 16.3 [228, 330, 331]. The noise sources, absent in GaAs, are resolved in the 2DEG channels at electric fields below the threshold for the intervalley transfer noise [the latter dominates in GaAs at fields over 2 kV/cm, (see Chapter 10)]. The threshold field for the noise specific to the 2DEG channels depends on the conduction band offset, which increases with Al mole ratio in AlGaAs (see Table 16.1).

A simple interpretation of the longitudinal hot-electron noise can be given in terms of field-controlled real-space transfer of hot electrons [330]. Figure 16.4 illustrates the subband energy spectra calculated by self-consistent solution of coupled Schrödinger–Poisson equations for single-heterojunction AlGaAs/GaAs channels A and B containing the subcritical densities of 2DEG: $n_0 < n_c$. At equilibrium all electrons occupy the right-hand lower well, where the electron mobility is high. The left-hand upper well is empty, unless the electrons are hot

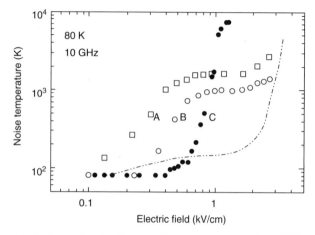

Figure 16.3. Hot-electron longitudinal noise temperature for AlGaAs-containing heterostructures with different Al mole ratio in the spacer (symbols): A, 25% [228]; B, 33% [330]; C, 100% [331]. For 2DEG parameters see Table 16.1. Dashed-dotted line is for lightly doped GaAs [192].

TABLE 16.1. AlGaAs/GaAs-Based Channel Parameters at 80 K [228, 330, 331]

	AlGaAs/GaAs		AlGaAs/GaAs/AlAs/GaAs
Channel	A	B	C
Al mole ratio in the spacer	0.25	0.33	1.0
Spacer thickness (nm)	3	40	2.5
Electron density, 10^{11} (cm^{-2})	5.7	1.9	13
Electron mobility (cm^2/(V·s))	75,000	103,000	35,000

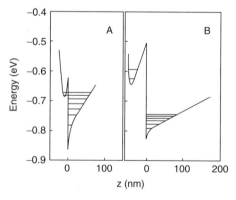

Figure 16.4. Subband energy spectra of selectively doped single-heterojunction AlGaAs/GaAs channels A and B [330] (see Table 16.1). At equilibrium, electrons occupy the right-hand quantum wells located in the undoped GaAs layer.

and the real-space transfer takes place between the undoped GaAs layer and the adjacent doped AlGaAs layer.

A comparison of the noise results (Fig. 16.3) for channels A and B having different energy separation between the wells (Fig. 16.4) confirm the natural expectation that the onset field for the real-space transfer should depend on the barrier responsible for the 2DEG confinement at equilibrium. This behavior cannot be accounted for by a different mobility in channels A and B at low electric fields (Table 16.1): The higher mobility μ_0 leads to the higher electron temperature T_e because of the higher power $P = e\mu_0 E^2$ received by an average electron from the applied electric field E.

Figure 16.5 presents the spectral intensity of hot-electron velocity fluctuations [330]. The maxima of S_V are resolved near the onset fields, and their position depends on the effective barrier height (cf. circles and squares in Fig. 16.5). The shape of the maxima is similar to that obtained by Monte Carlo simulation [328] of the real-space transfer fluctuations (triangles in Fig. 16.6). It is noteworthy that the real-space transfer in GaAs-based 2DEG channels suppresses the intervalley fluctuations of hot electrons dominating in GaAs at field over 2 kV/cm (cf. circles and squares for the QW channels and dashed–dotted line for GaAs in Fig. 16.5) [228, 330, 331].

Relaxation Time of Reversible Real-Space Transfer

By assuming the intrawell processes to be essentially faster as compared to the interwell transfer, the spectral intensity of velocity fluctuations in the direction of current can be decomposed into [cf. Eq. (9.5)]:

$$S_V = \frac{n_1}{n_0}S_1 + \frac{n_2}{n_0}S_2 + S_V^{\text{inter}}, \qquad (16.1)$$

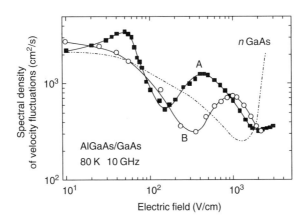

Figure 16.5. Field-dependent spectral intensity of hot-electron longitudinal velocity fluctuations for AlGaAs/GaAs channels A and B (symbols) [228, 330] at 80 K lattice temperature together with data for lightly doped GaAs (dashed–dotted line) [192]. Solid lines are to guide the eye. For 2DEG parameters see Table 16.1.

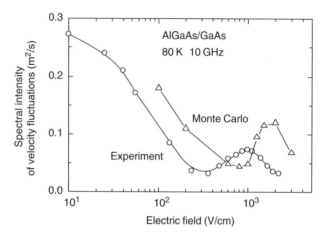

Figure 16.6. Field-dependent spectral intensity of hot-electron longitudinal velocity fluctuations for AlGaAs/GaAs channel at 80 K lattice temperature. ○ represents experiment, channel B (see Table 16.1) [330]; △ represents results of Monte Carlo simulation [328]. Solids curves are to guide the eye.

where n_1/n_0 and n_2/n_0 are the partial densities of electrons in the lower and the upper wells under steady state, $S_1 n_1/n_0$ and $S_2 n_2/n_0$ are the weighted spectral intensities of drift velocity fluctuations in the wells neglecting the well coupling, and S_v^{inter} is the contribution of the partition fluctuations resulting from the interwell transitions.

A simple formula can be applied to discuss the partition fluctuations caused by the reversible real-space transfer [cf. Eq. (9.6)]:

$$S_v^{\text{inter}} = 4\frac{n_1 n_2}{n_0^2}(v_1^{\text{dr}} - v_2^{\text{dr}})^2 \bar{\tau}^{\text{inter}}, \tag{16.2}$$

where v_1^{dr} and v_2^{dr} are the electron drift velocities in the wells, and $\bar{\tau}^{\text{inter}}$ is the real-space transfer relaxation time,

$$(\bar{\tau}^{\text{inter}})^{-1} = \tau_{12}^{-1} + \tau_{21}^{-1}, \tag{16.3}$$

determined by the real-space transfer rates from the first well and backwards, τ_{12}^{-1} and τ_{21}^{-1}, respectively.

Equations (16.1) and (16.2) show that the transverse real-space transfer contributes to the longitudinal velocity fluctuations, provided that the electron drift velocities in the wells differ. There is no contribution at thermal equilibrium, since $n_2 = 0$ and $v_1^{\text{dr}} = v_2^{\text{dr}} = 0$. A higher mobility causes a higher electron temperature; this leads to "evaporation" of the hot electrons from the high-mobility well and their "condensation" in the low-mobility well. A monotonous increase in the electron density ratio n_2/n_1 with an increase of electric field takes place in the channel containing nonequivalent wells (Fig. 16.4). The increase leads to formation of the maximum of S_{12} at $n_1 \approx n_2$ [see Eq. (16.2)].

The maximum associated with the real space transfer of hot electrons is observed [228] at 1 kV/cm field in channel B (Fig. 16.6). Under the assumption that $v_1^{\mathrm{dr}} - v_2^{\mathrm{dr}} \approx 10^7$ cm/s and $n_1 = n_2$, the real-space transfer time for channel B is estimated from Eq. (16.2) with the result $\bar{\tau}^{\mathrm{inter}} \sim 5$ ps [330]. The obtained time constant being short, the transfer rates, τ_{12}^{-1} and τ_{21}^{-1}, should be high at the electric field corresponding to $n_1 = n_2$. For the return rate not to be too small, efficient electron heating in the low-mobility well in the doped AlGaAs layer must be assumed [330].

In channel A (Table 16.1), the well in the undoped GaAs and that in the doped AlGaAs are separated by the thin barrier which is transparent for electrons at energies higher than the bottom energy of the upper well (cf. Fig. 16.4, channel A). The high-energy electrons are shared by the two wells, and they are subjected to impurity scattering, thereby resulting in a lower drift velocity. Thus, two groups of electrons with different drift velocities can be introduced, and, after certain modifications, the idea of real-space transfer fluctuations remains useful to interpret the maximum dominating at the intermediate fields (squares in Fig. 16.5).

Thermal Effects on Real-Space Transfer

Higher lattice temperatures enhance electron scattering by optical phonons and make electron heating less efficient. This reduces the number of electrons able to jump over the barrier. Thus, the transverse real-space transfer takes place at higher electric fields, as confirmed by the experiments (Fig. 16.7) [313, 323].

Figure 16.7 shows that the real-space transfer fluctuations reach the maximum at fields over 3 kV/cm at 175 K lattice temperature. Provided that the

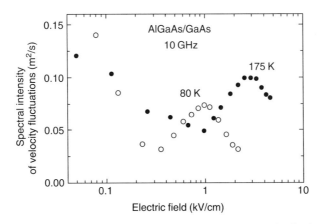

Figure 16.7. Field-dependent spectral intensity of hot-electron longitudinal velocity fluctuations for single-heterojunction AlGaAs/GaAs channel B (see Table 16.1) at different lattice temperatures [313, 323].

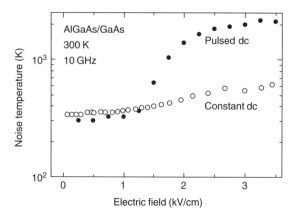

Figure 16.8. Hot-electron longitudinal noise temperature versus electric field for AlGaAs/GaAs channel (see channel B in Table 16.1) at room temperature [313]. ● represents pulsed bias (4 μs); ○ represents constant bias.

real-space transfer fluctuations were absent, the field would be strong enough for the intervalley noise in lightly doped GaAs (cf. Fig. 10.1). This overlap of the field ranges for the intervalley transfer and the real-space transfer makes it difficult to resolve the latter at room temperature [197], especially if the thermal walkout due to the Joule heat is not avoided [326, 327]. Figure 16.8 illustrates the interference of the thermal and the electronic processes: The electronic noise due to the real-space transfer in the 2DEG channel is strongly quenched by the channel overheat resulting from the Joule effect due to the DC current. Similar results are also available for channel A [197].

16.3 LONGITUDINAL FLUCTUATIONS DUE TO TRANSVERSE TUNNELING

Subband Structure

In order to avoid hot-electron jumps over the barrier, the 2DEG high-mobility channel can be separated from the ionized donors by a thin undoped layer of a wide-band-gap semiconductor constituting a high barrier for the confined electrons. A triple-heterojunction AlGaAs/δ-GaAs/AlAs/GaAs structure is designed [331] to have the δ-doped GaAs layer separated from the undoped GaAs layer by the thin layer of AlAs (Fig. 16.1d). Figure 16.9 presents the results of subband energy calculations; the occupied and the empty wells are treated independently [331]. At equilibrium, the electrons are confined in the right-hand high-mobility well.

Tunneling-Related Relaxation Time for Real-Space Transfer

At low and intermediate fields, the AlGaAs/δ-GaAs/AlAs/GaAs structure (Fig. 16.3, channel C) demonstrates a lower noise temperature [331] as compared

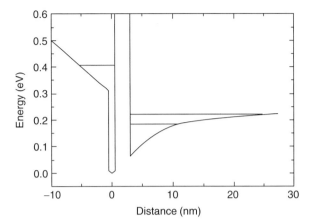

Figure 16.9. The lowest subbands in AlGaAs/δ-GaAs/AlAs/GaAs heterostructure (see Fig. 16.1d). The heterostructure consists (from right to left) of an undoped GaAs layer, a 2.5-nm undoped AlAs spacer, a 1-nm δ-doped GaAs layer, and an undoped AlGaAs layer [331].

to AlGaAs/GaAs channels without AlAs layer (channels A and B). As one could expect, the AlAs barrier is efficient in suppressing the electron jumps over the barrier and the associated real-space transfer fluctuations at electric fields below 700 V/cm at the considered lattice temperature.

However, strong excess noise onsets at electric fields exceeding 700 V/cm (Fig. 16.3, channel C). Estimations show that much higher fields are required for the electron jumps *over* the AlAs barrier and for the *intervalley transfer* in GaAs: The intervalley transfer noise in GaAs dominates at fields exceeding 2 kV/cm, and the AlAs barrier is higher than the intervalley separation gap in GaAs. On the other hand, the barrier is thin, and the electrons can penetrate it by tunneling. The energy required for the tunneling (cf. Fig. 16.9) is lower than the intervalley separation gap in GaAs. Therefore it has been concluded [331] that the transverse tunneling is responsible for the steep increase of the hot-electron noise temperature (Fig. 16.3, channel C).

Figure 16.10 presents the spectral intensity of velocity fluctuations for the same AlGaAs/δ-GaAs/AlAs/GaAs heterostructure (Fig. 16.1d). The maximum resolved at 1 kV/cm field (unfilled circles) provides a possibility to estimate the tunneling-related relaxation time. In the framework of Eq. (16.2) under assumption $v_1^{dr} - v_2^{dr} \sim 2 \times 10^7$ cm/s the relaxation time associated with the reversible transverse tunneling is estimated to be $\bar{\tau}^{inter} \sim 10$ ps [331].

Suppression of the Real-Space Transfer Noise

The obtained relatively long relaxation time estimated for the real-space transfer noise sources suggests a possibility for noise suppression in short channels (see Chapter 11). Indeed, the transverse-tunneling-related noise source (observed in

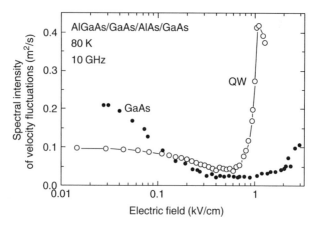

Figure 16.10. Field-dependent spectral intensity of hot-electron longitudinal velocity fluctuations: ○ represents AlGaAs/δ-GaAs/AlAs/GaAs heterostructure [331] (channel C, Table 16.1); ● represents lightly doped GaAs [192].

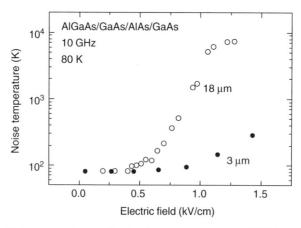

Figure 16.11. Hot-electron longitudinal noise temperature at 80 K lattice temperature for AlGaAs/δ-GaAs/AlAs/GaAs 2DEG channel of different length [331]. ○ represents $L = 18$ μm; ● represents $L = 3$ μm.

a long channel, Fig. 16.11, unfilled circles) is very weak in the 3-μm channel at fields below 1.5 kV/cm (filled circles). This strong dependence on the channel length can be interpreted in terms of electron transit time t^{tr}. Provided that the transit time is short as compared to the tunneling time constant $\bar{\tau}^{inter}$, the tunneling-related fluctuations cannot take place. By assuming $v^{dr} = 2 \times 10^7$ cm/s for the drift velocity in the 3-μm long channel, one obtains $t^{tr} = 15$ ps. Evidently, the time spent by the electrons at energies exceeding the threshold for the tunneling is essentially shorter than the transit time — the total time spent in the channel. Consequently, the result $\bar{\tau}^{inter} \sim t^{tr}$ [331] is in reasonable agreement with that obtained from the maximum of the spectral intensity of velocity fluctuations: $\bar{\tau}^{inter} \sim 10$ ps (Fig. 16.10). The tunneling time estimated from these two independent experiments do not contradict each other and are in a reasonably good agreement with the tunneling time available from luminescence data for the *resonant* tunneling in AlGaAs-containing structures [332]. The *nonresonant* tunneling time is essentially longer [50].

By the way, in the field range where the transverse-tunneling noise dominates at 80 K, the excess noise is essentially weaker at room temperature. At a fixed electric field, say 1.2 kV/cm, an increase in the lattice temperature leads to the reduction of the longitudinal noise temperature by a factor of 10. This is a nice illustration of the thermal quenching of the hot-electron noise related to the transverse tunneling.

16.4 INTERSUBBAND TRANSFER NOISE

Intersubband scattering can cause excess fluctuations provided electron mobility in the subbands differs. These partition-type fluctuations appear in the direction of the current. A δ-doped GaAs 2DEG channel (Fig. 15.7) is suited for demonstration of the related noise [329].

Figure 16.12 presents the experimental results on the hot-electron noise temperature for a δ-doped GaAs channel containing 2DEG ($n_0 = 1.5 \times 10^{12}$ cm^{-2}). The Shubnikov–de Haas measurements confirm [299] that the electron mobility in the channel subbands differs essentially: 600 cm^2/(V·s) in the lowest subband and 3000 cm^2/(V·s) in the upper one at 1.5 K. The excess noise in the 2DEG channel [313] is higher than the excess noise in a uniformly doped GaAs channel with a similar integral planar density of the electron gas. The noise specific to the

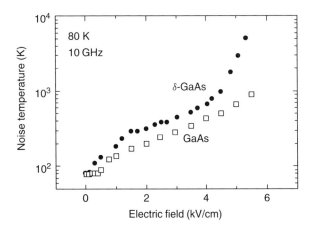

Figure 16.12. Hot-electron longitudinal noise temperature at 80 K lattice temperature for GaAs channels: ● represents δ-doped 2DEG channel (1.5×10^{12} cm^{-2} [313, 324]); ○ represents uniformly doped 3DEG channel (3×10^{17} cm^{-3} [228]).

2DEG channel results from the intersubband fluctuations. Since more than two subbands are involved at high electric fields, a simple two-part partition model [supporting a single maximum of S_V, cf. Eq. (16.2)] does not work in this case.

In conclusion, the sources of microwave noise specific to two-dimensional electron gas were excited and resolved in selectively doped heterostructure channels subjected to the electric field applied in the plane of electron confinement.

CHAPTER 17

HOT-ELECTRON NOISE IN InP-BASED 2DEG CHANNELS

High conduction-band offset at InAlAs/InGaAs and InP/InGaAs interfaces supports deep quantum wells in InGaAs 2DEG channels grown on InP substrates. This modifies the dominant intrawell and interwell kinetic processes; as a result, fluctuation pattern in InAlAs/InGaAs/InAlAs/InP and InP/InGaAs/InP 2DEG channels differs from that observed in AlGaAs/GaAs quantum-well structures [12, 324, 325]. This chapter discusses sources of hot-electron noise in lattice-matched and strained double-heterojunction InAlAs/InGaAs/InAlAs/InP 2DEG channels.

17.1 SUPPRESSION OF REAL-SPACE-TRANSFER NOISE

Onset Field

Let us consider longitudinal noise of hot electrons in subcritically doped InAlAs/InGaAs/InAlAs/InP and InP/InGaAs/InP channels containing 2DEG in the undoped InGaAs quantum well (see Fig. 16.1b,c). Table 17.1 gives the channel specification and some 2DEG properties. Structure 13Q has a lattice-matched $In_xGa_{1-x}As$ channel ($x = 53\%$), and structure 14Q contains a splitted channel with a 30-nm lattice-matched lower $In_{0.53}Ga_{0.47}As$ layer and a 12-nm-thick strained upper $In_{0.7}Ga_{0.3}As$ layer embedded into the $In_{0.52}Al_{0.48}As$ barriers (Fig. 16.1b). Structure 15Q is double delta-doped: Both electron-supplying layers are located in $In_{0.52}Al_{0.48}As$ layers at 3-nm distance from the 12-nm-thick strained $In_{0.7}Ga_{0.3}As$ layer (Fig. 16.1c). Structure 3-18 is MBE-grown and contains an InGaAs channel of graded composition. MOCVD-grown Al-free structure H42 has a strained InGaAs channel of uniform composition.

TABLE 17.1. Low-Field Properties of 2DEG at 300 K for Subcritically Doped $In_xGa_{1-x}As$ Channels [334]

No.	Structure	InGaAs Width (nm)	x (%)	$\Delta\mathcal{E}_c$ (eV)	μ_0 (cm²/(V·s))	n_0 (cm⁻²)
13Q	InAlAs/InGaAs/InAlAs	30	53	0.45	10560	2.4×10^{12}
14Q	InAlAs/InGaAs/InAlAs	12/30	70/53	0.55/0.45	11170	2.3×10^{12}
15Q	InAlAs/InGaAs/InAlAs	12	70	0.55	7260	4.1×10^{12}
3-18	InAlAs/InGaAs/InAlAs	35	80–53	0.6–0.45	8800	1.5×10^{12}
H42	InP/InGaAs/InP	12	67	0.32	7840	1.7×10^{12}

Heterostructures 13Q and 14Q have similar electron densities, similar zero-field mobility, and essentially different height of the barrier responsible for the electron confinement. According to Table 15.2 the conduction band offset at the heterojunction of $In_{0.53}Ga_{0.47}As$ and $In_{0.52}Al_{0.48}As$ is near 0.5 eV. The barrier height in the subcritically doped lattice-matched channel 13Q is around 0.25 eV (cf. curve 2 in Fig. 15.11). The splitted pseudomorphic channel 14Q has a deeper quantum well (the effective barrier is around 0.35 eV).

Figure 17.1 illustrates the field-dependent longitudinal noise temperature for the mentioned two heterostructures containing the 2DEG in the InGaAs layer [333]. Hot electrons behave in a similar way at low/moderate electric fields, but the curves spread apart at high fields: A strong source of noise onsets in the lattice-matched channel 13Q at electric field exceeding 2 kV/cm. Evidently, the same source tends to appear at 3.5 kV/cm (circles in Fig. 17.1) in channel 14Q.

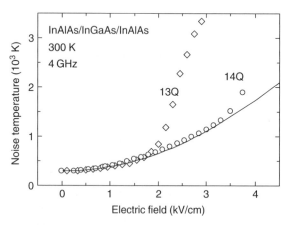

Figure 17.1. Longitudinal hot-electron noise temperature for InAlAs/InGaAs/InAlAs/InP channels [333]. Channel 13Q is lattice-matched ($In_xGa_{1-x}As$, $x = 0.53$), and structure 14Q contains a splitted pseudomorphic $In_xGa_{1-x}As$ channel ($x = 0.7$ and $x = 0.53$), for 2DEG parameters (see Table 17.1). Solid line is the electron temperature estimated according to $T_e = T_0 + e\mu_0 E^2 \bar{\tau}^{en}/k_B$.

Thus, two ranges of electric field are evident. The noise temperature T_n can be approximated by the electron temperature T_e at low/moderate electric fields until another source of noise onsets at high electric fields. The onset field correlates with the effective height of the confining barrier. This suggests that the noise source in question is associated with the reversible real-space transfer of hot electrons [333].

The dependence of the hot-electron noise on frequency is given in Fig. 17.2 for the same InAlAs/InGaAs/InAlAs structures biased with a high electric field [333]. The observed dependence on frequency in the frequency range from 200 MHz to 10 GHz can be interpreted in terms of the real-space transfer, assuming a rather long relaxation time $\tau^{\text{inter}} > 100$ ps. It is noteworthy that the typical relaxation time for the reversible real-space transfer in AlGaAs/GaAs channels is around 5 ps [330] (see Fig. 16.6).

Length Dependence of Real-Space-Transfer Noise

The obtained relatively long relaxation time estimated for the real-space transfer noise suggests a possibility for suppression of the associated source of noise in short channels (cf. Chapter 11 and Fig. 16.11). Figure 17.3 compares the experimental results on hot-electron noise in InAlAs/InGaAs/InAlAs/InP channels (13Q) of different length [333]. As mentioned, the hot-electron real-space transfer noise manifests itself in long channels at fields exceeding 2 kV/cm. This source is partially suppressed as the channel length is reduced (Fig. 17.3, filled diamonds).

Let us interpret this strong dependence on the channel length in terms of the electron transit time t^{tr}. The real-space transfer fluctuations cannot take place if the transit time is much shorter than the relaxation time $\bar{\tau}^{\text{inter}}$ of the associated

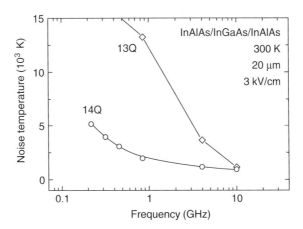

Figure 17.2. Frequency-dependent longitudinal noise temperature for InAlAs/InGaAs/InAlAs/InP channels at 3-kV/cm electric field at 300 K [333]. ◊ represents 13Q, ○ represents 14Q. (For 2DEG parameters see Table 17.1).

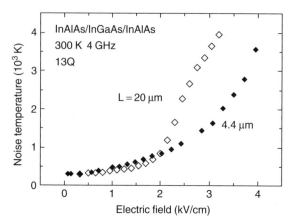

Figure 17.3. Suppression of the hot-electron real-space transfer noise in InAlAs/InGaAs/InAlAs/InP channel of reduced length [333]. ◊ represents $L = 20$ μm, ♦ represents $L = 4.4$ μm. (For 2DEG parameters see 13Q in Table 17.1).

kinetic process, $\bar{\tau}^{\text{inter}} > t^{\text{tr}}$. By assuming $v^{\text{dr}} = 10^7$ cm/s for the drift velocity in the 4.4-μm channel, one obtains $t^{\text{tr}} \sim 50$ ps. Thus, $\bar{\tau}^{\text{inter}} > 50$ ps, and the obtained estimate for the relaxation time does not contradict with the result $\bar{\tau}^{\text{inter}} \sim 100$ ps obtained from the frequency-dependent noise spectra (Fig. 17.2).

17.2 ENERGY RELAXATION TIME IN InGaAs QUANTUM WELL CHANNELS

Energy Relaxation and Hot-Electron Noise

Now let us consider hot-electron noise in the subcritically doped structures in the electric field range where the real-space transfer noise is not important (cf. Fig. 17.1). In this range of electric fields, two sources of noise remain active [see Eq. (9.5)]: (a) the velocity fluctuations determined by the electron temperature T_e (hot-electron "thermal" noise) and (b) the fluctuations in the electron temperature, δT_e, (convective noise). The convective noise is considered repeatedly in Sections 7.2, 9.3, 9.7, 11.1 and 12.1. This source is not observed unless the electron mobility depends on electric field. After Eq. (12.38), the correction term O_c due to the electron temperature fluctuations enters the relation between the longitudinal noise temperature and the electron temperature:

$$T_{n\|} = T_e(1 + O_c). \qquad (17.1)$$

The dimensionless term O_c accounts for the nonohmic behavior. The estimated values for O_c for channel 14Q (bars in Fig. 17.4) are below 20% at $E < 1.5$ kV/cm at room temperature [334]. In the range of electric fields where the correction term O_c is small, the excess longitudinal noise power (per unit

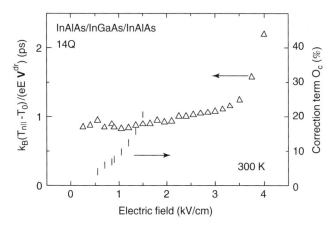

Figure 17.4. The emitted excess noise power (per unit bandwidth) over the consumed power (per electron) for the splitted pseudomorphic InGaAs 2DEG channel in the InAlAs/InGaAs/InAlAs/InP heterostructure ($\triangle$), and the estimated correction term O_c due to the electron temperature fluctuations (bars) [334]. (For 2DEG parameters see 14Q in Table 17.1).

bandwidth) divided by the consumed power (per electron) provides us with an estimate of the effective energy relaxation time $\bar{\tau}^{en}$ [cf. Eqs. (9.2) and (9.3)]. For a two-dimensional electron gas, one has

$$\bar{\tau}^{en}(\mathbf{E}) = \frac{k_B(T_{n\parallel} - T_0)}{e\mathbf{E} \cdot \mathbf{v}^{dr}}. \qquad (17.2)$$

Triangles in Fig. 17.4 show that the rhs of Eq. (17.2) is almost independent of the electric field in the range of low and moderate electric fields, where the contributions from the real-space transfer noise and from the fluctuations in electron temperature are not of primary importance. In this range of fields for the structure in question, one obtains $\bar{\tau}^{en} \approx 0.9$ ps.

It is noteworthy that the energy relaxation time is almost independent of electric field. This is in contrast to the results of Fig. 9.2 demonstrating a strong dependence of the energy relaxation time on the bias: The transition from the acoustic-phonon-controlled to the optical-phonon-controlled energy loss mechanisms takes place when electrons (holes) are heated up. In the considered case at room temperature (see Fig. 17.4), optical phonons dominate, and the experimental results give an experimental evidence that the energy loss on optical phonons is almost independent of the electron energy. The data at 80 K (Fig. 17.5) [325] demonstrate a different behavior and support the statement that the role of the optical phonons diminishes at low temperatures: The energy relaxation time becomes longer and tends to increase at lower electric fields, this being in agreement with the discussed transition from the optical-phonon-emission-dominated energy loss to the acoustic-phonon-controlled energy balance at zero electric field (see Section 9.1).

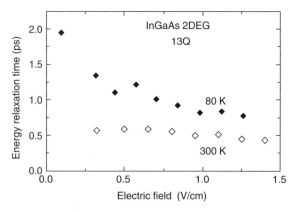

Figure 17.5. Energy relaxation time for lattice-matched InAlAs/InGaAs/InAlAs channel at 80 K and 300 K [325]. A transition from the optical-phonon-controlled to the acoustic-phonon-controlled energy relaxation is evidenced. (For 2DEG properties see 13Q in Table 17.1).

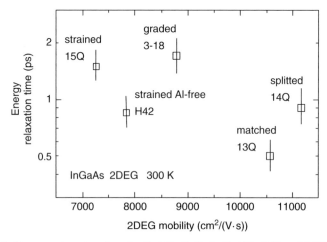

Figure 17.6. Energy relaxation time for InAlAs/InGaAs/InAlAs/InP and InP/InGaAs/InP heterostructure 2DEG channels demonstrating different mobility at room temperature [334]. (For heterostructure properties see Table 17.1).

Energy Relaxation in Subcritically Doped Heterostructures

The discussed technique to estimate the energy relaxation time from the hot-electron noise data provides us with a unique possibility to study interaction of the confined electrons with the confined phonons in 2DEG channels. Figure 17.6 presents the energy relaxation time determined for the 2DEG confined in the InGaAs channels at room temperature. No evident correlation of the time constant $\bar{\tau}^{en}$ and the electron mobility μ_0 is found [334]. This is a good experimental illustration that these two kinetic parameters are responsible for different aspects of the hot-electron behavior: (i) The energy relaxation is determined by the phonon scattering, and (ii) the mobility is determined by all scattering mechanisms (including elastic scattering on the residual and the remote impurities).

A reasonable correlation of the energy relaxation time constant and the conduction band offset is obtained only for the Al-containing channels — the result for the Al-free heterostructure H42 stays apart (Fig. 17.7). This is the experimental evidence for the electron confinement not being of the primary importance for the electron energy relaxation. Evidently, the phonon confinement must be taken into consideration for a better understanding of the experimental data on the hot-electron noise in the 2DEG structures.

Energy Relaxation and Phonon Confinement

Figure 17.8 relates the energy relaxation time and the InGaAs composition at the heterojunction. Now, the Al-containing structures and the Al-free structure tend to obey the same dependence. It gives an experimental evidence that the electron–phonon interaction (responsible for the energy relaxation and the hot-electron microwave noise at low/moderate electric fields) is sensitive to the lattice mismatch at the heterojunction where the 2DEG is located.

The following interpretation is possible. At room temperature, the main contribution to the energy loss comes for the spontaneous emission of the optical phonons by the hot electrons present in the high-energy tail of their distribution. The energy of the confined optical phonons, those confined inside the InGaAs layer, and those localized at the heterointerface tends to increase under the strain, and the "weight" of the high-energy electron tail decreases in the strained layer. This causes the reduced energy loss, which supports the enhanced hot-electron

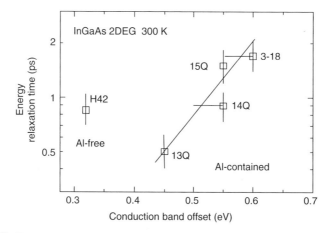

Figure 17.7. Energy relaxation time for InAlAs/InGaAs/InAlAs/InP and InP/InGaAs/InP heterostructure 2DEG channels versus the conduction band offset at the heterojunction [334]. (For 2DEG properties see Table 17.1).

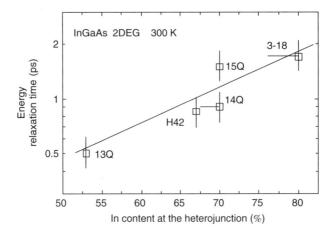

Figure 17.8. Energy relaxation time for InAlAs/InGaAs/InAlAs/InP and InP/InGaAs/InP heterostructure 2DEG channels versus the mole ratio of InAs in InGaAs at the heterojunction [334]. (For 2DEG properties see Table 17.1).

noise and the longer energy relaxation time reported for the strained InAs-rich InGaAs layers containing 2DEG [334].

The obtained dependence of the energy loss on the lattice mismatch suggests that we should reconsider the lattice-controlled electron mobility in the 2DEG channels at room temperature, where the optical phonon contribution prevails over that of the acoustic phonons. The lattice-controlled mobility is foreseen to exceed the value deduced without taking into account the observed weaker interaction of the two-dimensional electrons with the confined optical phonons in the pseudomorphic InAs-rich InGaAs quantum well channels.

17.3 REAL-SPACE TRANSFER NOISE AT LOW ELECTRIC FIELDS

Transverse Tunneling at Equilibrium

Figure 17.9 illustrates that the longitudinal noise temperature for a supercritically doped InAlAs/InGaAs/InAlAs/InP heterostructure (triangles) is essentially higher than that in a subcritically doped one (solid line) [336]. This doping-dependent behavior can be understood, taking into account two-channel conduction in the supercritically doped structure. An undoped quantum well can confine a limited density of 2DEG (see Fig. 15.10). At a high 2DEG density the two-channel conduction takes place at thermal equilibrium: The low-mobility channel is present in the δ-doped barrier layer (InAlAs in our case), and the high-mobility channel is present in the undoped narrow-band-gap semiconductor (InGaAs layer) as illustrated by curves 4 and 5 in Fig. 15.11. Supposing that the channels are separated by a low and thin barrier (see Fig. 15.11), the fluctuations of the electron population in the high-mobility/low-mobility channels can take place without electron heating.

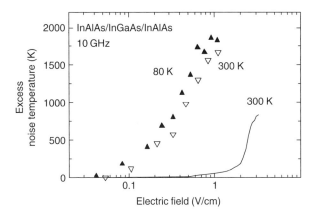

Figure 17.9. The excess noise temperature $(T_{n\parallel} - T_0)$ for the lattice-matched supercritically doped InAlAs/InGaAs/InAlAs/InP structures [335]. $\triangle$ represents $T_0 = 300$ K, $n_0 = 8.6 \times 10^{12}$ cm^{-2}, $\mu_0 = 4200$ cm^2/(V·s); $\blacktriangle$ represents $T_0 = 80$ K, $n_0 = 7.6 \times 10^{12}$ cm^{-2}, $\mu_0 = 15200$ cm^2/(V·s). Solid line stands for the subcritically doped 13Q channel at 300 K.

In agreement with Eq. (16.1) the longitudinal noise (in the direction of the applied electric field) is caused by the velocity fluctuations inside each channel and by the partition-type fluctuations due to the real-space transfer. The experimental results show no thermal activation of the excess noise dominating at low/intermediate electric fields in the supercritically doped channel (cf. unfilled and filled triangles in Fig. 17.9). The observed weak thermal quenching of the real-space transfer noise allows one to state that the electron tunneling rather than the jumps over the barrier dominate the excess noise.

Let us consider the spectral intensity of the velocity fluctuations due to the transverse tunneling at low electric fields. Because of essential frequency dependence of the real-space transfer in InP-based quantum-well channels at microwave frequencies, Eq. (16.2) is modified into

$$S_V^{\text{inter}} = 4 \frac{n_1 n_2}{n_0^2} (\mu_1 - \mu_2)^2 E^2 \bar{\tau}^{\text{inter}} \frac{1}{1 + (\omega \bar{\tau}^{\text{inter}})^2}. \qquad (17.3)$$

The squared dependence on electric field of the spectral intensity holds at low electric fields where the hot-electron effects are not important and the deviations from the Ohm's law are negligible. The observed E^2-dependence at the low fields supports the idea that no electric field is needed for the electron tunneling. (On the other hand, the deviations from Ohm's law develop at high electric fields, and the deviations from the E^2-dependence appear as the electron mobility in the channels changes considerably due to electron heating [336].)

Transverse Noise in the 2DEG Plane

Let us consider the field-induced in-plane anisotropy of microwave noise—the anisotropy in the plane of electron confinement induced by the electric field applied in the 2DEG plane. It is obvious that the directions perpendicular and parallel to the current are not equivalent: Many sources of hot-electron microwave fluctuations (convective, intervalley, and others) also need a DC current to manifest themselves (see Chapter 9). These sources are superimposed over the hot-electron velocity fluctuations due to electron kinetic energy referred to as the hot-electron "thermal" noise acting in all directions. The in-plane anisotropy experiments in a 2DEG are difficult to perform in the conventional configuration [151, 156], where the side walls of a bulk sample serve for convenient capacitative coupling of the noise source to the noise measuring device.

The first experiments on the transverse hot-electron noise temperature [337] were performed on test samples for the Hall effect; this solution of the problem of coupling of the 2DEG to the waveguide in the direction transverse to the current was successful. The anisotropy of noise was observed in supercritically doped InAlAs/InGaAs/InAlAs/InP heterostructures [338]. The experimental results are analyzed in terms of the two-channel model; the analysis allows one to estimate the real-space transfer relaxation time and the energy relaxation time.

Figure 17.10 compares the hot-electron noise temperature measured in the transverse and longitudinal directions for a supercritically doped InAlAs/InGaAs/InAlAs/InP heterostructure containing the low-mobility and the high-mobility 2DEG channels at equilibrium (cf. curves 4 and 5 in Fig. 15.11). The transverse noise is caused by the properly weighted hot-electron 'thermal' noise agitated in both channels. Equation (17.2) is applied with the following result: The effective energy relaxation time $\bar{\tau}^{en} \approx 1.9$ ps is estimated at room temperature in the field range up to 2 kV/cm [338]. The experimental results on the longitudinal and the transverse noise are used for an estimate of the interwell relaxation time. At low electric fields, say $E < 250$ V/cm, the deviations from Ohm's law are

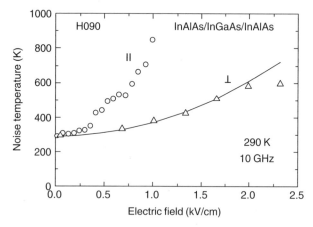

Figure 17.10. Hot-electron noise temperature, longitudinal $T_{n\parallel}$ (O), and transverse $T_{n\perp}$ (△) for lattice-matched supercritically doped InAlAs/InGaAs/InAlAs/InP structures $n_0 = 3 \times 10^{13}$ cm^{-2}, $\mu = 4500$ cm^2/V·s, containing low-mobility and high-mobility channels, $T_0 = 290$ K [337, 338].

insignificant, and the hot-electron 'thermal' noise can be assumed as isotropic. Now, the difference of the longitudinal and the transverse noise temperatures provide with the spectral intensity of velocity fluctuations due to the interwell transfer.

In conclusion, different sources of hot-electron noise were excited and resolved in InAlAs/InGaAs/InAlAs/InP and InP/InGaAs/InP channels. The microscopic origin of the hot-electron fluctuations is analyzed and interpreted in terms of the associated ultrafast intrawell and interwell kinetic processes specific to a two-dimensional electron gas. The values of the relaxation time for the processes are estimated.

CHAPTER 18

CUTOFF FREQUENCIES OF FAST AND ULTRAFAST PROCESSES

In the foregoing chapters, the experimental results on hot-electron noise in semiconductors and semiconductor structures were considered in terms of nonequilibrium processes of dissipation taking place inside the conduction band. The spectral pattern was often decomposed into sources of noise, with each source being approximated by an effective single Lorentzian of equivalent integral intensity. Within this simplified treatment, a reasonable agreement with the solution of the kinetic equation for fluctuations and with the detailed Monte Carlo simulation of fluctuations was obtained for different semiconductors and semiconductor structures.

The noise source of interest was resolved among the concurrent ones using techniques of hot-electron fluctuation spectroscopy. Dependence of noise on electric field, doping, channel length, and lattice temperature was used to suppress some sources of noise in favor of the others; spectral windows were cleaned for observation of the latter. So, different sources of noise such as hot-electron "thermal" noise, convective noise, intervalley noise, interwell transfer noise, and others were investigated in homogeneous bulk semiconductors, uniformly doped channels, and channels containing two-dimensional electron gas. This chapter considers cutoff frequencies of fast and ultrafast kinetic processes estimated during experimental investigation of hot-electron fluctuations.

18.1 HOT-ELECTRON ENERGY RELAXATION

Hot-electron energy relaxation is one of the most universal kinetic processes responsible for two essentially different sources of hot-electron noise: the convective noise and the hot-electron "thermal" noise. These two sources are basic in case of one-valley conduction. In the relaxation time approximation, the effective energy relaxation time determines the intensity and/or the cutoff frequency of these two sources of noise.

After Eq. (9.5), the effective single Lorentzian for a convective noise is as follows:

$$S^{en}_{\mathcal{V}\|}(\omega, \mathbf{E}) = \frac{S^{en}_{\mathcal{V}\|}(0, \mathbf{E})}{1 + (\omega \bar{\tau}^{en})^2}, \quad (18.1)$$

where $\bar{\tau}^{en}$ determines the cutoff frequency of the equivalent source:

$$f^{cutoff} = 1/(2\pi \bar{\tau}^{en}). \quad (18.2)$$

Equation (18.1) contains the effective energy relaxation time, $\bar{\tau}^{en}$, rather than the energy-dependent relaxation time τ^{en} (cf. Section 7.2); as discussed in Sections 9.1 and 9.2, the factor near unity appears, provided that the effective relaxation time is introduced. Within this uncertainty, the effective energy relaxation time gives an estimate of the cutoff frequency of the equivalent source for the hot-electron convective noise.

Hot-electron energy relaxation time enters hot-electron "thermal" noise as well. The spectral intensity of velocity fluctuations in this case is approximately given by [see Eqs. (9.1), (9.2), and (9.3), and (9.5)]

$$S^{\text{"therm"}}_{\mathcal{V}}(\omega, \mathbf{E}) = \frac{4}{m}(k_B T_0 + be \mathbf{E} \cdot \mathbf{v}^{dr} \bar{\tau}^{en}) \frac{\bar{\tau}}{1 + (\omega \bar{\tau})^2}, \quad (18.3)$$

where $b = 2/3$ stands for a three-dimensional gas and $b = 1$ stands for a two-dimensional electron gas. The effective energy relaxation time $\bar{\tau}^{en}$ together with the consumed power $e\mathbf{E} \cdot \mathbf{v}^{dr}$ determine the spectral intensity of the excess hot-electron noise [in excess over the Nyquist thermal noise $4k_B T_0 \mu_0 / e$ in Eq. (18.2)]. The effective momentum relaxation time $\bar{\tau}$ determines the cutoff frequency of the equivalent source. In an isotropic semiconductor, the hot-electron "thermal" noise is the only source of hot-electron noise manifesting itself in the transverse direction to the applied electric field (see Section 9.2). Of course, the hot-electron "thermal" noise acts in the direction of current as well [see Eq. (9.5)].

Under efficient interelectron collisions the convective noise is often referred to as noise due to hot-electron temperature fluctuations. In this case the noise temperature is given by Eqs. (12.32) and (12.33). The electron temperature relaxation time τ_T defined by Eq. (12.10) stands for the energy relaxation time $\bar{\tau}^{en}$ in Eqs. (18.1) and (18.3).

The investigation of hot-electron noise provides us with the effective electron energy relaxation time (see Chapters 9, 10, 11, and 17), among other characteristics. Table 18.1 summarizes the data for long samples on $\bar{\tau}^{en}$ estimated using noise techniques. The values ranging from tens of picoseconds to hundreds of femtoseconds were obtained for different semiconductors and semiconductor structures. The energy relaxation controls the cutoff frequency of the associated

TABLE 18.1. Effective Energy Relaxation Time in 2DEG Channels Estimated Using Noise Techniques [339][a]

Channel	Electron Density (cm^{-2}, cm^{-3})	Lattice Temperature (K)	Relaxation Time (ps)	Reference	Symbol Number in Fig. 18.1
InAlAs/In$_x$Ga$_{1-x}$As					
$x = 0.53$	2.4×10^{12}	300	0.5	Fig. 17.5	1
$x = 0.53$	2.4×10^{12}	80	1.9–0.8	Fig. 17.5	2
$x = 0.7/0.53$	2.3×10^{12}	300	0.9	Fig. 17.4	3
$x = 0.7$	4.1×10^{12}	300	1.5	340	4
$x = 0.8$–0.53	1.5×10^{12}	300	1.7	Fig. 17.6	5
$x = 0.53$	3.3×10^{13}	300	1.9	Fig. 17.10	6
InP/In$_{0.67}$Ga$_{0.33}$As	1.7×10^{12}	300	0.85	Fig. 17.6	7
AlGaAs/GaAs	5.7×10^{11}	80	40	330	8
AlGaAs/GaAs	3.8×10^{11}	17	50	341	9
n-type GaAs	3×10^{17}	300	0.25	Fig. 10.10	10
n-type InP	1×10^{17}	300	0.3	Fig. 10.9	11
p-type Ge	1.4×10^{14}	80	33–2.3	Fig. 9.2	12
p-type Ge	1.4×10^{14}	80	17	Fig. 9.3	13
p-type Si	5×10^{14}	80	18–0.6	Fig. 9.2	14
n-type GaAs	3×10^{15}	300	0.7	Fig. 10.2	15

[a] Data for lightly doped Ge, Si, GaAs, and InP are included for comparison.

source of noise [Eq. (18.2)] in the microwave and millimeter-wave range (from several gigahertz to a terahertz).

In general, the energy relaxation time changes under electron heating. This introduces the dependence of the cutoff frequency on the electric field as illustrated by Fig. 18.1, where a summary of the cutoff frequency of the equivalent noise source is shown as a function of electric field for 2DEG channels (unfilled symbols) and for GaAs, InP, Ge, and Si (centered symbols). The effect of lattice temperature and doping is evidenced as well. The dependence on the electric field varies between a nearly linear increase of the cutoff frequency with the field and a rather weak (if any) dependence. Figure 18.1 also gives the estimated range of electric field where the energy relaxation controls the hot-electron noise in different semiconductors and structures, with the range being dependent on semiconductor properties.

A weak (if any) dependence of the cutoff frequency on electric field is obtained for 2DEG channels at room temperature (Fig. 18.1, unfilled symbols). Indeed, the electron density per unit volume is very high in a quantum-well channel. This condition acting together with the high lattice temperature favors the field-independent cutoff frequency in the range of electric fields where the real-space transfer is not important.

A low density of ionized impurities inside the 2DEG channel has an indirect effect on the energy relaxation: The energy loss in 2DEG channels is not as heavy

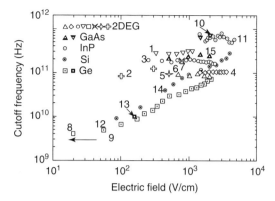

Figure 18.1. Electric-field-dependent cutoff frequency for the equivalent source of noise caused by hot-electron energy relaxation for long 2DEG channels (unfilled symbols) and 3DEG channels (centered symbols). Numbers by the symbols specify the channel (for details see Table 18.1).

as in the standard-doped GaAs and InP as indicated by the energy relaxation time data in Table 18.1 (cf. numbers 1–6 and 10, 11). Correspondingly, the associated cutoff frequency is lower (unfilled symbols in Fig. 18.1) than that given by centered down-triangles 10 and centered circles 11. In line with this statement, the data for the lightly doped GaAs (centered up-triangles 15) are comparable with those for the 2DEG channels. However, as discussed in Chapter 17, the energy relaxation time depends on phonon confinement in double-heterojunction channels containing two-dimensional electron gas, and this has an effect on the cutoff frequency: The cutoff frequency is lower in pseudomorphic channels (Fig. 18.1, diamonds 4). The effect of lattice temperature and doping on the cutoff frequency is evidenced through a comparison of the data for lattice-matched InGaAs channels (down-triangles 1 versus unfilled crosses 2, and down-triangles 1 versus up-triangles 6 in Fig. 18.1).

Unlike this, a strong dependence on electric field is observed at low temperatures in lightly doped p-type germanium and silicon (Fig. 18.1: squares 12, 13; circles 14). A possible interpretation is as follows. At low electric fields the energy relaxation is controlled by acoustic phonons. This leads to a slow energy relaxation and a relatively low cutoff frequency. Under heating by the applied electric field, the holes acquire energy needed for emission of optical phonons, and a greater number of holes is involved in the energy-loss mechanism — a more efficient energy loss on optical phonons sets in. As a result, the cutoff frequency increases with field. Centered squares (symbols 12 and 13 in Fig. 18.1) compare the results obtained for p-type Ge by two independent techniques. The technique based on measurement of the transverse noise temperature (see Fig. 9.2) leads to essentially the same result as that based on the cutoff of the longitudinal convective noise (see Fig. 9.3).

The energy-loss mechanism is essentially influenced by interelectron collisions. At high electron densities, as discussed in Chapter 13, the interelectron collisions control the energy loss enhancing emission of optical phonons in GaAs subjected to electric fields ranging from ~10 V/cm to ~300 V/cm at 80 K lattice temperature (see Fig. 13.4) — the parabolic dependence of the noise temperature on electric field holds (Fig. 13.3). This behavior is in agreement with the assumption that the energy relaxation time is independent of the electric field in the field range in question. The estimated effective energy relaxation time, ~4–5 ps, leads to (30- to 40-GHz cutoff frequency for the electron temperature fluctuations in standard-doped GaAs channels at 80 K temperature in the field range 10–300 V/cm. This is not the only case of almost field-independent energy relaxation time.

The field-dependent interplay of energy loss on acoustic and optical phonons fades away at room temperature — many electrons have enough energy for emission of optical phonons. Moreover, the spontaneous emission rate is almost independent of the electron energy, provided that it is high enough. Thus, the dependence of the cutoff frequency on electric field is very weak (if any) (centered up-triangles 15 in Fig. 18.1) in lightly doped GaAs at room temperature in the range of electric fields below the onset of the intervalley noise. This result is supported by the comparison of the field dependencies of the experimentally determined longitudinal noise temperature $T_{n\|}$ and of the electron temperature T_e determined by the average electron energy $\bar{\varepsilon}$ obtained through Monte Carlo simulation (see Fig. 10.2).

Centered circles 11 in Fig. 18.1 stand for the cutoff frequency obtained for standard doped n-type InP channels in the following way. Assuming that the electron temperature approximation is valid, the dependence of the electron temperature $T_e(E)$ on electric field is extracted from data on the electron mobility $\mu_\|(E)$ and the longitudinal noise temperature $T_{n\|}(E)$ (Fig. 10.9) according to Eq. (12.36). Then, Eq. (9.2) is applied to estimate the energy relaxation time. The results show a weak, if any, dependence on electric field in the (1.4- to 4.6-kV/cm range (centered circles 11 in Fig. 18.1). Similar results are obtained for the standard-doped GaAs channels at room temperature (Fig. 18.1, centered down-triangles 10).

18.2 INTERWELL AND INTERVALLEY TRANSFER

Weakly interacting groups of electrons are often present in heterostructure channels containing parallel wells separated by barriers (see Chapter 15). Electron exchange between the groups cause fluctuations of occupancies manifesting themselves as a source of noise similar to that due to the intervalley fluctuations. Table 18.2 gives the values of the effective relaxation time $\bar{\tau}^{\text{inter}}$ obtained using experimental data on the corresponding noise sources in 2DEG channels, in GaAs and InP channels, and in bulk crystals.

After Eqs. (9.6) and (16.2), the spectral intensity of velocity fluctuations for the noise source caused by the intervalley and/or interwell transfer is given by

$$S_{\mathcal{V}\|}^{\text{inter}}(\omega, \mathbf{E}) = 4 \frac{n_1 n_2}{n_0^2} (v_1^{\text{dr}} - v_2^{\text{dr}})^2 \frac{\bar{\tau}^{\text{inter}}}{1 + (\omega \bar{\tau}^{\text{inter}})^2}. \tag{18.4}$$

TABLE 18.2. Relaxation Time for Some Interwell/Intervalley Processes Estimated from Experimental Data on Hot-Electron Fluctuations [339]

Channel	Electron Density (cm^{-2},cm^{-3})	Lattice Temperature (K)	Electric Field (kV/cm)	Relaxation Time (ps)	Reference	Symbol Number in Fig. 18.2
InAlAs/ In$_{0.53}$Ga$_{0.47}$As	2.4×10^{12}	300	>2	100	339	16
InAlAs/ In$_{0.53}$Ga$_{0.47}$As	3.3×10^{13}	300	<1	70	Fig. 17.10	17
GaAs/AlAs/ GaAs	1.3×10^{12}	80	1	10	Fig. 16.10	18
AlGaAs/GaAs	1.9×10^{11}	80	1	5	Fig. 16.6	19
n-type Si	3×10^{13}	80	0.2	45	Fig. 9.15	20
n-type Ge	1.5×10^{14}	80	2	17	Fig. 9.17	21
n-type GaAs	3×10^{17}	300	2.5	0.05	257	22
n-type GaAs	3×10^{17}	300	20	0.4	Fig. 11.4	23
n-type GaAs	3×10^{17}	300	200	0.045	Fig. 11.14	24

The dissipation mechanism includes the kinetic process supported by (numbers in the last column): 16, 19 — hot-electron interwell jumps over barrier; 17 — interwell transverse tunneling at equilibrium; 18 — hot-electron interwell transverse tunneling; 20, 21 — intervalley transfer between equivalent valleys, 22 — resonant impurity scattering; 23 — Γ-L intervalley transfer at very high electric fields; 24 — Γ-X intervalley transfer at extremely high electric fields.

Figure 18.2 presents the cutoff frequencies $f^{\text{cutoff}} = 1/(2\pi \bar{\tau}^{\text{inter}})$ for the equivalent sources of intervalley/interwell noise associated with different processes of dissipation.

According to the definition, the noise source due to the interwell/intervalley transfer can be introduced, provided that the groups of electrons are separated and their interaction is weak. This also means that the effective relaxation time should exceed the interwell/intravalley relaxation time (usually the energy relaxation time inside the well and/or the valley) in the range of electric fields where the interwell/intervalley relaxation is important. A comparison of the experimental results presented in Tables 18.1 and 18.2 support this idea. In particular, the effective hot-electron interwell relaxation time in long 2DEG channels (numbers 16 and 17 in Table 18.2) exceeds the effective energy relaxation time determined for 2DEG channels (numbers 1 and 6 in Table 18.1). Since the effective energy relaxation time as well as the intervalley or the interwell relaxation time often depend on electric field (see Figs. 9.2, 9.14, and 9.15), care should be taken when the data obtained at essentially different electric field strength are under comparison. In particular, data on the energy relaxation time are often available in the field range below the onset of the noise sources due to the intervalley and interwell transfer. As a result, the energy relaxation time in AlGaAs/GaAs 2DEG channels at cryogenic temperatures at 20 V/cm electric field (numbers 8 in Table 18.1) exceeds the interwell transfer time at the same lattice temperature

226 CUTOFF FREQUENCIES OF FAST AND ULTRAFAST PROCESSES

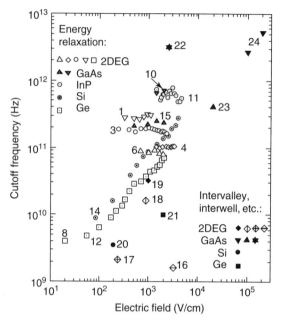

Figure 18.2. Cutoff frequency for the equivalent sources of noise caused by interwell and intervalley relaxation versus electric field for 2DEG channels (diamonds) and 3DEG channels (filled symbols) together with the data for the equivalent source of noise caused by hot-electron energy relaxation. Numbers by the symbols specify the channel (for the details see Tables 18.1 and 18.2).

estimated at an essentially higher electric field (near 1 kV/cm, see number 19 in Table 18.2).

As discussed in Chapter 17, a long relaxation time obtained for the interwell real-space transfer time in InAlAs/InGaAs/InAlAs/InP channels (lattice-matched and pseudomorphic) most probably results from trapping of the transferred electrons. A similar phenomenon seems to be less expressed in the investigated AlGaAs/GaAs channels: the relaxation time of the interwell transfer caused by the hot-electron jumps over the barrier (see number 19 in Table 18.2) is quite short as compared to the similar time in InP-based heterostructure channels (numbers 16 and 17). Correspondingly, the cutoff frequency for the real-space transfer is higher in AlGaAs/GaAs channels (cf. numbers 18, 19 and 16, 17 in Fig. 18.2).

So far in this chapter we dealt with long channels. According to Eq. (18.4), a longer relaxation time means that the associated source of noise has a higher intensity. Such a source tends to dominate over the other sources. On the other hand, the long-lasting relaxation process can be eliminated, provided that the channel is made short enough: The hot-electron velocity correlation function is cutoff at the electron transit time as discussed in Chapter 11. This opens a possibility to get rid of the relatively slow dissipation process and the associated source of noise. Thus, weaker sources of noise caused by faster kinetic processes can be resolved. The condition for the suppression of the mth relaxation processes is as follows: $\tau_m > t_{tr}$. Tables 18.1 and 18.2 provide with the data on the relaxation time for estimation of the channel length for the suppression of the source of noise in question. Of course, dependence of the relaxation time on electric field should be taken into account.

Naturally enough, at very high electric fields applied to short channels, ultrafast relaxation processes tend to dominate over the fast ones. The values on the relaxation time falling into the range of hundreds and tens of femtoseconds are given in Table 18.2 (numbers 22, 23, and 24). In particular, the relaxation time for Γ–X transfer noise is estimated to be 45 ± 15 fs at 200 kV/cm in short standard-doped GaAs channels at room temperature (number 24 in Table 18.2). In the range of extremely high electric fields the cutoff frequency of the noise source due to the Γ–X intervalley transfer exceeds 2 THz (black down-triangles 24 in Fig. 18.2). In short 0.2-μm channels, Γ–L coupling seems to be essentially weaker — the associated source of noise is overshadowed by Γ–X transfer noise at high electric fields ($E > 25$ kV/cm) (see Fig. 11.14). Because of at least ten times lower noise intensity, the relaxation time for Γ–L transfer is expected to be at least ten times longer: $\tau_{\Gamma L} \sim 0.4$ ps at $E \sim 20$ kV/cm (see Section 11.6).

Consequently, each equivalent source of noise can be associated with the corresponding kinetic process inside the conduction band. The spectrum of a single equivalent source of noise can be approximated by a Lorentzian, with its cutoff frequency being determined by the effective relaxation time. Methods based on fluctuation measurements provide with the electric-field-dependent effective relaxation time and intensity of the source. The source of interest is resolved by changing the conditions such as channel length, electric field, lattice temperature, and so on. In the case of several concurrent sources of noise, the experimental investigation supported by the adequate simulation of the fluctuations by Monte Carlo technique is quite helpful.

18.3 NOISE SOURCES FOR EQUIVALENT CIRCUITS

The discussed equivalent sources of noise (together with the reactive impedances) can be used in the physics-based equivalent circuit in order to predict the spectral properties of its noise. The noise characterization begins with the determination of the circuit elements (ideal resistances, capacitances, and inductances) during the procedure of fitting the static and the small-signal-response characteristics. Once the equivalent circuit elements are determined, the sources of noise (associated with the resistances) are introduced, and the noise figure of the circuit is determined. Ideal capacitors and inductances introduce no noise, but they influence the resultant spectral properties of the circuit noise.

Figure 18.3 presents a simplified equivalent circuit for a field-effect transistor (a MESFET, a HEMT, see Figs. 1.2 and 1.3). Semiconducting part of the transistor (intrinsic transistor) is treated as a combination of a channel and a diode. The main element of the gate circuit of a MESFET (and HEMT) is the Schottky barrier

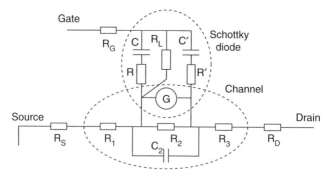

Figure 18.3. A lumped-element equivalent circuit of a field-effect transistor in a common-source configuration. Semiconducting part of the transistor (dashed ellipses) is represented by a combination of a channel and a Schottky diode. Solid circle (G) accounts for the gain. Resistors are associated with the equivalent sources of noise. Inductances due to metallic electrodes of source, gate, and drain and parasitic capacitances (for example, source–gate, gate–drain capacitances) are not shown.

represented by lumped ideal capacitances, C and G', and resistances in series, R and R'; the barrier leakage resistance R_L is connected in parallel. The latter is usually associated with the shot noise, while the resistances in series produce hot-electron noise. The channel is represented by several resistances connected in series (see R_1, R_2, R_3 located inside the horizontal ellipse in Fig. 18.3); the associated sources of noise are due to hot electrons. The source of hot-electron noise located closer to the source (R_1) differs essentially from those located under the gate and closer to the drain (R_2, R_3) because the electric field is essentially higher in the latter two parts of the channel. Moreover, the noise due to R_1 is amplified (solid circle G in Fig. 18.3 stands for the gain effect). Resistances R_G, R_S, and R_D are due to the metal electrodes—they introduce the Nyquist sources of noise. (Inductances and parasitic capacitances due to metallic source, gate, and drain electrodes are not shown in Fig. 18.3 for the sake of simplicity.) Equivalent sources of noise can be introduced in different ways.

The approach originated by Fukui [342] relates the minimum noise figure $NF_{\min}$ at a given frequency with the resistance of the gate to source circuit ($R_S + R_1 + R_G + R$ in Fig. 18.3), the transconductance, and the gate to source capacitance (C in Fig. 18.3). A quantitative agreement can be obtained after a choice of the fitting factor. This semiempiric approach is attractive because of its simple form, and it is widely used. However, the physical meaning of the fitting factor remains unclear. Evidently, the dominant source of noise is associated with the resistances R_S, R_1, R_G, and R. In the approach, the resistances are treated on equal rights despite of the fact that the Nyquist theorem holds for the noise originated in the metallic electrodes (R_S and R_G) and does not hold for the biased channel (resistance R_1). Thus, the hot-electron fluctuations in the channel are ignored explicitly (they are taken in account implicitly, through the fitting factor).

The approach developed by Pospieszalski [36] takes into account hot-electron effects in the channel explicitly. The noise parameters of the intrinsic transistor is calculated, assuming two equivalent noise temperatures, T_n^G and T_n^D, associated with the intrinsic source to gate resistance $R_1 + R$ and the drain conductance $1/(R_2 + R_3)$, respectively in addition to other elements of the equivalent circuit treated as noiseless. Within the accuracy of the experiments on MESFET and HEMT transistors, the equivalent temperatures assume the values $T_n^G \sim T_0$ and $T_n^D \sim 10^4$ K. The latter value is in a satisfactory agreement with the data of Fig. 11.14, where the saturated value of the noise temperature $\sim 10^4$ K is measured in the range of electric fields from 50 kV/cm to 200 kV/cm for 0.2-μm GaAs channels. Figure 11.14 also supports the assumption $T_n^G \sim T_0$ in the channel near the source, provided that the field in this part of the channel does not exceed 10 kV/cm. Indeed, these values of field are typical for high-speed transistors with the channel length falling into the sub-micrometer range. Analysis of numerous experimental data on high-speed transistors shows [35] that the Pospieszalski [36] model is the best among the simple ones.

In the physics-based equivalent circuit approach, the noise characteristics of the lumped sources can be taken from the experimental investigation of hot-electron fluctuations performed on two-terminal test samples of similar geometry (gateless channels) as well as from theory and Monte Carlo simulation as discussed in the foregoing chapters. In order to apply the available experimental results [Figs. 18.1 and 18.2, and Eqs. (18.1), (18.3), and (18.4)], the estimate of the average field in the lumped element is necessary; fitting the static characteristics of the model and the device provides us with the required estimate of the DC voltages on the lumped elements.

Of course, short-channel effects come into play (see Chapter 11). The long sample data on the relaxation time of the kinetic processes (Tables 18.1 and 18.2) are useful to consider the short-channel effects: The data on the relaxation time help to estimate at what channel length the source of noise in question is suppressed due to the short-channel effects. On the other hand, the experimental results on equivalent noise temperature are directly available for short channels from experiments on test samples.

Nevertheless, the lumped-equivalent-circuit approach has serious drawbacks. For example, the shot noise in the barrier depends on the actual space charge distribution in the barrier and its vicinity, but the required information is beyond the lumped-element equivalent circuit approach. In a real transistor, there is a strong correlation of the sources of hot-electron noise in the channel represented by several lumped resistors connected in series. The correlation of the fluctuations in the barrier and in the channel is essential as well. A physics-based solution of these problems calls for consideration of spatially inhomogeneous fluctuations. These problems are treated self-consistently inside the microscopic approach where the kinetic equation for fluctuations is coupled to the Poisson equation for the electric potential.

CHAPTER 19

SPATIALLY INHOMOGENEOUS FLUCTUATIONS

So far we have dealt only with the spatially homogeneous fluctuations, or fluctuations integrated over the volume of the channel. This type of fluctuations reveals itself as a channel noise measured in the circuit (see Chapter 8). Of course, this is not the only type of fluctuations occurring in the system. Spatially inhomogeneous fluctuations of different scale also take place even in the spatially homogeneous system. The sufficiently large-scale (long-wavelength) spatially inhomogeneous fluctuations may be directly detected and measured by a probe method (see, e.g., reference 343). Such an important phenomenon as scattering of electromagnetic waves from plasma (including solid-state plasma) is caused by fluctuations of charge density (in the simplest case); experiments on light scattering in solids may give a deep insight into spatially inhomogeneous fluctuation phenomena (see references 53–62).

On the microscopic level, the spatial homogeneity of fluctuations meant that we considered fluctuations of distribution function only in the quasi-momentum space (see Chapter 3). Now we shall take into consideration also a possible coordinate dependence of the "actual" ("momentary") distribution of particles. The kinetic theory of fluctuations in $(\mathbf{p}, \mathbf{r})$-space for nonequilibrium systems was developed by efforts of many authors (see references 37, 38, 42, 70, 71, 83, 85–88, 91, 92, 95, 96, 133, 134, 344–350).

19.1 SPATIALLY INHOMOGENEOUS FLUCTUATIONS OF CARRIER DISTRIBUTION

The theory of spatially inhomogeneous fluctuations on the microscopic level in fact is quite similar [70] to that of space-independent fluctuations as presented in Chapters 3–5. For simplicity, we limit ourselves to the case when the stationary particle distribution function is coordinate-independent, and we introduce coordinate-dependent fluctuations of the distribution function,

$$\delta F_{\mathbf{p}}(\mathbf{r}, t) = F_{\mathbf{p}}(\mathbf{r}, t) - \overline{F}_{\mathbf{p}}, \tag{19.1}$$

where $F_{\mathbf{p}}(\mathbf{r}, t)$ is the "actual" distribution of particles in the momentum-coordinate space ("μ-space") at a moment t. We introduce the correlation function

$$\overline{\delta F_{\mathbf{p}}(\mathbf{r}, t) \delta F_{\mathbf{p}_1}(\mathbf{r}_1, t_1)}, \tag{19.2}$$

the time-displaced two-point correlation function, which for fluctuations from a homogeneous steady state depends only on the differences $\mathbf{r} - \mathbf{r}_1$ and $t - t_1$. It is convenient to introduce the Fourier transforms in these differences

$$(\delta F_{\mathbf{p}} \delta F_{\mathbf{p}_1})_{\mathbf{q}\omega} \equiv \frac{1}{\mathcal{V}_0} \int \overline{\delta F_{\mathbf{p}}(\mathbf{r}, t) \delta F_{\mathbf{p}_1}(\mathbf{r}_1, t_1)} \exp[i\omega(t - t_1) - i\mathbf{q}(\mathbf{r} - \mathbf{r}_1)] \\ \times d(t - t_1) d(\mathbf{r} - \mathbf{r}_1) \tag{19.3}$$

and

$$(\delta F_{\mathbf{p}} \delta F_{\mathbf{p}_1})_{\mathbf{q}} \equiv \frac{1}{\mathcal{V}_0} \int \overline{\delta F_{\mathbf{p}}(\mathbf{r}) \delta F_{\mathbf{p}_1}(\mathbf{r}_1)} \exp[-i\mathbf{q}(\mathbf{r} - \mathbf{r}_1)] \, d(\mathbf{r} - \mathbf{r}_1), \tag{19.4}$$

where

$$\overline{\delta F_{\mathbf{p}}(\mathbf{r}) \delta F_{\mathbf{p}_1}(\mathbf{r}_1)} \equiv \overline{\delta F_{\mathbf{p}}(\mathbf{r}, t) \delta F_{\mathbf{p}_1}(\mathbf{r}_1, t_1)}|_{t=t_1} \tag{19.5}$$

is the equal-time two-point correlation function. The obvious relation

$$(\delta F_{\mathbf{p}} \delta F_{\mathbf{p}_1})_{\mathbf{q}} = \frac{1}{2\pi} \int_{-\infty}^{+\infty} (\delta F_{\mathbf{p}} \delta F_{\mathbf{p}_1})_{\mathbf{q}\omega} \, d\omega \tag{19.6}$$

connects the Fourier transforms (19.3) and (19.4).

Basic Formulae

This section is aimed at generalizing expression (4.10) for the spectral intensity of distribution function fluctuations in order to cover the spatially inhomogeneous fluctuations. Expression (4.10) contains (i) a fluctuation "source"

$$(I_{\mathbf{p}} + I_{\mathbf{p}_1}) \overline{F}_{\mathbf{p}} \delta_{\mathbf{p}\mathbf{p}_1} - I_{\mathbf{p}\mathbf{p}_1}^{ee} \{\overline{F}, \overline{F}\}, \tag{19.7}$$

and (ii) a pair of inverse response operators that describe an evolution of fluctuations already born. For spatially inhomogeneous fluctuations with wavelengths much larger than the electron's de Broglie wavelength ($q \ll \varepsilon_{\mathbf{p}}/\hbar v_{\mathbf{p}}$), the fluctuation source is obviously the *same* as for spatially homogeneous fluctuations, with the source as well as the collision operator being formed on the time and space scale of the order of $\hbar/\varepsilon_\mathbf{p}$ and $\hbar v_\mathbf{p}/\varepsilon_\mathbf{p}$ (the scale on which the collision probabilities are formed). On the other hand, the further evolution of created (due to stochastic character of collisions) fluctuation in the distribution of particles is governed by the linearized Boltzmann equation. This means that, in comparison with time and space scale of collision itself, the evolution of the fluctuation is a comparatively slow and spatially smooth process being described by the response operator which generates a comparatively large time and space scale conditioned by the mean free time τ and mean free path $l = v_\mathbf{p}\tau$. Thus a generalization of Eq. (4.10) is straightforward: The spatially homogeneous response operators should be replaced by the operators of response to spatially inhomogeneous perturbation, while expression (19.7) for the fluctuation source should not be altered.

To find the response operator in question, let us write the Boltzmann equation for a response, $\Delta \overline{F}_\mathbf{p}(\mathbf{r}, t)$, of the spatially uniform stable nonequilibrium system to a weak perturbation varying in space and time, $x_\mathbf{p}(\mathbf{r}, t)$. For the Fourier transform $\Delta \overline{F}_\mathbf{p}(\mathbf{q}, \omega)$ we have

$$\mathcal{B}_\mathbf{p}(\mathbf{q}, \omega)\Delta \overline{F}_\mathbf{p}(\mathbf{q}, \omega) \equiv (-i\omega + i\mathbf{q}\mathbf{v}_\mathbf{p} + I_\mathbf{p})\Delta \overline{F}_\mathbf{p}(\mathbf{q}, \omega)$$
$$- i\mathbf{q}U(\mathbf{q})\frac{\partial \overline{F}_\mathbf{p}}{\partial \mathbf{p}}\sum_{\mathbf{p}'}\Delta \overline{F}_{\mathbf{p}'}(\mathbf{q}, \omega)$$
$$= x_\mathbf{p}(\mathbf{q}, \omega), \tag{19.8}$$

where $x_\mathbf{p}(\mathbf{q}, \omega)$ is the Fourier transform of an external perturbation [cf. Eq. (2.28)]. Two new terms arise in the operator of the spatially inhomogeneous response, denoted as $\mathcal{B}_\mathbf{p}(\mathbf{q}, \omega)$, in comparison to the operator of the spatially homogeneous response, $(-i\omega + I_\mathbf{p})$. The first new term is the trivial convective one, $i\mathbf{q}\mathbf{v}_\mathbf{p}$ (or, in coordinate representation, $\mathbf{v}_\mathbf{p}\partial/\partial \mathbf{p}$). The second new term is characteristic for the gas of charged particles. It describes the action of the self-consistent electric field $\Delta \mathbf{E}(\mathbf{r}, t)$ arising from the space charge caused by the spatial redistribution of electrons. The self-consistent electric field satisfies the Poisson equation

$$\varepsilon_{0\alpha\beta}\frac{\partial}{\partial r_\alpha}\Delta E_\beta(\mathbf{r}, t) = \frac{4\pi e}{\mathcal{V}_0}\sum_\mathbf{p}\Delta \overline{F}_\mathbf{p}(\mathbf{r}, t), \tag{19.9}$$

so that

$$U(\mathbf{q}) = \frac{4\pi e^2}{\varepsilon_0^\mathbf{q} q^2 \mathcal{V}_0}, \tag{19.10}$$

where $\varepsilon_{0\alpha\beta}$ is the lattice static dielectric constant tensor, and the superscript $\mathbf{q}$ denotes the contraction of the corresponding tensor:

$$\varepsilon_0^\mathbf{q} \equiv \varepsilon_{0\alpha\beta}q_\alpha q_\beta/q^2. \tag{19.11}$$

It is supposed that $e^2 n_0^{1/3}/\varepsilon_0^{\mathbf{q}} \ll \bar{\varepsilon}_{\mathbf{p}}$ and $e^2 q/\varepsilon_0^{\mathbf{q}} \ll \bar{\varepsilon}_{\mathbf{p}}$, where $\bar{\varepsilon}_{\mathbf{p}}$ is the average energy of electron.

Once the response operator is known, we come to the following expression for the spectral intensity of spatially inhomogeneous fluctuations of the distribution function in the nonequilibrium gas of charged particles:

$$(\delta F_{\mathbf{p}} \delta F_{\mathbf{p}_1})_{\mathbf{q}\omega} = \frac{1}{\mathcal{B}_{\mathbf{p}}(\mathbf{q},\omega)\mathcal{B}_{\mathbf{p}_1}^*(\mathbf{q},\omega)}[(I_{\mathbf{p}} + I_{\mathbf{p}_1})\overline{F}_{\mathbf{p}}\delta_{\mathbf{p}\mathbf{p}_1} - I_{\mathbf{p}\mathbf{p}_1}^{ee}\{\overline{F},\overline{F}\}]. \quad (19.12)$$

This formula was obtained in reference 83 as a formal solution of the equations of the kinetics of spatially inhomogeneous fluctuations, derived in reference 37.

Expression (19.12) can also be understood as a result of the averaging of the formal solutions of the Boltzmann–Langevin equation [42],

$$\mathcal{B}_{\mathbf{p}}(\mathbf{q},\omega)\delta F_{\mathbf{p}}^{\mathbf{q}\omega} = \delta J_{\mathbf{p}}, \quad (19.13)$$

or

$$\delta F_{\mathbf{p}}^{\mathbf{q}\omega} = \mathcal{B}_{\mathbf{p}}^{-1}(\mathbf{q},\omega)\delta J_{\mathbf{p}}, \quad (19.14)$$

with the spectral intensity of "extraneous" random fluxes $\delta J_{\mathbf{p}}$ given by the expression

$$(\delta J_{\mathbf{p}}\delta J_{\mathbf{p}_1})_{\mathbf{q}\omega} = (I_{\mathbf{p}} + I_{\mathbf{p}_1})\overline{F}_{\mathbf{p}}\delta_{\mathbf{p}\mathbf{p}_1} - I_{\mathbf{p}\mathbf{p}_1}^{ee}\{\overline{F},\overline{F}\} \quad (19.15)$$

[cf. Eq. (5.4)].

Fluctuations of Carrier Density

From formula (19.12), we can get the expression for the spectral intensity of electron density fluctuations

$$(\delta n^2)_{\mathbf{q}\omega} = (1/\mathcal{V}_0)^2 \sum_{\mathbf{p}\mathbf{p}_1}(\delta F_{\mathbf{p}}\delta F_{\mathbf{p}_1})_{\mathbf{q}\omega}. \quad (19.16)$$

We shall see below that this quantity describes light scattering from electrons as well as some other important physical phenomena.

Summation of expressions containing the inverse response operator $\mathcal{B}_{\mathbf{p}}^{-1}(\mathbf{q},\omega)$ are facilitated thanks to the identity

$$\sum_{\mathbf{p}} \mathcal{B}_{\mathbf{p}}^{-1}(\mathbf{q},\omega) x_{\mathbf{p}} = \frac{1}{\varepsilon(\mathbf{q},\omega)} \sum_{\mathbf{p}} \frac{1}{-i\omega + i\mathbf{q}\mathbf{v}_{\mathbf{p}} + I_{\mathbf{p}}} x_{\mathbf{p}}, \quad (19.17)$$

which can be obtained by using Eq. (19.8). Here

$$\varepsilon(\mathbf{q},\omega) = 1 - i\mathbf{q}\frac{4\pi e^2}{\varepsilon_0^{\mathbf{q}} q^2 \mathcal{V}_0} \sum_{\mathbf{p}} \frac{1}{-i\omega + i\mathbf{q}\mathbf{v}_{\mathbf{p}} + I_{\mathbf{p}}} \frac{\partial \overline{F}_{\mathbf{p}}}{\partial \mathbf{p}}. \quad (19.18)$$

The quantity $\varepsilon(\mathbf{q}, \omega)$ may be called the *longitudinal dielectric constant of the electron gas* in a crystal.

Using identity (19.17) and Eq. (19.12), we transform Eq. (19.16) into

$$(\delta n^2)_{\mathbf{q}\omega} = \frac{1}{V_0^2 |\varepsilon(\mathbf{q}, \omega)|^2} \sum_{\mathbf{p}\mathbf{p}_1} \frac{1}{(-i\omega + i\mathbf{q}\mathbf{v}_\mathbf{p} + I_\mathbf{p})(i\omega - i\mathbf{q}\mathbf{v}_{\mathbf{p}_1} + I_{\mathbf{p}_1})}$$
$$\cdot [(I_\mathbf{p} + I_{\mathbf{p}_1})\overline{F}_\mathbf{p} \delta_{\mathbf{p}\mathbf{p}_1} - I^{ee}_{\mathbf{p}\mathbf{p}_1}\{\overline{F}, \overline{F}\}]. \qquad (19.19)$$

The first term in the source can be rewritten as

$$(I_\mathbf{p} + I_{\mathbf{p}_1})\overline{F}_\mathbf{p} \delta_{\mathbf{p}\mathbf{p}_1} = [(-i\omega + i\mathbf{q}\mathbf{v}_\mathbf{p} + I_\mathbf{p}) + (i\omega - i\mathbf{q}\mathbf{v}_{\mathbf{p}_1} + I_{\mathbf{p}_1})]\overline{F}_\mathbf{p} \delta_{\mathbf{p}\mathbf{p}_1}, \qquad (19.20)$$

so that

$$(\delta n^2)_{\mathbf{q}\omega} = \frac{1}{V_0^2 |\varepsilon(\mathbf{q}, \omega)|^2} \left[2\operatorname{Re} \sum_\mathbf{p} \frac{1}{-i\omega + i\mathbf{q}\mathbf{v}_\mathbf{p} + I_\mathbf{p}} \overline{F}_\mathbf{p} \right.$$
$$\left. - \sum_{\mathbf{p}\mathbf{p}_1} \frac{1}{(-i\omega + i\mathbf{q}\mathbf{v}_\mathbf{p} + I_\mathbf{p})(i\omega - i\mathbf{q}\mathbf{v}_{\mathbf{p}_1} + I_{\mathbf{p}_1})} I^{ee}_{\mathbf{p}\mathbf{p}_1}\{\overline{F}, \overline{F}\} \right]. \qquad (19.21)$$

The second term appears only in a nonequilibrium state and only if electron–electron collisions are essential.

If $ql \gg 1$, $\omega\tau \gg 1$, we come to the well-known formula for the concentration fluctuations in the collisionless limit

$$(\delta n^2)_{\mathbf{q}\omega} = \frac{2\pi}{V_0^2 |\varepsilon(\mathbf{q}, \omega)|^2} \sum_\mathbf{p} \delta(\omega - \mathbf{q}\mathbf{v}_\mathbf{p}) \overline{F}_\mathbf{p}. \qquad (19.22)$$

In the thermal-equilibrium state the spectral intensity of electron density fluctuations is expressible through the longitudinal dielectric constant $\varepsilon(\mathbf{q}, \omega)$. Indeed, in this case

$$I_\mathbf{p} \to I_\mathbf{p}^{eq}, \qquad I_\mathbf{p}^{eq} F_\mathbf{p}^{eq} = 0, \qquad I_{\mathbf{p}\mathbf{p}_1}^{eq}\{F^{eq}, F^{eq}\} = 0 \qquad (19.23)$$

[see Eqs. (3.37) and (4.7)]. Thus,

$$(\delta n^2)_{\mathbf{q}\omega}^{eq} = \frac{2}{V_0^2 |\varepsilon^{eq}(\mathbf{q}, \omega)|^2} \operatorname{Re} \sum_\mathbf{p} \frac{1}{-i\omega + i\mathbf{q}\mathbf{v}_\mathbf{p} + I_\mathbf{p}^{eq}} F_\mathbf{p}^{eq}. \qquad (19.24)$$

The latter expression can be transformed (see reference 70) into

$$(\delta n^2)_{\mathbf{q}\omega}^{eq} = \frac{2}{V_0^2 |\varepsilon^{eq}(\mathbf{q}, \omega)|^2} \operatorname{Re} \left[\frac{\mathbf{q}}{\omega} \sum_\mathbf{p} \frac{1}{-i\omega + i\mathbf{q}\mathbf{v}_\mathbf{p} + I_\mathbf{p}^{eq}} \mathbf{v}_\mathbf{p} F_\mathbf{p}^{eq} \right]. \qquad (19.25)$$

Since

$$F_{\mathbf{p}}^{eq} \propto \exp[-\varepsilon_{\mathbf{p}}/T_0], \qquad \frac{\partial F_{\mathbf{p}}^{eq}}{\partial \mathbf{p}} = -\frac{\mathbf{v}_{\mathbf{p}} F_{\mathbf{p}}^{eq}}{T_0}, \qquad (19.26)$$

we have from Eq. (19.18)

$$\operatorname{Im} \varepsilon^{eq}(\mathbf{q},\omega) = \frac{4\pi e^2}{\varepsilon_0^{\mathbf{q}} q^2 \mathcal{V}_0 T_0} \operatorname{Re}\left[\mathbf{q} \sum_{\mathbf{p}} \frac{1}{-i\omega + i\mathbf{q}\mathbf{v}_{\mathbf{p}} + I_{\mathbf{p}}^{eq}} \mathbf{v}_{\mathbf{p}} F_{\mathbf{p}}^{eq}\right], \qquad (19.27)$$

and finally

$$(\delta n^2)_{\mathbf{q}\omega}^{eq} = \frac{2}{\mathcal{V}_0^2 |\varepsilon^{eq}(\mathbf{q},\omega)|^2} \frac{q^2 \varepsilon_0^{\mathbf{q}} T_0}{4\pi e^2 \omega} \operatorname{Im} \varepsilon^{eq}(\mathbf{q},\omega), \qquad (19.28)$$

or

$$(\delta n^2)_{\mathbf{q}\omega}^{eq} = \frac{2 n_0 q^2 r_D^2}{\omega \mathcal{V}_0} \operatorname{Im} \frac{1}{\varepsilon^{eq}(\mathbf{q},\omega)}, \qquad (19.29)$$

where

$$r_D \equiv \left(\frac{\varepsilon_0^{\mathbf{q}} T_0}{4\pi e^2 n_0}\right)^{1/2} \qquad (19.30)$$

is the *Debye screening length*. The quantities $\varepsilon(\mathbf{q}, \omega)$ and $(\delta n^2)_{\mathbf{q}\omega}$ for equilibrium as well as some nonequilibrium situations, have been calculated by many authors (see, e.g., references 351 and 352).

Equal-Time Two-Point Correlation Function

In this subsection we consider in detail a correlation between the occupancies in the momentum and coordinate space at the same moment, characterized by the equal-time two-point correlation function $\overline{\delta F_{\mathbf{p}}(\mathbf{r}) \delta F_{\mathbf{p}_1}(\mathbf{r}_1)}$ defined by (19.5). We decompose the Fourier transform of this function, Eq. (19.4), into two parts:

$$(\delta F_{\mathbf{p}} \delta F_{\mathbf{p}_1})_{\mathbf{q}} = \overline{F}_{\mathbf{p}} \delta_{\mathbf{p}\mathbf{p}_1} + \Phi_{\mathbf{p}\mathbf{p}_1}^{\mathbf{q}}. \qquad (19.31)$$

The equation for $\Phi_{\mathbf{p}\mathbf{p}_1}^{\mathbf{q}}$ can be obtained from the basic formula (19.12) by using Eq. (19.6) and the identity

$$\frac{1}{2\pi} \int_{-\infty}^{+\infty} \frac{d\omega}{(-i\omega + a)(i\omega + b)} = \frac{1}{a+b}. \qquad (19.32)$$

Thus, after the integration of (19.12) over ω we have

$$\overline{F}_{\mathbf{p}} \delta_{\mathbf{p}\mathbf{p}_1} + \Phi_{\mathbf{p}\mathbf{p}_1}^{\mathbf{q}} = \mathcal{B}_{\mathbf{p}\mathbf{p}_1}^{-1}(\mathbf{q})[(I_{\mathbf{p}} + I_{\mathbf{p}_1}) \overline{F}_{\mathbf{p}} \delta_{\mathbf{p}\mathbf{p}_1} - I_{\mathbf{p}\mathbf{p}_1}^{ee} \{\overline{F}, \overline{F}\}], \qquad (19.33)$$

where we denoted as $\mathcal{B}_{\mathbf{pp}_1}(\mathbf{q})$ the sum of the operators $\mathcal{B}_{\mathbf{p}}(\mathbf{q}, \omega)$ and $\mathcal{B}_{\mathbf{p}_1}^*(\mathbf{q}, \omega)$:

$$\mathcal{B}_{\mathbf{pp}_1}(\mathbf{q}) \equiv \mathcal{B}_{\mathbf{p}}(\mathbf{q}, \omega) + \mathcal{B}_{\mathbf{p}_1}^*(\mathbf{q}, \omega) \equiv [i\mathbf{q}(\mathbf{v} - \mathbf{v}_1) + I_{\mathbf{p}} + I_{\mathbf{p}_1}]x_{\mathbf{pp}_1}$$
$$- i\frac{4\pi e^2}{\varepsilon_0^q q^2 V_0} \mathbf{q} \cdot \left[\frac{\partial \overline{F}_{\mathbf{p}}}{\partial \mathbf{p}} \sum_{\mathbf{p}'} x_{\mathbf{p}'\mathbf{p}_1} - \frac{\partial \overline{F}_{\mathbf{p}_1}}{\partial \mathbf{p}_1} \sum_{\mathbf{p}'} x_{\mathbf{pp}'} \right]. \quad (19.34)$$

Thus we come to the following equation for $\Phi_{\mathbf{pp}_1}^{\mathbf{q}}$:

$$\mathcal{B}_{\mathbf{pp}_1}(\mathbf{q})\Phi_{\mathbf{pp}_1}^{\mathbf{q}} = i\frac{4\pi e^2}{\varepsilon_0^q q^2 V_0} \mathbf{q} \cdot \left(\overline{F}_{\mathbf{p}_1} \frac{\partial \overline{F}_{\mathbf{p}}}{\partial \mathbf{p}} - \overline{F}_{\mathbf{p}} \frac{\partial \overline{F}_{\mathbf{p}_1}}{\partial \mathbf{p}_1} \right) - I_{\mathbf{pp}_1}^{ee}\{\overline{F}, \overline{F}\}. \quad (19.35)$$

This equation is easy to solve in the case of thermal equilibrium, where Eqs. (19.23) and (19.26) hold. Seeking the solution of Eq. (19.35) in the form const $\cdot F_{\mathbf{p}}^{eq} F_{\mathbf{p}_1}^{eq}$, we find that

$$\Phi_{\mathbf{pp}_1}^{\mathbf{q}} = -\frac{F_{\mathbf{p}}^{eq} F_{\mathbf{p}_1}^{eq}}{N(1 + q^2 r_D^2)}, \quad (19.36)$$

where r_D is the Debye screening length defined by Eq. (19.30). Accordingly, the Fourier transform (19.31) of the equal-time two-point correlation function in the thermal equilibrium is

$$(\delta F_{\mathbf{p}} \delta F_{\mathbf{p}_1})_{\mathbf{q}}^{eq} = F_{\mathbf{p}}^{eq} \delta_{\mathbf{pp}_1} - \frac{F_{\mathbf{p}}^{eq} F_{\mathbf{p}_1}^{eq}}{N(1 + q^2 r_D^2)}. \quad (19.37)$$

The inverse Fourier transformation of Eq. (19.37) leads to the following expression for the equal-time two-point correlation function at thermal equilibrium:

$$\overline{\delta F_{\mathbf{p}}(\mathbf{r})\delta F_{\mathbf{p}_1}(\mathbf{r}_1)}^{eq} = F_{\mathbf{p}}^{eq} \delta_{\mathbf{pp}_1} V_0 \delta(\mathbf{r} - \mathbf{r}_1) - \frac{F_{\mathbf{p}}^{eq} F_{\mathbf{p}_1}^{eq}}{N} \cdot \frac{V_0 \exp(-|\mathbf{r} - \mathbf{r}_1|/r_D)}{4\pi r_D^2 |\mathbf{r} - \mathbf{r}_1|}, \quad (19.38)$$

valid for distances $|\mathbf{r} - \mathbf{r}_1| \gg e^2/\varepsilon_0^q T_0$.

At $qr_D \ll 1$, expression (19.37) coincides with that for the equal-time correlation function for spatially homogeneous fluctuations of the distribution function in the system with the fixed number of particles [see Eq. (3.11)]:

$$(\delta F_{\mathbf{p}} \delta F_{\mathbf{p}_1})_{\mathbf{q}}^{eq} = F_{\mathbf{p}}^{eq} \delta_{\mathbf{pp}_1} - F_{\mathbf{p}}^{eq} F_{\mathbf{p}_1}^{eq}/N, \quad qr_D \ll 1. \quad (19.39)$$

The equal-time fluctuations of carrier density are characterized by the function

$$(\delta n^2)_{\mathbf{q}} = \frac{1}{V_0^2} \sum_{\mathbf{pp}_1} (\delta F_{\mathbf{p}} \delta F_{\mathbf{p}_1})_{\mathbf{q}}, \quad (19.40)$$

which in the thermal equilibrium, according to Eqs. (19.31) and (19.36), is

$$(\delta n^2)_{\mathbf{q}}^{\text{eq}} = \frac{n_0 q^2 r_{\text{D}}^2}{\mathcal{V}_0 (1 + q^2 r_{\text{D}}^2)}. \tag{19.41}$$

At $qr_{\text{D}} \gg 1$ we obtain

$$(\delta n^2)_{\mathbf{q}}^{\text{eq}} = \frac{n_0}{\mathcal{V}_0}. \tag{19.42}$$

The fluctuations of the number of electrons in small parts of the system — in volumes small as compared to r_{D}^3 — are practically unaffected by the screening. Contrary to it, at $qr_{\text{D}} \ll 1$ we have

$$(\delta n^2)_{\mathbf{q}}^{\text{eq}} = \frac{n_0}{\mathcal{V}_0} q^2 r_{\text{D}}^2. \tag{19.43}$$

In volumes larger than r_{D}^3 the fluctuations of the number of electrons are effectively suppressed by the Coulomb interaction between them.

19.2 DRIFT–DIFFUSION EQUATION WITH FLUCTUATIONS

The formal expression for $(\delta F_{\mathbf{p}} \delta F_{\mathbf{p}_1})_{\mathbf{q}\omega}$ we have obtained in the preceding section contains the inverse response operators of the spatially inhomogeneous Boltzmann equation [see Eqs. (19.12) and (19.19)]. In spite of a formal character of these expressions, they can be readily analyzed and will serve as a starting point for the application of the Chapman–Enskog procedure (see Chapter 6), leading to an adequate description of long-range low-frequency fluctuations. The expressions for spectral intensities of such "hydrodynamic" fluctuations (smooth enough fluctuational redistributions of carrier density in space, the time evolution of which is rather slow) will be entered by the diffusion coefficients (which have been defined in Chapter 6) as well as by the differential conductivities and the spectral intensities of spatially homogeneous current fluctuations (which have been analyzed beginning with Chapter 3).

The procedure leading from the microscopic (kinetic) description of fluctuations in $(\mathbf{p}, \mathbf{r})$-space to macroscopic, hydrodynamic-like description of long-wavelength low-frequency fluctuations is quite similar to that used in Section 6.3 while deriving the drift–diffusion equation from the kinetic equation. Let the fluctuational redistribution in space be smooth enough; that is, let the microscopic dimension (scale) — the "mean free path" $l \equiv v\tau$ — be small as compared to the wavelength $1/q$, which is the characteristic scale of fluctuational inhomogeneity in the system of carriers. Now we are interested only in frequencies ω low enough as compared to the characteristic frequency of collisions, $1/\tau$. In other words, the term *long-wavelength low-frequency fluctuations* will mean the fluctuations whose characteristic lengths and frequencies obey the inequalities

$$ql \ll 1, \qquad \omega\tau \ll 1. \tag{19.44}$$

238 SPATIALLY INHOMOGENEOUS FLUCTUATIONS

For a detailed calculation of fluctuations by means of Eq. (19.12), it is necessary to know the *inverse* operator of the linearized Boltzmann equation, $\mathcal{B}_{\mathbf{p}}^{-1}(\mathbf{q}, \omega)$ — that is, to be able to solve the response equation. This problem is a difficult one and generally cannot be resolved analytically. However, we have seen in Chapter 6 how this can be done in the case (19.44). In this important case of small q and ω, the gas of electrons reacts upon a perturbation as a continuous medium, and its response can be described by a set of macroscopic equations discussed in Chapter 6. To derive the macroscopic equations for long-wavelength low-frequency fluctuations, one starts with the Boltzmann–Langevin equation (19.13) and applies the Chapman–Enskog iterative procedure described in detail in Section 6.3.

The idea of the Chapman–Enskog method is to reduce the solution of the space- and time-dependent Boltzmann equation to the solution of a number of differential equations with the coefficients expressed through the solutions of the Boltzmann equations for space- and time-independent response. Consecutively applied, the procedure leads to the *drift–diffusion equation with fluctuations*, valid as far as inequalities (19.44) hold:

$$\delta j_\alpha(\mathbf{r}, t) = eV'_\alpha \delta n(\mathbf{r}, t) - eD_{\alpha\beta}\frac{\partial \delta n(\mathbf{r}, t)}{\partial r_\beta} + \delta g_\alpha(\mathbf{r}, t), \tag{19.45}$$

where the differential drift velocity $\mathbf{V}'$ is defined by Eq. (6.7) while the carrier diffusion tensor $D_{\alpha\beta}$ is defined by Eq. (6.6). For a moment we omit in Eq. (19.45) the term with self-consistent field in the operator $\mathcal{B}_{\mathbf{p}}(\mathbf{q}, \omega)$, as we used to do in Chapter 6. The "sources" provoking long-range low-frequency fluctuations — the "extraneous" *local random currents* $\delta \mathbf{g}(\mathbf{r}, t)$ — in the "macroscopic" time-space scale outlined by inequalities (19.44) have no memory of themselves and are correlated in space only at the same "point":

$$\overline{\delta g_\alpha(\mathbf{r}, t)\delta g_\beta(\mathbf{r}', t')} \propto \delta(t - t')\delta(\mathbf{r} - \mathbf{r}'), \qquad |\mathbf{r} - \mathbf{r}'| \gg l, \quad |t - t'| \gg \tau. \tag{19.46}$$

This means that the Fourier transforms, with respect to time and space differences $t - t'$ and $\mathbf{r} - \mathbf{r}'$, of the correlation function of the extraneous local random currents are independent of the corresponding Fourier variables — of frequency ω and wave vector $\mathbf{q}$. In other words, in the "macroscopic" time-space scale defined by Eq. (19.44), the impetus of fluctuations — the extraneous local random currents $\delta \mathbf{g}(\mathbf{r}, t)$ — *behave as a white noise*. Being independent of frequency ω and wave vector $\mathbf{q}$, the spectral intensity of the extraneous local random currents is obviously equal to its long-wave low-frequency limit, with the latter, in turn, being none other than the *low-frequency limit of the spectral intensity of spatially homogeneous current density fluctuations*, investigated in Chapters 3 and 4 [see Eq. (4.12)]:

$$(\delta g_\alpha \delta g_\beta)_{\mathbf{q}\omega} = (\delta j_\alpha \delta j_\beta)_{\omega\tau \ll 1} \tag{19.47}$$

or, explicitly,

$$(\delta g_\alpha \delta g_\beta)_{\mathbf{q}\omega} = (e/\mathcal{V}_0)^2 \sum_{\mathbf{p}\mathbf{p}_1} v_{\mathbf{p}\alpha} v_{\mathbf{p}_1\beta} I_{\mathbf{p}}^{-1} I_{\mathbf{p}_1}^{-1}[(I_{\mathbf{p}} + I_{\mathbf{p}_1})\overline{F}_{\mathbf{p}}\delta_{\mathbf{p}\mathbf{p}_1} - I_{\mathbf{p}\mathbf{p}_1}^{ee}\{\overline{F}, \overline{F}\}]. \tag{19.48}$$

The drift–diffusion equation with fluctuations, Eq. (19.45), and the relations (19.47) and (19.48) were derived for electron gas in a nonequilibrium state in references 37 and 83 (see also [38]). Relation (19.47) sometimes is called the *second fluctuation–dissipation theorem* (cf. reference 72) since it relates the spectral intensity of "macroscopic" Langevin forces, entering the fluctuational drift–diffusion equation (19.45), to an independently measurable quantity, namely, the spectral intensity of space-independent current fluctuations. Due to Eq. (6.8) we rewrite Eq. (19.47) as

$$(\delta g_\alpha \delta g_\beta)_{\mathbf{q}\omega} = (e^2 n_0/\mathcal{V}_0)(D_{\alpha\beta} + D_{\beta\alpha} - \Delta_{\alpha\beta}), \tag{19.49}$$

where $\Delta_{\alpha\beta}$ is the low-frequency limit of the tensor of additional correlation defined by Eq. (6.9):

$$\Delta_{\alpha\beta} = \frac{1}{N} \sum_{\mathbf{pp}_1} v_{\mathbf{p}\alpha} v_{\mathbf{p}_1\beta} I_{\mathbf{p}}^{-1} I_{\mathbf{p}_1}^{-1} [I_{\mathbf{pp}_1}^{ee}\{\overline{F},\overline{F}\} - (I_{\mathbf{p}} + I_{\mathbf{p}_1})(\overline{F}_{\mathbf{p}} - N\partial\overline{F}_{\mathbf{p}}/\partial N)\delta_{\mathbf{pp}_1}], \tag{19.50}$$

In Section 6.1 we found that in the case of nonintercolliding carriers the Price relation holds; that is, the low-frequency limit of the spectral intensity of spatially homogeneous current density fluctuations is expressible in terms of carrier diffusion coefficients [see Eq. (6.4)]. It follows that for nonintercolliding carriers the spectral intensity of Langevin forces entering the fluctuational drift–diffusion equation (19.45) is expressible in terms of kinetic coefficients, namely, the carrier diffusion coefficients:

$$(\delta g_\alpha \delta g_\beta)_{\mathbf{q}\omega} = (e^2 n_0/\mathcal{V}_0)(D_{\alpha\beta} + D_{\beta\alpha}). \tag{19.51}$$

Of course, the diffusion coefficients can in principle be measured from the average response of the current to an inhomogeneous carrier density perturbation. Hence, the diffusion coefficients are related to transport, while the left-hand side of Eq. (19.51) is clearly a fluctuational characteristic. So, relation (19.51), when valid, has every reason to be referred to as the second fluctuation–dissipation theorem: It expresses the spectrum of Langevin forces in the drift–diffusion equation in terms of the diffusion coefficients entering the expression for the diffusion current.

The temporarily omitted term with self-consistent field of fluctuations in the long-wavelength low-frequency limit (19.44) would lead to the appearance, in Eq. (19.45) for the fluctuational current, of the local currents created by the local fluctuational fields. The correct expression for the Fourier transform of the current fluctuation is

$$\delta j_\alpha^{\mathbf{q}\omega} = e\left(V'_\alpha - iD_{\alpha\beta}q_\beta - i\frac{4\pi\sigma_{\alpha\beta}q_\beta}{\varepsilon_0^{\mathbf{q}} q^2}\right)\delta n^{\mathbf{q}\omega} + \delta g_\alpha^{\mathbf{q}\omega}, \tag{19.52}$$

with the fluctuations of current and electron density being connected by the continuity equation

$$-i\omega \delta n^{\mathbf{q}\omega} + i\mathbf{q}\delta\mathbf{j}^{\mathbf{q}\omega} = 0. \tag{19.53}$$

240 SPATIALLY INHOMOGENEOUS FLUCTUATIONS

Returning from the Fourier transforms to the space- and time-dependent functions, we conclude that macroscopic fluctuations in the system of carriers obey the following set of equations:

$$\delta j_\alpha(\mathbf{r}, t) = eV'_\alpha \delta n(\mathbf{r}, t) - eD_{\alpha\beta}\frac{\partial \delta n(\mathbf{r}, t)}{\partial r_\beta} + \sigma_{\alpha\beta}\delta E_\beta(\mathbf{r}, t) + \delta g_\alpha(\mathbf{r}, t), \quad (19.54)$$

$$\varepsilon_{0\alpha\beta}\frac{\partial \delta E_\beta(\mathbf{r}, t)}{\partial r_\alpha} = 4\pi e \delta n(\mathbf{r}, t), \quad (19.55)$$

$$e\frac{\partial}{\partial t}\delta n(\mathbf{r}, t) + \frac{\partial}{\partial r_\alpha}\delta j_\alpha(\mathbf{r}, t) = 0. \quad (19.56)$$

The latter two equations are the Poisson equation and the continuity equation that the fluctuations should satisfy. The first one — the expression for the local fluctuational current density $\delta \mathbf{j}(\mathbf{r}, t)$ — includes the drift of the density fluctuation caused by the external field, the diffusion current arising due to gradients of the fluctuating density, the term caused by the local fluctuating field, and the impetus to all these fluctuation phenomena — the "extraneous" random current $\mathbf{g}(\mathbf{r}, t)$.

The set of equations (19.54)–(19.56) corresponds to the following physical picture. There are fluctuations of electron distribution in quasi-momentum space accompanied by local currents $\delta \mathbf{g}(\mathbf{r}, t)$. Since $\omega\tau \ll 1$ and $ql \ll 1$, the decay of fluctuations in momentum space is almost instantaneous from a macroscopic point of view. From this point of view, local currents taking origin from these fast-developing fluctuations are instantaneous and uncorrelated in space and time [see Eq. (19.46)]. Furthermore, the local current correlation function is none other than the current correlation function considered in Chapter 3.

The random currents $\delta \mathbf{g}(\mathbf{r}, t)$ give rise to a redistribution of electrons in coordinate space which then slowly develops according to the macroscopic equations (19.54)–(19.56). By using the macroscopic equations just obtained, we can calculate, for instance, the spectral intensity of long-range low-frequency fluctuations of the electron density:

$$(\delta n^2)_{\mathbf{q}\omega} = \frac{2n_0}{V_0}\frac{q^2(D^\mathbf{q} - \Delta^\mathbf{q}/2)}{(\omega - \mathbf{q}\mathbf{V}')^2 + (4\pi\sigma^\mathbf{q}/\varepsilon_0^\mathbf{q} + q^2 D^\mathbf{q})^2}, \quad (19.57)$$

where $D^\mathbf{q}$, $\Delta^\mathbf{q}$, and $\sigma^\mathbf{q}$ stand for the contractions of the corresponding tensors [cf. Eq. (19.11)].

For equal-time long-range electron-density fluctuations characterized by Eq. (19.40) we obtain

$$(\delta n^2)_\mathbf{q} = \frac{n_0}{V_0}\frac{q^2 r_\mathbf{q}^2(1 - \Delta^\mathbf{q}/2D^\mathbf{q})}{1 + q^2 r_\mathbf{q}^2}, \quad (19.58)$$

where the quantity

$$r_{\mathbf{q}} \equiv \left(\frac{\varepsilon_0^{\mathbf{q}} q_\alpha q_\beta D_{\alpha\beta}}{4\pi q_\alpha q_\beta \sigma_{\alpha\beta}} \right)^{1/2} \tag{19.59}$$

plays the role of a screening length [cf. Eq. (19.30)] in the nonequilibrium system depending on the direction of the vector $\mathbf{q}$ not only through the contracted static lattice dielectric tensor $\varepsilon_0^{\mathbf{q}}$, but also through the ratio of the contracted differential conductivity and diffusion tensors. If $qr_{\mathbf{q}} \to 0$, fluctuations in the coordinate space vanish: Strong Coulomb forces prevent the charge redistribution between sufficiently large volumes.

To resume the section, let us stress that the drift–diffusion equation with fluctuations, Eq. (19.54), together with the Poisson and continuity equations, (19.55) and (19.56), and the correlation property of Langevin currents given by Eq. (19.49) [or, in special cases, by Eq. (19.51)], exhausts the theory of long-wavelength low-frequency fluctuations in the gas of electrons in a semiconductor or semiconductor structure. Of course, the prerequisite for calculation of spectra of spatially inhomogeneous fluctuations in the concrete situation is the knowledge of the kinetic coefficients and the characteristics of space-independent fluctuations of the system (sometimes, but by no means always, being expressible in terms of the kinetic coefficients).

19.3 EXPERIMENTAL DETECTION OF SPATIALLY INHOMOGENEOUS FLUCTUATIONS

Nowadays, due to the advanced laser technique, the light scattering from plasma (including solid-state plasma) [53–62] has become a powerful tool of experimental investigation. The experiments on the subject may give a deep insight into fluctuation phenomena.

The differential cross section for electromagnetic wave scattering is a function of the change in the wave vector, $\mathbf{q}$, and of that in the frequency, ω; we assume the latter change to be small in comparison with the frequency of the electromagnetic wave. Then the *differential cross section for scattering of electromagnetic wave by an isotropic single-component electron plasma*, with the electron having the dispersion relation $\varepsilon_{\mathbf{p}} = p^2/2m$, is given by the well-known expression

$$\frac{dS}{do\,d\omega} = r_0^2 (\mathbf{e}_i \cdot \mathbf{e}_s)^2 (\delta n^2)_{\mathbf{q}\omega} \frac{\mathcal{V}_0^2}{2\pi}. \tag{19.60}$$

Here $\mathbf{e}_i$ and $\mathbf{e}_s$ are the *polarization vectors of the incident and scattered waves*, respectively, do is the element of solid angle into which the radiation is scattered, and

$$r_0 = \frac{e^2}{mc^2} \tag{19.61}$$

is the so-called classical or Thomson radius of electron (c is the speed of light), characterizing the coupling strength of electromagnetic radiation with the electron.

242 SPATIALLY INHOMOGENEOUS FLUCTUATIONS

For q and ω large enough,

$$ql \gg 1, \qquad \omega\tau \gg 1, \tag{19.62}$$

the density–density correlation function in the *collisionless limit* can be used. By inserting (19.22) into (19.60) we get

$$\frac{dS}{do\,d\omega} = r_0^2(\mathbf{e}_i \cdot \mathbf{e}_s)^2 \frac{1}{|\varepsilon(\mathbf{q},\omega)|^2} \sum_\mathbf{p} \delta(\omega - \mathbf{q}\mathbf{v}_\mathbf{p})\overline{F}_\mathbf{p}. \tag{19.63}$$

Under conditions (19.62) each electron scatters light independently. Expression (19.63) provides a basis for the determination of nonequilibrium distribution function from light scattering experiments [52].

However, the conditions (19.62) frequently do not hold in practice, and the opposite limiting case

$$ql \ll 1, \qquad \omega\tau \ll 1,$$

that is, the collision-controlled scattering of electromagnetic radiation, is also quite realistic in semiconductors. In this case, the electromagnetic radiation is scattered by long-wavelength low-frequency stochastic excitations of a system of mobile electrons, against the background of its steady state. By using (19.57) we have in this case [37, 63]

$$\frac{dS}{do\,d\omega} = r_0^2(\mathbf{e}_i \cdot \mathbf{e}_s)^2 n_0 \frac{q^2(D^\mathbf{q} - \Delta^\mathbf{q}/2)}{(\omega - \mathbf{q}\mathbf{V}')^2 + (4\pi\sigma^\mathbf{q}/\varepsilon_0^\mathbf{q} + q^2 D^\mathbf{q})^2} \frac{V_0}{\pi}. \tag{19.64}$$

The electromagnetic wave is scattered from fluctuational "charge density waves," and the line of scattered light is of Lorentz form, with the half-width determined by the conductivity and diffusion and the center displaced from $\omega = 0$ due to Doppler-shift caused by the drift of density fluctuation in a DC field.

Correspondingly, the integral scattering cross section—the scattering cross section integrated over the change of frequency, ω—is

$$\frac{dS}{do} = r_0^2(\mathbf{e}_i \cdot \mathbf{e}_s)^2 (\delta n^2)_\mathbf{q} V_0^2. \tag{19.65}$$

For $ql \ll 1$ we obtain [see (19.58)]

$$\frac{dS}{do} = r_0^2(\mathbf{e}_i \cdot \mathbf{e}_s)^2 n_0 V_0 \frac{q^2 R_\mathbf{q}^2 (1 - \Delta^\mathbf{q}/2D^\mathbf{q})}{1 + q^2 r_\mathbf{q}^2}. \tag{19.66}$$

Because the ratio $(1 - \Delta/2D)$ is always of the order of unity, the main dependence of dS/do upon the concentration n_0 and the DC field $\mathbf{E}$ is due to the factor $n_0 q^2 r_\mathbf{q}^2/(1 + q^2 r_\mathbf{q}^2)$. At low electron concentrations, $q^2 r_\mathbf{q}^2 \gg 1$, this factor is proportional to n_0, but at high concentrations ($q^2 r_\mathbf{q}^2 \ll 1$) it ceases to depend on n_0 because $r_\mathbf{q}^2$ is approximately inversely proportional to n_0 [see Eq. (19.59)].

The saturation value of this factor is $q^2 n_0 \varepsilon_0^q D^q / 4\pi \sigma^q$, and its field dependence is determined by the ratio D^q/σ^q. This ratio increases with the growth of the mean electron energy $\bar{\varepsilon}_\mathbf{p}$ (by a rough estimate D/σ is proportional to $\bar{\varepsilon}$). Therefore, the electromagnetic wave scattering intensity grows with the heating of electrons. Moreover, there are some cases where the scattering cross section in certain directions grows even more rapidly than $\bar{\varepsilon}$. Indeed, if the current in a semiconductor saturates with heating of electrons — that is, the differential conductivity in the DC current direction at high fields tends to zero — the ratio D^q/σ^q for $\mathbf{q}$ parallel to DC current grows more rapidly than $\bar{\varepsilon}$. In such a case, one should expect not only a general enhancement, but also a sharp angular dependence of the electromagnetic wave scattering intensity. In other words, in the case of DC current saturation the charge density fluctuations with $\mathbf{q}$ parallel to the DC current are practically unscreened, thereby making the observation conditions for electromagnetic wave scattering more favorable.

Thus, the long-wavelength carrier-density fluctuations can be directly observed in electromagnetic wave scattering experiments, provided that the absolute value of the change of the wave vector is small as compared to the inverse mean free path of carriers. The half-width of the line of the electromagnetic wave scattered within such a collision-controlled regime is determined by the carrier diffusion coefficient and differential conductivity. The collision-controlled scattering of light by free electrons in CdS was studied in reference 145. A scattered-light line of Lorentz form, Doppler-shifted in an applied electric field, was observed, supplying us with the knowledge of the carrier diffusion coefficient.

The scattering of light by condensed media has been studied comprehensively during recent decades [55–62]. The first experimental detection of light scattering from electronic excitations in GaAs, in n-Ge, and in n-Si was performed in references 53 and 54, in references 146 and 147, and in references 353 and 354, respectively. Various mechanisms and sources for the scattering of light by carriers were studied. It was shown that the electromagnetic radiation can be scattered in semiconductors not only by carrier density fluctuations, but also by other types of carrier fluctuations: unscreened fluctuations due to intervalley transitions in many-valley semiconductors, or due to electron energy fluctuations in the case of nonparabolic electron energy dependence on quasi-momentum (see reference 55); spin density fluctuations also are "seen" by electromagnetic waves and can be detected in experiments with scattering of polarized light (see reference 55). An approach to these problems, based on the Chapman–Enskog procedure, was used in references 63, 138–141, and 355.

19.4 SPACE-DEPENDENT FLUCTUATIONS OF EFFECTIVE ELECTRON TEMPERATURE

The drift–diffusion equation (19.54) describes only long-wavelength low-frequency fluctuations, the wave vectors and frequencies of which satisfy inequalities $ql \ll 1$, $\omega \tau \ll 1$, where l and τ are microscopic parameters. In this case the fluctuations are those of carrier density, the total number of carriers being the only integral of motion.

In some cases, however, the kinetics of the system of carriers is characterized by more than one, say, relaxation time; some types of microscopic excitations can decay slower than others. Then the region of frequencies ω and/or wave vectors **q** can exist where for the largest values of microscopic parameters one or both inequalities (19.44) are not fulfilled. Moreover, for special physical reasons the "slowest" kinetic process can happen to be none other than the slow relaxation of a few variables or even one variable. Then the additional "quasi-macroscopic" equation can be derived governing birth and evolution of fluctuations of such a "quasi-macroscopic" variable.

An example is given by fluctuations of effective electron temperature (see Chapter 12) — the slowly changing characteristic of the energy of the system of intensively intercolliding hot electrons. Such a gas has two fluctuational degrees of freedom, which correspond to excitation of stochastic damped waves of the electron density and of the effective electron temperature at *intermediate* frequencies and wave vectors. The "quasi-macroscopic" equations describing fluctuations of density and of effective temperature in the system of hot electrons were derived by Shulman and Kogan [150]. For a special case the intensity of electron temperature fluctuations integrated over frequency was obtained and used to calculate the cross section of electromagnetic wave scattering by electron plasma [142]. The most complete investigation of peculiarities of spectra of electron-density and electron-temperature fluctuations as well as their cross-correlations was performed in references 143 and 144. In particular, it was shown that in nonequilibrium the mutual effects of fluctuations in electron density and electron temperature are directly related to the additional kinetic correlation arising from interelectron collisions (described in Chapter 4 and Section 6.2). It was shown that the additional kinetic correlation may either intensify or suppress the scattering of electromagnetic waves by electron plasma. The additional kinetic correlation also was shown to lead to a shift of the peak of the spectral line of the scattered light, unrelated to the drift of electrons.

To conclude, let us summarize the main results of the chapter. Long-wavelength spatially inhomogeneous fluctuations can be effectively investigated using so-called drift–diffusion equation with fluctuations, which is of Langevin type. The approach is analogous to the well-known method of investigation of hydrodynamic fluctuations in gases and fluids [69, 134]. In the case of the carrier drift–diffusion equation, the spectra of corresponding Langevin forces often are expressible in terms of diffusion coefficients of hot-carriers. This property is referred to as the second fluctuation–dissipation theorem. The quasi-hydrodynamical description of fluctuations is possible in even wider regions of wavelengths and frequencies in the so-called effective electron temperature approximation applicable at higher carrier densities. The spatially inhomogeneous fluctuations can be observed in experiments on electromagnetic wave scattering from solid-state plasma.

CHAPTER 20

MONTE CARLO APPROACH TO MICROWAVE NOISE IN DEVICES

In this chapter, we are going to illustrate how the microscopic approach to noise problems in devices works under conditions of nonuniform electric field and doping. We will give two examples to demonstrate a good fit of the experimental results and the simulation based on self-consistently coupled Monte Carlo procedure and Poisson solver.

20.1 MICROSCOPIC APPROACHES

The theoretic basis for a microscopic approach to problems of microwave noise is the kinetic equation for fluctuations. Its solution provides us with correlation functions required to obtain the desired spectral characteristics of the excess noise arising under deviations from the thermal equilibrium. Analytic solutions of the kinetic equation (3.5) for correlation functions available for a number of spatially uniform models (see Chapter 7) give a profound microscopic understanding of electric-field-dependent spectral features of hot-electron fluctuations — the main source of noise at microwave and millimeter wave frequencies in biased semiconductors.

Naturally enough, the analytic solutions are available only for a limited number of models. Therefore, numerical techniques are necessary for a quantitative microscopic treatment of microwave noise without limitations imposed onto the number and complexity of electron scattering mechanisms. The most versatile technique for solution of the kinetic equation in semiconductors — Monte Carlo technique — was modified to treat microwave noise in terms of electron velocity correlation functions [206, 225]. The first simulations considered the applied electric field as uniform, and no interaction between electrons was taken into account. Nevertheless, this approach was fruitful in the case of lightly doped uniform bulk crystals and led to important results in physics of hot-electron fluctuations in semiconductors.

However, the approach of uniform electric field is of limited applicability in devices where electrostatic and other barriers are used to control currents. The nonuniformity of the electric field becomes of primary importance for device performance. Consequently, a self-consistent solution of the kinetic equation coupled to that for the electric field is needed for adequate description of the transport and noise phenomena. An important step forward was made when the Monte Carlo technique was applied for a microscopic simulation of the intervalley transfer of hot electrons leading to microwave current oscillations in a Gunn diode caused by the high-field domain formation and motion [356]. The simulation was equivalent to a self-consistent solution of the kinetic equation coupled to the one-dimensional Poisson equation. The fluctuations of current appeared as a "by-product" in a natural way. Essentially the same microscopic approach was used [270] to describe the observed suppression of intervalley fluctuations in short samples of GaAs [261]; the performed Monte Carlo simulation took into account the nonuniformity of the electric field and space charge density fluctuations in a self-consistent way (see Chapter 11).

Quite a high density of electrons is desired for the optimal matching of device-building blocks contained in a high-speed device or a circuit. Interaction among electrons in a high-density electron gas calls for procedures incorporating interelectron collisions into the general scheme of Monte Carlo simulation [233, 235, 236] (see Chapter 12). Efficient time-saving procedures are needed for simulation of fluctuations assuming a realistic model of dissipation [136, 238, 246, 247]. The results of Monte Carlo simulation in ungated channels (see Chapter 13) demonstrate that the electron temperature approach is acceptable under conditions where the interelectron collisions are essential (Chapters 13 and 14). This is important from an applied point of view, since the electron temperature approach allows analytic solutions for fluctuations, provided that the small-signal response problem is solved [131]. Evidently, the microscopic approach based on the electron temperature is also applicable to quantum-well channels (electron confinement leads to extremely high densities of electrons, and electron gas degeneracy takes place). Under conditions defined in Chapter 19, the approach based on drift–diffusion equation is also acceptable for treatment of spatially nonuniform fluctuations. A quantitative microscopic description of high-frequency noise in microwave devices tends to combine the foregoing approaches.

Monte Carlo procedure coupled self-consistently to the Poisson equation is sufficient for treatment of high-frequency noise in a many-terminal device without addressing to the concepts of impedance field and local sources of noise. In a diode, where the electric field and doping are not uniform in the direction of current, the one-dimensional Poisson equation is applicable (there is no doping profile in the transverse direction). However, this approach is of a limited applicability to three-terminal devices like a field-effect transistor. In this type of transistor, a nonuniformity of electric field and space charge fluctuations are important in two directions: along the channel and in the transverse direction. The Poisson equation in two dimensions coupled to the kinetic equations for fluctuations is needed for transistors.

20.2 PLANAR-DOPED-BARRIER DIODE

Transition from Shot Noise to Hot-Electron Noise

Schottky diodes and planar-doped-barrier diodes are nonlinear two-terminal devices where the current is controlled by the barrier. The devices are unipolar, usually n-type, and the barrier for the electrons is formed by the negative charge accumulated by the localized states. In a Schottky diode, the occupied surface (interface) states in the metal–semiconductor contact are responsible for the barrier and for the current rectification by the diode. The states in a planar-doped-barrier diode are located in the bulk, under the surface; the barrier is formed by a plane of acceptors inserted into a lightly doped n-type layer during its growth [357]. Despite of the different origin and location of the localized states (see Fig. 20.1), the electrostatic potential inside the planar-doped-barrier diode resembles that in a Schottky diode. It is advantageous that the acceptor sheet density is under perfect technological control in the planar-doped-barrier diode while the density of the surface states in a Schottky diode is difficult to control precisely. Moreover, in a planar-doped-barrier diode, the barrier height and the diode asymmetry can be controlled independently [303]; the optimal combination of the barrier height and the diode asymmetry (needed for mixer applications at microwave frequencies) can be achieved. At a high forward bias, the importance of the barrier in a planar-doped-barrier diode fades away [358, 359]: A transition from the barrier-controlled nonohmic transport to the hot-electron-dominated current takes place.

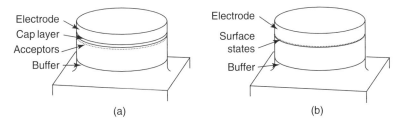

Figure 20.1. Schematic views of (a) a GaAs planar-doped-barrier diode and (b) a Schottky diode. The mesa structure on the n^{++} substrate consists of (from bottom to top) the n^+-type buffer layer, the n-type layer, and the metal top electrode (the bottom electrode is not shown). In the planar-doped-barrier diode (a) the n-type layer contains an n^--type barrier formed by the acceptor plane, and the n^{++}-type cap layer is inserted between the n-type layer and the metal electrode to facilitate formation of perfect ohmic contact. In the Schottky diode (b) the rectifying metal–semiconductor contact is formed; its properties are controlled by the surface (interface) states.

Figure 20.2 shows that there is a striking similarity between the noise properties of a Schottky diode and a planar-doped-barrier diode [360]. Shot noise dominates at low current densities, and the transition to hot-electron noise takes place at high currents. Let us consider the transition from the shot noise to the hot-electron noise from a microscopic point of view. Our choice falls on the planar-doped-barrier diode; its barrier and other properties are well-controlled, and no adjustable parameters are needed for simulation.

Device Model

Electron transport and fluctuations in the planar-doped GaAs diode are simulated [361] by the Monte Carlo particle technique within the three-valley Γ–L–X conduction band model [244], assuming intermediate Γ–L intervalley coupling (for details of the technique see references 241 and 362). Figure 20.1 gives a schematic view of the diode under simulation — a sandwich-type structure $n^{++}-n^{+}-n-n^{-}-n-n^{++}$ containing the layers of the following thickness, respectively (40–300–758–2–40–60) nm. The 2-nm-thick "plane" contains acceptors ($N_A(n^-) = 4 \times 10^{18}$ cm^{-3}). Their negative charge forms a quasi-triangular barrier, with its height being around 0.5 eV at equilibrium. The donor densities in the donor-doped layers are $N_D(n^{++}) = 5 \times 10^{18}$ cm^{-3}, $N_D(n^+) = 2 \times 10^{17}$ cm^{-3}, and $N_D(n) = 10^{14}$ cm^{-3}. The 40-nm and 60-nm n^{++}-layers are considered to have equal donor densities, and this eliminates the contact potential difference at the ends of the structure. The buffer n^+-layer thickness is 300 nm, in contrast to 600 nm in the experimentally investigated diodes (the reduced thickness saves computer time without any loss of physically important information). Free holes are not taken into account since the Fermi level is located in the upper half of the forbidden gap everywhere in the diode. Electron motion is simulated taking into account nonuniform electric field in a self-consistent

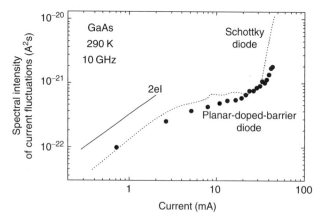

Figure 20.2. Spectral intensity of current fluctuations at 10-GHz frequency for forward-biased diodes [360]: a GaAs planar-doped-barrier diode (●) and a Schottky diode (dotted curve). Solid line stands for the ideal shot noise, $S_I = 2eI$.

way. Fluctuations of the space charge and electric field are taken into account using "charge sheets" [241]. In the procedure, the grid lines of the adaptive mesh are associated with the individual electrons under simulation. As a result, the mesh density is high where the field changes rapidly—the computer resources are used very efficiently. The results of simulation [361] at voltages $U > k_B T_0/e$ are presented in Fig. 20.3 (unfilled symbols).

Microwave Noise Measurements

The experimentally investigated GaAs planar-doped-barrier diode is a 14-μm-diameter mesa structure (see Fig. 20.1a) formed onto a n^{++}-type GaAs substratum covered by a 600-nm donor-doped buffer layer ($n^+ \approx 2 \times 10^{17}$ cm^{-3}). The mesa contains an 800-nm lightly doped n-type layer ($n \approx 10^{14}$ cm^{-3}) with a 2-nm acceptor plane inserted at a 40-nm distance from the n^{++}-type cap layer ($n^{++} \approx 5 \times 10^{18}$ cm^{-3}). The acceptor plane has a sheet density $N_A = 8 \times 10^{11}$ cm^{-2}, and the acceptors are totally ionized. The forward bias corresponds to a positive potential on the cap layer with respect to the substratum. In this direction, the current–voltage characteristics is almost exponential at voltages below 0.5 V; it tends to become linear at voltages over 2 V, and the resistance in this bias range is 127 ohms [360]. The voltage on the active layers of the diode is determined, assuming 15-ohm contact resistance.

The noise measurements are performed [360, 361] at 10-GHz frequency at room temperature using a high-sensitivity gated radiometric setup (see Chapter 8). Short pulses of bias voltage are applied—the pulsed technique helps to avoid thermal walkout, diode overheat, and thermal breakdown. The diode is mounted into a waveguide and matched, and the matching is controlled at any bias by a standing-wave-ratio meter. The equivalent noise temperature T_n is determined in the direction of current from a comparison of the noise power

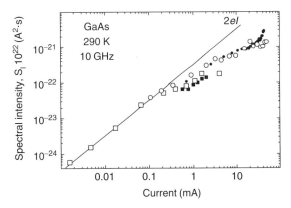

Figure 20.3. Spectral intensity of current fluctuations at 10-GHz frequency for the GaAs planar-doped-barrier diode at forward bias (circles) and reverse bias (squares) [361]: experimental data (filled symbols) and Monte Carlo simulation (unfilled symbols). Solid line stands for the ideal shot noise, $S_I = 2eI$.

emitted by the matched diode to that of the "black body" noise source kept at a known temperature. The spectral intensity of current fluctuations S_I is obtained according to Eq. (8.2) from the measured noise temperature T_n and the real part of the small-signal admittance determined at each bias. Figure 20.3 presents the spectral intensity of current fluctuations S_I for the forward-biased (circles) and the reverse-biased (squares) planar-doped-barrier diode.

Results of Experiment and Simulation

The results of simulation (Fig. 20.3, unfilled symbols) are in a reasonably good agreement with the experimental data (filled symbols) in the wide range of currents except for some discrepancy appearing at the highest forward current [361].

At a low current, $I < 0.2$ mA, the results of simulation show no dependence on the current direction and fit the Schottky formula [4]:

$$S_I = 2eI, \qquad (20.1)$$

valid for the ideal shot noise corresponding to uncorrelated passage of individual electrons through the barrier region (cf. solid line and unfilled symbols in Fig. 20.3).

At intermediate currents, the sublinear deviations develop, indicating an onset of correlation between the electrons. The shot noise becomes partially suppressed, with the suppression becoming stronger at a higher current. The experiments and the simulation show a systematic dependence on the current direction: The suppression is less efficient in the forward-biased diode (Fig. 20.3, circles). The mechanism of suppression is as follows. The negative charge of the electrons drifting across the barrier repels the electrons waiting for their turn to enter the barrier region; the probability to enter depends on the electron number present in the barrier region. A temporary decrease in the number causes an increase of the probability, leading to an enhanced injection of new electrons; the original fluctuation is reduced. In other words, the fluctuation is partially suppressed; the shot noise is suppressed as well.

Let us apply the same arguments for the noise suppression at the same current in the reverse and the forward directions. Since the forward-biased barrier is thin (the electrons drift across the 40-nm n-type layer), the electrons spend less time in the barrier region, and the total number of the drifting electrons is lower at a given forward current as compared to the reverse one. Thus, a higher forward current is needed to accumulate enough electrons for manifestation of the noise suppression, as evidenced by the simulation and the experimental data [361].

At high bias, the barrier almost disappears and stops controlling the current. The transition from the partially suppressed shot noise to the hot-electron noise takes place. Figure 20.4 compares the results on equivalent noise temperature for the GaAs planar-doped-barrier diode [363] and for GaAs n^+-n-n^+ structure. The latter structure contains the uniformly doped n-type region, with its length ($L = 1$ μm, Fig. 11.6) being nearly the same as that of the n-type region in

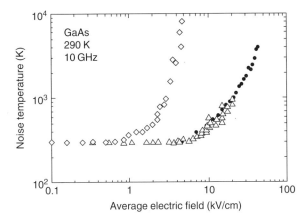

Figure 20.4. Dependence of equivalent noise temperature on average electric field for the GaAs planar-doped-barrier diode (● [363]) and for GaAs n^+–n–n^+ structures (unfilled symbols [261]). Data for the n^+–n–n^+ structures correspond to different length of the n-type region: △ represents $L = 1$ μm, ◊ represents $L = 7.5$ μm. The length of the diode n-type region (containing the acceptor plane) is $L = 0.8$ μm.

the planar doped-barrier diode ($L = 0.8$ μm). The equivalent noise temperature of the structure and the diode is approximately the same in the range of high electric fields ($\overline{E} > 10$ kV/cm in Fig. 20.4 corresponds to $U > 1$ V). In this range of voltages the barrier in the diode does not control the current and is of minor importance for noise properties (cf. circles and triangles in Fig. 20.4). The results of simulation [361] demonstrate an onset of occupation of the upper valleys by hot electrons at voltages near 1 V (Fig. 20.5, dashed curve). As discussed in Chapter 11 (see Fig. 11.6), partially suppressed intervalley fluctuations cause the main source of noise in this range of electric fields.

Impact Ionization

The results of simulation show the tendency for saturation of the intervalley noise at high voltages (Fig. 20.5, unfilled circles). The experiment is in agreement with the model at voltages below 2 V, but the tendency for saturation following from the model (Fig. 20.5, unfilled circles) is not observed; the experimental data [361] demonstrate a transition from the sublinear dependence of S_I on voltage to the superlinear one dominating at the highest voltages (Fig. 20.5, filled circles). The simulation shows that this behavior cannot be ascribed to the intervalley transfer, and another mechanism has to be assumed to interpret the excess noise at the highest bias.

Figure 20.6 illustrates the distribution of electric field simulated in a self-consistent way for the forward-biased diode. Electric field strength of 170 kV/cm at 2 V and 200 kV/cm at 3 V is reached in the 40-nm n-type layer located between the acceptor plane and the cap layer. The step of electric field at the acceptor plane (Fig. 20.6, see n^- arrow in the inset) favors ballistic acceleration

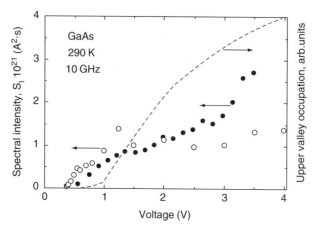

Figure 20.5. Voltage-dependent spectral intensity of current fluctuations at 10-GHz frequency for the forward-biased GaAs planar-doped-barrier diode [361]: ● represents experimental data, ○ represents results of and Monte Carlo simulation. Dashed line stands for the electron density in the upper valleys integrated over the n-type region.

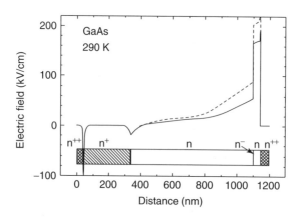

Figure 20.6. Electric field distribution in the forward-biased $n^{++}-n^+-n-n^--n-n^{++}$ diode at 2 V (solid curve) and at 3 V (dashes) [361]. The left-hand-side n^{++} layer is negative (see inset).

of the electrons entering the high-field region and causes an effective energy gain by them. For example, at 3-V bias, more than 1.3% of the electrons reach the n^{++} layer having accumulated energy exceeding 2 eV, with this energy being sufficient for impact ionization. The experimental data [201] available for submicrometer n^+-n-n^+ structures (see Fig. 11.14) demonstrate an onset of impact ionization noise at around 200 kV/cm fields, with this noise being superimposed onto the intervalley noise. The Monte Carlo model [241] neglects impact ionization, and the results of simulation [361] do not fit the experimental results at the highest bias (Fig. 20.5, circles). The misfit supports the idea that the observed superlinear growth of S_I with the bias results from impact ionization.

The Monte Carlo simulation of microwave noise gives a satisfactory interpretation of the experimental results for a GaAs planar-doped-barrier diode over a wide range of bias. The transitions from the ideal shot noise to the partially-suppressed shot noise and to the hot-electron noise are evidenced. Hot-electron intervalley-transfer noise is found to dominate at high forward bias until the impact ionization noise onsets at the highest forward bias.

20.3 ULTRAFAST FIELD-EFFECT TRANSISTORS

Two-Dimensional Problem

A high-frequency noise problem in a field-effect transistor is a typical two-dimensional problem. The electron transport takes place mainly in the two directions: along the channel and in the direction normal to the semiconductor surface. Usually the fringing effects at the channel side edges are negligible, provided that the channel is relatively wide as compared to its length and thickness. The electric field pattern in the plane normal to the surface controls the channel current; the solution in two dimensions is sufficient. A semiclassical Monte Carlo simulation of electron scattering and motion (equivalent to the solution of the time-dependent kinetic equation) can be used for a microscopic consideration of hot-electron effects. The simulation provides us with the electron velocities and the positions at any moment of time. In this way, the information on electron distribution in two-dimensional space required for the Poisson equation is obtained; simultaneously the correlation functions become available for noise treatment.

For numerical solution of the Poisson equation coupled to the Monte Carlo procedure the two-dimensional region of a transistor is subdivided into small sections by grid lines. Uniformly distributed rectangular grids can be used, but they are not the optimal ones. A restriction on computational resources suggests that we should apply an adaptive mesh with variable grid spacing to allow sufficient accuracy in the locations of rapid change and conserveing memory resources in the areas where the variables are slowly varying. Typical grid spacing ranges from 1 nm to 10 nm in a high-speed transistor with 100-nm gate length [25, 364]. Time steps are generally limited to very small values (often below 10 fs).

A semiclassical ensemble Monte Carlo procedure is essentially the same as for one-dimensional models. For example, a three-valley (Γ, L, and X) GaAs model is used for simulation of noise properties of a MESFET and a HEMT [364]. The valleys are treated as spherical and nonparabolic. Ionized impurity, alloy, polar and nonpolar optical phonon, acoustic phonon, and intervalley scattering mechanisms are taken into account. Electron gas degeneracy and screening effects are described in terms of the local electron temperature estimated during simulation.

The solution is defined by the boundary conditions. For an intrinsic transistor model, the metal electrodes (the source, the gate, and the drain) are excluded together with a part of the heavily doped semiconductor, if present in the vicinity of the source and the drain electrodes. In the latter case, the boundary condition of the virtual ohmic contact is set at a certain distance from the source and the drain. The boundary condition for the electric field at the surface takes into account the surface charge, and the boundary condition at the blocking gate electrode is determined by the gate potential.

Noise Simulation for Intrinsic MESFET and HEMT

Microscopic approach leads to an interesting comparison of noise properties of a GaAs MESFET and an AlGaAs/GaAs HEMT having the same electrode layout. In order to demonstrate the effect of semiconductor properties on noise rather than those due to electron confinement in 2DEG, the subband structure and the electron tunneling are excluded from the simulation [364]. No attempt is made to optimize the design for the best possible performance. Nevertheless, the simulation provides with important information on the noise behavior.

The first step is calculation of the transistor drain current–voltage characteristics in the common-source configuration [364]. The doping is chosen so that the high-mobility effect in the HEMT tends to override the effect of a slightly higher number of the electrons present in the channel in the MESFET. Thus, the small-signal response analysis shows that the HEMT transconductance is higher than that of the MESFET unless the drain bias is too high. At the high bias, the high-mobility effect becomes of minor importance because the velocity saturation effect takes place. The small-signal response provides us with the parameters of the equivalent circuit of the intrinsic transistor.

Once the static behavior and small-signal response characteristics are known, the following step is noise characterization. The autocorrelation functions of the gate current and the drain current and their cross-correlation functions are calculated, and the spectral intensities of current fluctuations are determined [364]. At microwave frequencies, the intensity of the gate current fluctuations and that due to the mutual correlation of the gate and the drain currents are found to be essentially lower than the spectral intensity of the drain current fluctuations. The difference decreases in the millimeter-wave range; in particular, all three intensities tend to become important at 100-GHz frequency. The spectral intensity of the drain current fluctuations is almost frequency-independent; a higher value is obtained for the MESFET, but one should be careful because its steady-state drain current is also higher.

The noise figure is the most widely used noise parameter characterizing a two-port low-noise device (see Chapter 1). The minimum noise figure at a 100-GHz frequency obtained for the intrinsic MESFET and HEMT through the microscopic Monte Carlo simulation [364], coupled with the Poission equation, is shown in Fig. 20.7. For a fixed drain–source voltage ($V_{DS} = 2$ V), the minimum is reached at intermediate gate voltages; the noise figure increases at both low and high drain currents. The results of microscopic simulation show (Fig. 20.7) that the considered HEMT is superior over the MESFET of the same layout, but this conclusion has to be confirmed for other electrode layouts and doping profiles.

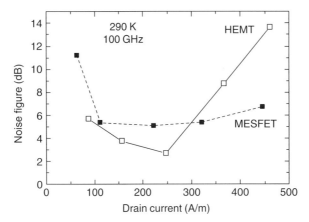

Figure 20.7. Noise figure at 100-GHz frequency for 0.1-μm gate intrinsic GaAs MESFET (■) and AlGaAs/GaAs HEMT (□) calculated using a two-dimensional Poisson solver coupled to Monte Carlo simulator [364].

The minimum noise figure at 10-GHz frequency is 0.7 dB and 0.3 dB for the simulated intrinsic MESFET and HEMT, respectively [364] .

20.4 COMPARISON TO EXPERIMENT: InGaAs-CHANNEL T-GATE HEMT

As discussed, the microscopic approach to high-frequency noise using the Monte Carlo technique provides us with the minimum noise figure of an intrinsic field-effect transistor. In order to make a comparison to the experimental results, the noise figure for the extrinsic transistor is to be determined. Once the intrinsic noise figure is known and the extrinsic circuit parameters (parasitic capacitances, inductances, and electrode/contact resistances) are available, it is straightforward to obtain the extrinsic noise figure for the device.

This problem has been solved for an InAlAs/InGaAs/InAlAs short-channel δ-doped T-gate HEMT [25] using the procedure discussed in the foregoing subsection. An excellent fit of the simulated and measured static drain current–voltage characteristics is obtained over wide range of bias. The parameters of the equivalent circuit of the intrinsic transistor are determined through small-signal-response simulation and experimentally. The highest experimental cutoff frequency is near 270 GHz, the same value is estimated through simulation.

The perfect fitting of the model and the device is a good basis for considering noise figure [25]. The model leads to the minimum intrinsic noise figure below 1.5 dB at 94 GHz (Fig. 20.8, solid circles). For a comparison with the results of experiment (Fig. 20.8, solid curve), the calculated extrinsic noise figure is presented by circles in Fig. 20.8. The calculated extrinsic noise figure exceeds the intrinsic one approximately by 1 dB. An excellent fit of the calculated and the measured noise figure is obtained at drain currents ranging from 50 A/m

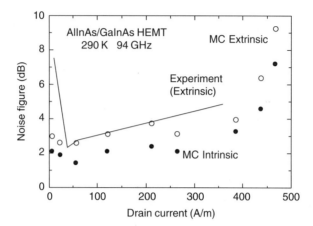

Figure 20.8. Calculated (symbols) and measured (solid curve) noise figure at 94-GHz frequency for 0.1-μm T-gate AlInAs/GaInAs HEMT [25]. Calculations are performed using Monte Carlo simulator coupled to a two-dimensional Poisson solver. ● represents intrinsic transistor, ○ represents extrinsic transistor.

to 200 A/m. The minimum extrinsic noise figure is determined to be 2.6 dB at 94 GHz. A strong effect of the T-gate is also obtained: The calculated values of the minimum noise figure reach 6 dB for the transistor without a T-gate.

This example demonstrates that the microscopic approach based on the Monte Carlo simulator coupled with the two-dimensional Poisson solver can take into account the layout and parameters of a real high-speed device. The simulation of small-signal response and noise characteristics at microwave and millimeter-wave frequencies is performed. An impressive fit of the results of simulation to the experimental data is obtained at 94-GHz frequency over a rather wide range of bias conditions. The model can consider subtle details such as the effect of the electrode shape and others. Despite some misfit remaining outside the optimum bias conditions, the approach has demonstrated its high potentiality.

20.5 CONCLUDING REMARKS

This now brings our book on fluctuations and microwave noise in semiconductors, semiconductor structures and devices to its end by completing our proposed discussion of microscopic explanation of noise in nonequilibrium systems. It is pleasing to have come to this end with the above illustrations, which show so clearly the deep and powerful character of combined analytic/computational approach to the problem in close connection with experimental study.

On the other hand, much has been omitted from the book in the way of considering important applications of nonequilibrium fluctuation theory to problems of interest for up-to-date technology and device engineering. In particular, we may regret the omission of further and more specific consideration of the applications to studies of noise performance of many high-speed devices operating under far-from-equilibrium conditions. This is a field in which much remains to be done and in which the methods of microscopic theory of fluctuations are specially needed.

It is hoped, however, that our exposition of theory and its applications, as well as presentation of available experimental results, have been sufficiently sound and complete to solve our main task of giving a true insight into the methods that must be employed when we wish to predict the noise behavior of structures and devices designed for high-speed electronics.

BIBLIOGRAPHY

[1] A. van der Ziel, *Noise in Solid State Devices and Circuits*, John Wiley & Sons, New York, 1986.

[2] J. B. Johnson, "Thermal agitation of electricity in conductors," *Phys. Rev.* **32**(1), 97–109 (1928).

[3] H. Nyquist, "Thermal agitation of electric charge in conductors," *Phys. Rev.* **32**(1), 110–113 (1928).

[4] W. Schottky, "Über spontane Stromschwankungen in verschiedenen Elektrizitätsleitern," *Ann. Phys.* **57**, 541–567 (1918).

[5] F. N. Hooge, "$1/f$ noise sources," *IEEE Trans. Electron Devices* **ED–41**(11), 1926–1935 (1994).

[6] Sh. Kogan, *Electronic Noise and Fluctuations in Solids*, Cambridge University Press, Cambridge, 1996.

[7] N. B. Lukyanchikova, *Noise Research in Semiconductor Physics*, Gordon & Breach Science Publishers, 1997.

[8] C. M. Van Vliet, "Forty-five years of generation–recombination noise in electronic and photonic materials and devices," in: *15th International Conference on Noise in Physical Systems and 1/f Fluctuations, Hong Kong 1999*, C. Surya, ed., Bentham Press, London, 1999, pp. 3–9.

[9] V. Bareikis, R. Katilius, J. Pozhela, S. Gantsevich, and V. Gurevich, "Fluctuation spectroscopy of hot electrons in semiconductors," in: *Spectroscopy of Nonequilibrium Electrons and Phonons*, C. V. Shank and B. P. Zakharchenya, eds., *Modern Problems in Condensed Matter Sciences*, Vol. 35, North-Holland, Amsterdam, 1992, pp. 327–396.

[10] J. P. Nougier, "Fluctuations and noise of hot carriers in semiconductor materials and devices," *IEEE Trans. Electron Devices* **ED–41**(11), 2034–2049 (1994).

[11] V. Bareikis, J. Liberis, I. Matulionienė, A. Matulionis, and P. Sakalas, "Experiments on hot electron noise in semiconductor materials for high-speed devices," *IEEE Trans. Electron Devices* **ED–41**(11), 2050–2060 (1994).

[12] A. Matulionis, "Noise, hot carrier effects," in: *Wiley Encyclopedia of Electrical and Electronics Engineering*, John G. Webster, ed., Vol. 14, John Wiley & Sons, 1999, pp. 410–428.

[13] S. M. Sze, "Introduction," in: *High-Speed Semiconductor Devices*, S. M. Sze, ed., John Wiley & Sons, New York, 1990, pp. 1–10.

[14] S. M. Sze, *Modern Semiconductor Device Physics*, John Wiley & Sons, New York, 1998.

[15] D. A. Bell, *Noise in the Solid State*, Pentech Press, London, 1985.

[16] M. J. Buckingham, *Noise in Electronic Devices and Systems*, John Wiley & Sons, New York, 1983.

[17] T. G. van de Roer, *Microwave Electronic Devices*, Chapman & Hall, London, 1994.

[18] M. Agethen, R. Reuter, T. Breder, R. M. Bertenburg, W. Brockerhoff, and F. J. Tegude, "Noise, high-frequency," in: *Wiley Encyclopedia of Electrical and*

Electronics Engineering, John G. Webster, ed., Vol. 14, John Wiley & Sons, New York, 1999, pp. 392–410.

[19] U. Erben, H. Schumacher, A. Schüppen, and J. Arndt, "Application of SiGe heterojunction bipolar transistors in 5.8 and 10 GHz low-noise amplifiers," *Electron. Lett.* **34**(15), 1498–1500 (1998).

[20] H. Dodo, Y. Amamiya, T. Niwa, M. Mamada, N. Goto, and H. Shimawaki, "Microwave low-noise AlGaAs/InGaAs HBT's with p^+-regrown base contacts," *IEEE Electron Device Lett.* **19**(4), 121–123 (1998).

[21] H. Hsia, Z. Tang, D. Caruth, D. Becher, and M. Feng, "Direct ion-implanted 0.12-μm GaAs MESFET with f_t of 121 GHz and f_{max} of 160 GHz," *IEEE Electron Device Lett.* **20**(5), 245–247 (1999).

[22] T. E. Kazior, "GaAs MHEMT: the evolution of GaAs-based FETs for high performance microwave and millimeter wave low noise applications," in: *23rd Workshop on Compound Semiconductor Devices and Integrated Cicuits—WOCSDICE'99, Chantilly, 1999*, pp. 71–74.

[23] T. Sonoda, Y. Yamamoto, N. Hayafuji, H. Yoshida, H. Sasaki, T. Kitano, S. Takamiya, and M. Ostubo, "Manufacturability and realiability of InP HEMTs," *Solid-State Electronics* **41**(10), 1621–1626 (1997).

[24] U. K. Mishra, A. S. Brown, S. E. Rosenbaum, C. E. Hooper, M. W. Pierce, M. J. Delaney, S. Vaughn, and K. White, "Microwave performance of AlInAs–GaInAs HEMT's with 0.2- and 0.1-μm gate length," *IEEE Electron Device Lett.* **9**(12), 647–648 (1988).

[25] J. Mateos, T. González, D. Pardo, V. Hoel, and A. Cappy, "Monte Carlo simulation of electronic noise in short channel δ-doped AlInAs/GaInAs HEMTs," in: *15th International Conference on Noise in Physical Systems and 1/f Fluctuations, Hong Kong 1999*, C. Surya, ed., Bentham Press, London, 1999, pp. 279–282.

[26] K. C. Hwang, P.C. Chao, C. Creamer, K. B. Nichols, S. Wang, D. Tu, W. Kong, D. Dugas, and G. Patton, "Very high gain millimeter-wave InAlAs/InGaAs/GaAs metamorphic HEMT's," *IEEE Electron Device Lett.* **20**(11), 551–553 (1999).

[27] G. R. Olbrich, "Characterization of low and high frequency noise in active devices," in: *15th International Conference on Noise in Physical Systems and 1/f Fluctuations, Hong Kong 1999*, C. Surya, ed., Bentham Press, London, 1999, pp. 25–30.

[28] G. Massobrio and P. Antognetti, *Semiconductor Device Modeling with SPICE*, McGraw-Hill, New York, 1988.

[29] M. Shur, *Physics of Semiconductor Devices*, Prentice Hall, Englewood Cliffs, New Jersey, 1990.

[30] J. Požela, *Physics of High-Speed Transistors*, Plenum, New York, 1993.

[31] C. Jacoboni and P. Lugli, *The Monte Carlo Method for Semiconductor Device Simulation*, Springer, Vienna, 1989.

[32] S. E. Laux and M. V. Fischetti, *Monte Carlo Device Simulation: Full Band and Beyond*, Kluwer, Boston, 1991.

[33] C. Moglestue, *Monte Carlo Simulation of Semiconductor Devices*, Chapman and Hall, London, 1993.

[34] G. H. Wannier, "Motion of gaseous ions in strong electric fields," *Bell Syst. Techn. J.* **32**(1), 170–254 (1953).

[35] P. Heymann, M. Rudolph, H. Prinzler, R. Doerner, L. Klapproth, G. Böck, "Experimental evaluation of microwave field-effect-transistor noise models," *IEEE Trans. Microwave Theory and Techniques* **MTT–47**(2), 156–163 (1999).

[36] M. W. Pospieszalski, "Modeling of noise parameters of MESFET's and MODFET's and their frequency and temperature dependence," *IEEE Trans. Microwave Theory and Techniques* **MTT–37**(9), 1340–1350 (1989).

[37] S. V. Gantsevich, V. L. Gurevich, and R. Katilius, "Fluctuations in semiconductor in strong electric field and light scattering from hot electrons," *Zh. Eksp. Teor. Fiz.* **57**(2), 503–519 (1969) [Sov. Phys. — JETP **30**(2), 276 (1970)].

[38] K. M. Van Vliet, "Markov approach to density fluctuations due to transport and scattering. II. Applications," *J. Math. Phys.* **12**(9), 1998–2012 (1971).

[39] B. B. Kadomtsev, "On fluctuations in gas," *Zh. Eksp. Teor. Fiz.* **32**(4), 943–944 (1957) [*Sov. Phys. — JETP* **5**(4), 771 (1957)].

[40] M. Lax, "Fluctuations from the nonequilibrium steady state," *Rev. Mod. Phys.* **32**(1), 25–64 (1960).

[41] P. J. Price, "Fluctuations of hot electrons," in: *Fluctuation Phenomena in Solids*, R. E. Burgess, ed., Academic Press, New York, 1965, pp. 355–380.

[42] Sh. M. Kogan and A. Ya. Shulman, "Theory of fluctuations in a nonequilibrium electron gas," *Zh. Eksp. Teor. Fiz.* **56**(3), 862–876 (1969) [*Sov. Phys. — JETP* **29**(3), 467 (1969)].

[43] E. M. Lifshitz and L. P. Pitaevskii, *Physical Kinetics*, Pergamon, New York, 1981.

[44] E. C. G. Stueckelberg, "Théorème H et unitarité de S," *Helv. Phys. Acta* **25**(5), 577–580 (1952).

[45] E. M. Conwell, *High Field Transport in Semiconductors*, Academic Press, New York, 1967.

[46] B. R. Nag, *Theory of Electrical Transport in Semiconductors*, Pergamon, London, 1972.

[47] B. R. Nag, *Electron Transport in Compound Semiconductors*, Springer, Berlin, 1980.

[48] *Hot-Electron Transport in Semiconductors*, L. Reggiani, ed., *Topics in Applied Physics*, Vol. 58, Springer, Berlin, 1985.

[49] V. F. Gantmakher and I. B. Levinson, *Carrier Scattering in Metals and Semiconductors*, North-Holland, Amsterdam, 1987.

[50] *Hot Carriers in Semiconductor Nanostructures: Physics and Applications*, J. Shah, ed., Academic Press, New York, 1992.

[51] *Spectroscopy of Nonequilibrium Electrons and Phonons*, C. V. Shank and B. P. Zakharchenya, eds., *Modern Problems in Condensed Matter Sciences*, Vol. 35, North-Holland, Amsterdam, 1992.

[52] D. Healey, T. P. McLean, "The determination of free carrier distribution functions by light scattering," *Phys. Lett.* **29A**(10), 607–608 (1969).

[53] A. Mooradian, "Light scattering from single-particle electron excitations in semiconductors," *Phys. Rev. Lett.* **20**(20), 1102–1104 (1968).

[54] A. Mooradian, "Light scattering in semiconductors," in: *Advances in Solid State Physics*, O. E. Madelung, ed., Vol. 9, Pergamon Press, New York, 1969, pp. 73–98.

[55] P. M. Platzman and P. A. Wolff, *Waves and Interactions in Solid-State Plasmas*, Academic Press, New York, 1973.

[56] *Light Scattering in Semiconductor Structures and Superlatices*, D. J. Lockwood and J. F. Young, eds., Plenum, New York, 1991.

[57] *Light Scattering in Solids I: Introduction Concepts*, M. Cardona, ed., 2nd ed., *Topics in Applied Physics*, Vol. 8, Springer, Berlin, 1982.

[58] *Light Scattering in Solids II: Basic Concepts and Instrumentation*, M. Cardona and G. Guntherodt, eds., *Topics in Applied Physics*, Vol. 50, Springer, Berlin, 1982.

[59] *Light Scattering in Solids III: Recent Results*, M. Cardona and G. Guntherodt, eds., *Topics in Applied Physics*, Vol. 51, Springer, Berlin, 1982.

[60] *Light Scattering in Solids IV: Electronic Scattering, Spin Effects, SERS, and Morphic Effects*, M. Cardona and G. Guntherodt, eds., *Topics in Applied Physics*, Vol. 54, Springer, Berlin, 1984.

[61] *Light Scattering in Solids V: Superlattices and other Microstructures*, M. Cardona and G. Guntherodt, eds., *Topics in Applied Physics*, Vol. 66, Springer, Berlin, 1989.

[62] *Light Scattering in Solids VI*, M. Cardona and G. Guntherodt, eds., Springer, Berlin, 1991.

[63] S. V. Gantsevich, V. L. Gurevich, V. D. Kagan, and R. Katilius, "Collision-controlled light scattering from hot solid state plasma," *Proceedings of the Second International Conference on Light Scattering in Solids, Paris, 1971*, M. Balkanski, ed., Flammarion Sciences, Paris, 1971, pp. 94–97.

[64] A. Einstein, *Investigations on the Theory of the Brownian Movement*, Dover, New York, 1965.

[65] M. von Smoluchowski, "Zur kinetischen Theorie der Brownschen Molekularbewegung und der Suspensionen," *Ann. Phys.* **21**(14), 756–780 (1906).

[66] Lord Rayleigh, "Remarks upon the law of complete radiation," *Philos. Mag.* **49**(301), 539–540 (1900).

[67] H. B. Callen and T. A. Welton, "Irreversibility and generalized noise," *Phys. Rev.* **83**(1), 34–40 (1951).

[68] L. D Landau and E. M. Lifshitz, *Statistical Physics, Part 1*, 3rd ed., Pergamon, New York, 1985.

[69] E. M. Lifshitz and L. P. Pitaevskii, *Statistical Physics, Part 2*, Pergamon, New York, 1980.

[70] S. V. Gantsevich, V. L. Gurevich, and R. Katilius, "Theory of fluctuations in nonequilibrium electron gas," *Rivista del Nuovo Cimento* **2**(5), 1–87 (1979).

[71] M. H. Ernst and E. G. D. Cohen, "Nonequilibrium fluctuations in μ space," *Journal of Statistical Physics* **25**(1), 153–180 (1981).

[72] A.-M. S. Tremblay, "Theories of fluctuations in nonequilibrium systems," in: *Recent Developments in Nonequilibrium Thermodynamics*, J. Casas-Vasquez, D. Jou and G. Lebon, eds., *Lecture Notes in Physics*, Vol. 199, Springer, Berlin, 1984, pp. 267–316.

[73] L. Reggiani, "General theory," in: *Hot-Electron Transport in Semiconductors*, L. Reggiani, ed., *Topics in Applied Physics*, Vol. 58, Springer, Berlin, 1985, pp. 7–86.

[74] J. Keizer, *Statistical Thermodynamics of Nonequilibrium Processes*, Springer, New York, 1987.

[75] V. Bareikis, R. Katilius, and R. Miliušytė, *Fluctuation Phenomena in Semiconductors in Nonequilibrium State* (in Russian), *Electrons in Semiconductors*, Vol. 8, Mokslas, Vilnius, 1989.

[76] N. A. Hashitsume, "A statistical theory of linear dissipative systems, II," *Prog. Theor. Phys.* **15**(4), 369–413 (1956).

[77] L. Onsager, "Reciprocal relations in irreversible processes. I," *Phys. Rev.* **37**(4), 405–426 (1931).

[78] L. Onsager, "Reciprocal relations in irreversible processes. II," *Phys. Rev.* **38**(12), 2265–2279 (1931).

[79] S. V. Gantsevich, V. L. Gurevich, and R. Katilius, "Current fluctuations in a semiconductor in high electric field," *Fiz. Tverd. Tela (Leningrad)* **11**(2), 308–315 (1969) [*Sov. Phys.—Solid State* **11**(2), 247 (1969)].

[80] M. A. Leontovich, "Main equations of kinetic theory of gas from the point of view of the theory of random processes," *Zh. Eksp. Teor. Fiz.* **5**(3/4), 211–231 (1935). Reprinted in: M.A. Leontovich, *Selected Papers. Theoretical Physics* (in Russian), Nauka, Moscow, 1985, p. 151.

[81] M. S. Green, "The nonequilibrium pair distribution function at low densities," *Physica* (Utrecht) **24**(6), 393–403, (1958).

[82] G. Ludwig, "An equation for description of fluctuation phenomena and turbulence in gases," *Physica* (Utrecht) **28**(9), 841–860 (1962).

[83] S. V. Gantsevich, V. L. Gurevich and R. Katilius, "On fluctuations in a nonequilibrium stationary state," *Zh. Eksp. Teor. Fiz.* **59**(2), 533–541 (1970) [*Sov. Phys.—JETP* **32**(2), 291 (1971)].

[84] Sh. M. Kogan and A. Ya. Shulman, "Extraneous random forces and equations for correlation functions in the theory of nonequilibrium fluctuations," *Fiz. Tverd. Tela (Leningrad)* **12**(4), 1119–1123 (1970) [*Sov. Phys.—Solid State* **12**(4), 874 (1970)].

[85] Yu. L. Klimontovich, "On nonequilibrium fluctuations in gas," *Teoret. Matemat. Fiz.* **9**(1), 109–123 (1971).

[86] K. M. Van Vliet, "Markov approach to density fluctuations due to transport and scattering. I. Mathematical formalism," *J. Math. Phys.* **12**(9), 1981–1998 (1971).

[87] A. G. Aronov and E. L. Ivchenko, "The theory of generation-recombination fluctuations in semiconductors in nonequilibrium conditions" *Fiz. Tverd. Tela (Leningrad)*, **13**(9), 2550–2557 (1971) [*Sov. Phys.—Solid State* **13**(9), 2142 (1972)].

[88] M. H. Ernst and J. R. Dorfman, "Nonanalytic dispersion relations in classical fluids. I. The hard-sphere gas," *Physica (Utrecht)* **61**(2), 157–181 (1972).

[89] H. K. Janssen, "Fluktuationen von Besetzungszahlen," *Z. Physik* **258**(3), 243–250 (1973).

[90] N. G. Van Kampen, "Fluctuations in Boltzmann's equation," *Phys. Lett.* **50A**(4), 237–238 (1974).

[91] V. D. Kagan, "Fluctuations in a system of charged particles" *Fiz. Tverd. Tela (Leningrad)* **17**(7), 1969–1977 (1975) [*Sov. Phys.—Solid State* **17**(7), 1289 (1975)].

[92] J. Keizer, "A theory of spontaneous fluctuations in macroscopic systems," *J. Chem. Phys.* **63**(1), 398–403 (1975).

[93] J. Logan and M. Kac, "Fluctuations and the Boltzmann equation. I," *Phys. Rev.* **A13**(1), 458–470 (1976).

[94] V. V. Bely and Yu. L. Klimontovich, "Kinetic fluctuations in a partially ionized plasma and in chemically reacting gases," *Zh. Eksp. Teor. Fiz.* **74**(5), 1660–1667 (1978) [*Sov. Phys.—JETP* **47**(5), 866–869 (1978)].

[95] A. Onuki, "On fluctuations in μ space," *Journal of Statistical Physics* **18**(5), 475–499 (1978).

[96] M. C. Marchetti and J. W. Dufty, "Kinetic and hydrodynamic theories of nonequilibrium fluctuations," *Physica* **118A**, 205–216 (1983).

[97] S. V. Gantsevich, V. L. Gurevich, and R. Katilius, "Comment on 'Fluctuations of the hot-carrier state-occupancy function in homogeneous semiconductors'," *Phys. Rev.* **B40**(17), 11958–11960 (1989).

[98] Sh. M. Kogan, "Equations for the correlation functions using a generalized Keldysh technique," *Phys. Rev.* **A44**(12), 8072–8083 (1991).

[99] K. G. Moh, H. S. Min, and Y. J. Park, "Equivalent noise source for Boltzmann transport equation with relaxation-time approximation in nondegenerate semiconductors," *J. Appl. Phys.* **74**(10), 6217–6221 (1993).

[100] C. M. Van Vliet and P. Vasilopoulos, "Master hierarchy of kinetic equations for binary collisions II: Addenda," *Can. J. Phys.* **75**(1), 1–6 (1997).

[101] P. Langevin, "Sur la théorie du mouvement Brownien," *Compt. Rend. Acad. des Sci. (Paris)*, **146**(10), 530 (1908).

[102] M. Bixon and R. Zwanzig, "Boltzmann-Langevin equation and hydrodynamic fluctuations," *Phys. Rev.* **187**(1), 267–272 (1969).

[103] R. F. Fox and G. E. Uhlenbeck, "Contributions to nonequilibrium thermodynamics. II. Fluctuation theory for the Boltzmann equation," *Phys. Fluids* **13**(12), 2881–2890 (1970).

[104] M. Lax, "Classical noise IV: Langevin methods," *Rev. Mod. Phys.* **38**(3), 541–566 (1966).

[105] S. V. Gantsevich, V. L. Gurevich, and R. Katilius, "Diffusion near nonequilibrium steady state," *Phys. Condensed Matter* **18**(3), 165–178 (1974).

[106] L. Huxley and R. Crompton, *The Diffusion and Drift of Electrons in Gases*, John Wiley & Sons, New York, 1974.

[107] S. Chapman and T. G. Cowling, *The Mathematical Theory of Non-uniform Gases*, 3rd edition, Cambridge University Press, Cambridge, 1970.

[108] W. V. Lovitt, *Linear Integral Equations*, McGraw-Hill, New York, 1924, Chapter III.

[109] N. Dunford and J. T. Schwartz, *Linear Operators, Part I: General Theory*, Interscience Publishers, New York, 1958, Chapter VI, Section 6.

[110] V. L. Gurevich and R. Katilius, "Theory of hot electrons in an anisotropic semiconductor," *Zh. Eksp. Teor. Fiz.* **49**(4), 1145–1156 (1965) [*Sov. Phys.—JETP* **22**(4), 796 (1966)].

[111] B. Davydov, "On the theory of the motion of electrons in gases and semiconductors," *Physik. Zeits. Sowjetunion* **12**(3), 269–300 (1937).

[112] H. R. Skullerud, "Longitudinal diffusion of electrons in electrostatic fields in gases," *J. Phys. B*, Ser. 2, **2**(6), 696–705 (1969).

[113] R. Katilius and R. Miliušytė, "Anisotropy of current fluctuations and of the electron diffusion in a strong electric field in the case of quasielastic scattering," *Zh. Eksp. Teor. Fiz.* **79**(2), 631–640 (1980) [*Sov. Phys.—JETP* **52**(2), 320–324 (1980)].

[114] R. I. Rabinovich, "On galvanomagnetic phenomena under hot-electron energy scattering on optical phonons," *Fiz. Tekhn. Poluprovodn.* **3**(7), 996–1004 (1969) [*Sov. Phys.—Semicond.* **3**(7), 839 (1969)].

[115] N. A. Zakhleniuk, V. A. Kochelap and V. V. Mitin, "Nonequilibrium fluctuations in semiconductors under scattering of energy of charge carriers on optical vibrations of the lattice," *Zh. Eksp. Teor. Fiz.* **95**(4), 1495–1512 (1989) [*Sov. Phys. — JETP* **68**(4), 863 (1989)].

[116] Z. S. Gribnikov and V. A. Kochelap, "Cooling of current carriers under scattering of energy on optical vibrations of the lattice," *Zh. Eksp. Teor. Fiz.* **58** (3), 1046–1056 (1970) [*Sov. Phys. — JETP* **31**(3), 562 (1970)].

[117] W. Shockley, "Hot electrons in germanium and Ohm's law," *Bell System Tech. J.* **30**(4), 990–1040 (1951).

[118] W. E. Pinson and R. Bray, "Experimental determination of the energy distribution functions and analysis of the energy-loss mechanisms of hot carriers in p-type germanium," *Phys. Rev.* **136**(5A), 1449–1466 (1964).

[119] I. Vosilyus and I. B. Levinson, "Generation of optical phonons and galvanomagnetic effects in the case of highly anisotropic electron distribution," *Zh. Eksp. Teor. Fiz.* **50**(6), 1660–1665 (1966) [*Sov. Phys. — JETP* **23**(6), 1104 (1966)].

[120] S. Komiyama, "Streaming motion and population inversion of hot carriers in crossed electric and magnetic fields," *Adv. Phys.* **31**(3), 255–297 (1982).

[121] S. Komiyama, T. Kurosawa, and T. Masumi, "Streaming motion of carriers in crossed electric and magnetic fields," in: *Hot-Electron Transport in Semiconductors*, L. Reggiani, ed., *Topics in Applied Physics*, Vol. 58, Springer, Berlin, 1985, pp. 177–199.

[122] A. A. Andronov, "The highly non-equilibrium hot-hole distributions in germanium," in: *Spectroscopy of Nonequilibrium Electrons and Phonons*, C. V. Shank and B. P. Zakharchenya, eds., *Modern Problems in Condensed Matter Sciences*, Vol. 35, North-Holland, Amsterdam, 1992, pp. 169–214.

[123] P. J. Price, "Noise theory of hot electrons," *IBM J. Res. Dev.* **3**(2), 191–193 (1959).

[124] I. B. Levinson and A. Yu. Matulis, "Current fluctuations in semiconductors in a high electric field," *Zh. Eksp. Teor. Fiz.* **54**(5), 1466–1478 (1968) [*Sov. Phys. — JETP* **27**(5), 786 (1968)].

[125] A. A. Andronov and V. A. Kozlov, "Low-temperature negative differential RF conductivity of semiconductors under inelastic scattering of electrons," *Pis'ma Zh. Eksp. Teor. Fiz.* **17**(2), 124–127 (1973) [*Sov. Phys. — JETP Lett.* **17**(2), 87 (1973)].

[126] Yu. V. Gulyaev and I. I. Chusov, "High-frequency effects in semiconductors at strongly anisotropic distribution of electrons," *Fiz. Tverd. Tela* (Leningrad) **20**(9), 2637–2644 (1978) [*Sov. Phys. — Solid State* **20**(9), 1524 (1978)].

[127] A. Matulis and A. Chenis, "Differential conductivity in semiconductors under inelastic scattering of electrons," *Zh. Eksp. Teor. Fiz.* **77**(3), 1134–1143 (1979) [*Sov. Phys. — JETP* **50**, 572 (1979)].

[128] P. J. Price, "Intervalley noise," *J. Appl. Phys.* **31**(6), 949–953 (1960).

[129] V. L. Gurevich, "On current fluctuations in semiconductors near nonequilibrium stedy state," *Zh. Eksp. Teor. Fiz.* **43**(5), 1771–1781, (1962) [*Sov. Phys. — JETP* **16**(5), 1252 (1963)].

[130] Sh. M. Kogan, "Electron temperature fluctuations and noise created by them," *Fiz. Tverd. Tela* (Leningrad) **5**(1), 224–228 (1963) [*Sov. Phys. — Solid State* **5**(1), 162 (1963)].

[131] Sh. M. Kogan and A. Ya. Shulman, "Electrical fluctuations in solid state plasma in high electric field," *Fiz. Tverd. Tela* (Leningrad) **9**(8), 2259–2264 (1967) [*Sov. Phys. — Solid State* **9**(8), 1771 (1968)].

[132] A. Ya. Shulman, "Correlation functions of nonequilibrium fluctuations in semiconductors in the electron-temperature approximation," *Fiz. Tverd. Tela* (Leningrad) **12**(4), 1181–1186 (1970) [*Sov. Phys.—Solid State* **12**(4), 922–925 (1970)].

[133] T. Kirkpatrick, E. G. D. Cohen, and J. R. Dorfman, "Kinetic theory of light scattering from a fluid not in equilibrium," *Phys. Rev. Lett.* **42**(14), 862–865 (1979).

[134] T. R. Kirkpatrick, E. G. D. Cohen, and J. R. Dorfman, "Fluctuations in a nonequilibrium steady state: Basic equations," *Phys. Rev.* **A26**(2), 950–971 (1982).

[135] R. Barkauskas and R. Katilius, "Current fluctuations in a strong electric field under conditions of frequent interelectron collisions ensuring a drifted Maxwellian distribution," *Zh. Eksp. Teor. Fiz.* **77**(3), 1144–1156 (1979) [*Sov. Phys.—JETP* **50**(3), 576–582 (1979)].

[136] S. Dedulevich, Ž. Kancleris, and A. Matulis, "Fluctuations and diffusion in a weakly heated electron gas," *Zh. Eksp. Teor. Fiz.* **95**(5), 1701–1710 (1989) [*Sov. Phys.—JETP* **68**(5), 982–987 (1989)].

[137] I. Ya. Korenblit and B. G. Tankhilevich, "Fluctuations in a nonequilibrium system of electrons and magnons in ferromagnetic semiconductors," *Phys. Stat. Sol. (b)* **124**(1), 79–86 (1984).

[138] S. V. Gantsevich, R. Katilius, and N. G. Ustinov, "Light scattering by nonequilibrium free carriers in a multivalley semiconductor," *Fiz. Tverd. Tela (Leningrad)* **16**(4), 1106–1113 (1974) [*Sov. Phys.—Solid State* **16**(4), 711–715 (1974)].

[139] S. V. Gantsevich, R. Katilius, and N. G. Ustinov, "Collision-controlled light scattering in a multivalley semiconductor," *Fiz. Tverd. Tela (Leningrad)* **16**(4), 1114–1121 (1974) [*Sov. Phys.—Solid State* **16**(4), 716–720 (1974)].

[140] R. Barkauskas, S. V. Gantsevich, and R. Katilius, "Scattering of light by an electron-hole plasma in a semiconductor in the collisional regime," *Fiz. Tverd. Tela (Leningrad)* **30**(10), 3030–3035 (1988) [*Sov. Phys.—Solid State* **30**(10), 1743–1746 (1988)].

[141] R. Barkauskas, S. V. Gantsevich, and R. Katilius, "Scattering of electromagnetic waves from nonequilibrium collisional multicomponent plasma in semiconductors," *Fiz. Tverd. Tela (Leningrad)* **31**(10), 157–165 (1989) [*Sov. Phys.—Solid State* **31**(10), 1741–1745 (1989)].

[142] Sh. M. Kogan and V. D. Shadrin, "Scattering of infrared radiation in semiconductors in strong electric fields," *Fiz. Tekhn. Poluprovodn.* **5**(11), 2224–2226 (1971) [*Sov. Phys.—Semicond.* **5**(11), 1942 (1972)].

[143] N. A. Zakhlenyuk and V. A. Kochelap, "Hydrodynamic fluctuations and scattering of electromagnetic waves in semiconductors in strong electric fields," *Zh. Eksp. Teor. Fiz.* **102**(3), 934–962 (1992) [*Sov. Phys.—JETP* **75**(3), 510–524 (1992)].

[144] V. A. Kochelap and N. A. Zakhleniuk, "Hydrodynamic fluctuations of a hot-electron gas," *Phys. Rev.* **B50**(12), 8325–8341 (1994).

[145] L. L. Akatov, S. V. Gantsevich, R. Katilius, and V. M. Rysakov, "Collision-controlled scattering of light by free electrons in semiconductors and the influence exerted on it by an electric field," *Pis'ma Zh. Eksp. Teor. Fiz.* **27**(11), 633–636 (1978) [*Sov. Phys.—JETP Lett.* **27**(11), 597–601 (1978)].

[146] J. Doehler, P. J. Colwell, and S. A. Solin, "Raman scattering from semiconducting and metallic Ge(As)," *Phys. Rev. Lett.* **34**(10), 584–587 (1975).

[147] J. Doehler, "Raman scattering from electronic excitations in Ge," *Phys. Rev.* **B12**(8), 2917–2931 (1975).

[148] T. L. Andreeva and A. V. Malyugin, "Spectrum of Rayleigh light scattering in a nonequilibrium gas," *Zh. Eksp. Teor. Fiz.* **92**(5), 1549–1563 (1987) [*Sov. Phys.—JETP* **65**(5), 869–876 (1987)].

[149] C. R. Oberman and E. A. Williams, "Theory of fluctuations in plasma," in: *Basic Plasma Physics—1, Handbook of Plasma Physics*, Vol. 1, A. Galeev and R. N. Sudan, eds., North-Holland, Amsterdam, 1983, pp. 279–334.

[150] A. Ya. Shulman and Sh. M. Kogan, "Theory of fluctuations in nonequilibrium electron gas, II, Space-inhomogeneous fluctuations" *Zh. Eksp. Teor. Fiz.* **57**(6), 2112–2119 (1969) [*Sov. Phys.—JETP* **30**(6), 1146 (1970)].

[151] E. Erlbach and J. B. Gunn, "Noise temperature of hot electrons in germanium," *Phys. Rev. Lett.* **8**(7), 280–282 (1962).

[152] C. A. Bryant, "Noise temperature of hot electrons in gallium-arsenide," *Bull. Am. Phys. Soc.* **9**(1), 62 (1964).

[153] V. Bareikis, R. Šaltis, and J. Požela, "On the electrical fluctuations in germanium in high electric field" (in Russian), *Lietuvos fizikos rinkinys* **6**(1), 99–104 (1966).

[154] V. Bareikis, I. Vaitkevičiūtė, and J. Požela, "Fluctuations of hot current carriers in germanium" (in Russian), *Lietuvos fizikos rinkinys* **6**(3), 437–440 (1966).

[155] E. B. Wagner, F. J. Davis, and G. S. Hurst, "Time-of-flight investigations of electron transport in some atomic and molecular gases," *J. Chem. Phys.* **47**(9), 3138–3147 (1967).

[156] V. Bareikis, J. Pozhela, and I. Matulionienė, "Noise and diffusion of hot holes in p-type germanium," in: *IX International Conference on the Physics of Semiconductors, Moscow, 1968, Proceedings*, S. M. Ryvkin, ed., Nauka, Leningrad, 1968, Vol. 2, pp. 760–765.

[157] V. Bareikis, I. Matulionienė, and J. Požela, "Noise of hot current carriers in p-type germanium" (in Russian), *Lietuvos fizikos rinkinys* **7**(2), 381–385 (1967).

[158] J. K. Požela, V. A. Bareikis, and I. B. Matulionienė, "Electric fluctuations of hot electrons in silicon," *Fiz. Tekhn. Poluprovodn.* **2**(4), 606–607 (1968) [*Sov. Phys.—Semicond.* **2**(4), 503–504 (1968)].

[159] J. P. Nougier and M. Rolland, "Mobility, noise temperature, and diffusivity of hot holes in germanium," *Phys Rev.* **B8**(12), 5728–5737 (1973).

[160] J. P. Nougier, J. Comallonga, and M. Rolland, "Pulsed technique for noise temperature measurement," *J. Phys. E: Scientific Instruments* **7**(4), 287–290 (1974).

[161] J. G. Ruch and G. S. Kino, "Transport properties of GaAs," *Phys Rev.* **174**(4), 921–931 (1968).

[162] L. Reggiani, C. Canali, F. Nava, and A. Alberigi-Quaranta, "Diffusion coefficient of holes in Ge," *J. Appl. Phys.* **49**(8), 4446–4452 (1978).

[163] R. Brunetti, C. Jacoboni, F. Nava, L. Reggiani, G. Bosman, and R. J. J. Zijlstra, "Diffusion coefficient of electrons in silicon," *J. Appl. Phys.* **52**(11), 6713–6722 (1981).

[164] V. Bareikis, A. Galdikas, R. Miliušytė, and V. Viktoravičius, "Noise of hot holes in Ge due to predominant inelastic scattering," in: *Sixth International Conference on Noise in Physical Systems, Gaithersburg, 1981*, Proceedings, P. H. E. Meijer, R. D. Mountain, and R. J. Soulen, Jr., eds., National Bureau of Standards, Washington, D.C., 1981, pp. 406–408.

[165] V. Bareikis, V. Viktoravičius, and A. Galdikas, "Hot-hole noise in germanium at liquid-helium temperatures," *Lietuvos Fizikos Rinkinys* **21**(2), 73–79 (1981) [*Sov. Phys.—Collection*, **21**(2), 51–55 (1981)].

[166] V. Bareikis, A. Galdikas, R. Miliušytė, J. Pozhela, and V. Viktoravičius, "Noise and diffusivity of hot electrons in n-type InSb," *J. de Phys. Colloq. C7*, suppl. **42**(10), C7-215–C7-220 (1981).

[167] V. Bareikis, V. Viktoravičius, A. Galdikas, R. Miliušytė, and J. Požela, "Noise and diffusion of hot electrons in n-type InSb," *Fiz. Tekhn. Poluprovodn.* **16**(10), 1816–1819 (1982) [*Sov. Phys.—Semicond.* **16**(10), 1165 (1982)].

[168] V. Bareikis, A. Galdikas, and J. Požela, "Noise, energy relaxation time and diffusion of hot holes in p-Ge in a magnetic field," *Fiz. Tekhn. Poluprovodn.* **11**(2), 365–372 (1977) [*Sov. Phys.—Semicond.* **11**(2), 210 (1977)].

[169] V. Bareikis, V. Viktoravičius, A. Galdikas, and R. Miliušytė, "Noise and diffusion of hot carriers in n-type Ge and n-type Si in a magnetic field," *Fiz. Tekhn. Poluprovodn.* **12**(1), 156–160 (1978) [*Sov. Phys.—Semicond.* **12**(1), 89 (1978)].

[170] E. Erlbach, "New hot-electron negative resistance effect," *Phys. Rev.* **132**(5), 1976–1979 (1963).

[171] T. W. Sigmon and J. F. Gibbons, "Diffusivity of electrons and holes in silicon," *Appl. Phys. Lett.* **15**(10) 320–322 (1969).

[172] L. G. Hart, "High-field current fluctuations in n-type germanium," *Can. J. Phys.* **48**(5), 531–542 (1970).

[173] D. J. Bartelink and G. Persky, "Diffusion of electrons in silicon transverse to a high electric field," *Appl. Phys. Lett.* **16**(5), 191–194 (1970).

[174] G. Persky and D. J. Bartelink, "High-field diffusivity of electrons in silicon," *J. Appl. Phys.* **42**(11), 4414–4421 (1971).

[175] C. Canali, C. Jacoboni, G. Ottaviani, and A. Alberigi-Quaranta, "High-field diffusion of electrons in silicon," *Appl. Phys. Lett.* **27**(5), 278–280 (1975).

[176] V. Bareikis, A. Galdikas, I. Matulionienė, and R. Miliušytė, "Anisotropy of hot electron noise in n-Ge in SHF range," *Fiz. Tekhn. Poluprovodn.* **9**(6), 1180–1182 (1975) [*Sov. Phys.—Semicond.* **9**(6), 784 (1975)].

[177] V. Bareikis, V. Viktoravičius, A. Galdikas, and R. Miliušytė, "Noise, small-signal conductivity and diffusion of charge carriers in n-type Ge and n-type Si in a high electric field," *Fiz. Tekhn. Poluprovodn.* **12**(1), 151–155 (1978) [*Sov. Phys.—Semicond.* **12**(1), 85 (1978)].

[178] D. Gasquet and J. P. Nougier, "Anisotropy of the differential conductivity and of the transverse diffusion coefficient in n-type silicon," *Appl. Phys. Lett.* **33**(1), 89–91 (1978).

[179] C. Canali, C. Jacoboni, and F. Nava, "Intervalley diffusion of hot electons in germanium," *Solid State Commun.* **26**(12), 889–892 (1978).

[180] V. Bareikis, V. Viktoravičius, and A. Galdikas, "Frequency dependence of noise in n-type Si in high electric fields," *Fiz. Tekhn. Poluprovodn.* **16**(10), 1868–1870 (1982) [*Sov. Phys.—Semicond.* **16**(10), 1202 (1982)].

[181] V. Bareikis, V. Viktoravičius, A. Galdikas, and R. Miliušytė, "Microwave noise and the coupling constant of Γ and L valleys in the three-valley model of GaAs,"

Fiz. Tekhn. Poluprovodn. **14**(7), 1427–1429 (1980) [*Sov. Phys.—Semicond.* **14**(7), 847–849 (1980)].

[182] D. Gasquet, M. Fadel, and J. P. Nougier, "Noise of hot electrons in indium phosphide," in: *Noise in Physical Systems and 1/f Noise, Proceedings of the 7th International Conference on Noise in Physical Systems and the 3rd International Conference on 1/f Noise*, Montpellier, 1983, M. Savelli, G. Lecoy, and J. P. Nougier, eds., North-Holland, Amsterdam, 1983, pp. 169–171.

[183] D. Gasquet, M. de Murcia, J. P. Nougier, and C. Gontrand, "Transport parameters of hot electrons in GaAs at 300 K," *Physica* **134 B&C**(1–3), 264–268 (1985).

[184] V. Bareikis, Ž. Bilkis, J. Liberis, A. Matulionis, R. Miliušytė, J. Požela, P. Sakalas, and R. Šaltis, "Hot electron noise and diffusion in short n^+-n-n^+ GaAs and InP structures," *Lietuvos Fizikos Rinkinys* **27**(5), 511–521 (1987) [*Sov. Phys.—Collection* **27**(5), 1–9 (1987)].

[185] J. P. Nougier, "Survey of noise of hot carriers: Some experimental and theoretical aspects," in: *Noise in Physical Systems, Proceedings of the 5th International Conference on Noise*, Bad Nauheim, 1978, D. Wolf, ed., Springer, Berlin, 1978, pp. 72–89.

[186] V. Bareikis, A. Galdikas, and R. Miliušytė, "Fluctuations and diffusion coefficient" (in Russian), in: *Hot Electron Diffusion*, J. Požela, ed., *Electrons in Semiconductors*, Vol. 3, Mokslas, Vilnius, 1981, pp. 127–161.

[187] V. Bareikis, R. Katilius, A. Matulionis, R. Šaltis, S. Gantsevich, V. Gurevich, Sh. Kogan, and A. Shulman, "Fluctuations and noise of hot electrons in semiconductors (review)," *Lithuanian Phys. J.* **31**(1), 3–25 (1991).

[188] Y. K. Pozhela, "Transport parameters from microwave conductivity and noise measurements," in: *Hot-Electron Transport in Semiconductors*, L. Reggiani, ed., *Topics in Applied Physics*, Vol. 58, Springer, Berlin, 1985, pp. 113–147.

[189] J. P. Nougier and M. Rolland, "Differential relaxation times and diffusivities of hot carriers in isotropic semiconductors," *J. Appl. Phys.* **48**(4), 1683–1687 (1977).

[190] M. Rolland and J. P. Nougier, "Hot electron longitudinal diffusion in silicon at 300 K," in: *Physics of Semiconductors, Proceedings of the 13th International Conference*, Roma, 1976, F. G. Fumi, ed., Roma, 1976, pp. 1227–1230.

[191] D. Gasquet, H. Tijani, J. P. Nougier, and A. Van der Ziel, "Diffusion and generation recombination of hot electrons in silicon at 77 K," in: *Noise in Physical Systems and 1/f Noise, Proceedings of the 7th International Conference on Noise in Physical Systems and the 3rd International Conference on 1/f Noise*, Montpellier, 1983, M. Savelli, G. Lecoy, and J. P. Nougier, eds., North-Holland, Amsterdam, 1983, pp. 165–167.

[192] V. Bareikis, J. Liberis, A. Matulionis, R. Miliušytė, and P. Sakalas, "Noise and diffusivity of hot electrons in GaAs and InP at 80 K," in: *Ninth International Conference on Noise in Physical Systems, Montréal, 1987*, C. M. VanVliet, ed., World Scientific, Singapore, 1987, pp. 109–112.

[193] M. de Murcia, D. Gasquet, A. Elamri, J. P. Nougier, and J. Vanbremeersch, "Diffusion and noise in GaAs material and devices," *IEEE Trans. Electron Devices* **ED-38**(11), 2531–2539 (1991).

[194] V. Bareikis, Z. Bilkis, J. Liberis, P. Sakalas, and R. Šaltis, "Noise and diffusivity in short n^+-n-n^+ structures of InP," *Fiz. Tekhn. Poluprovodn.* **22**(6), 1040–1044 (1988) [*Sov. Phys.—Semicond.* **22**(6), 656 (1988)].

[195] M. de Murcia, E. Richard, D. Gasquet, J. P. Dubuc, J. Vanbremeersch, and J. Zimmermann, "High frequency noise and diffusion coefficient of hot electrons in bulk $Al_{0.25}Ga_{0.75}As$," *Solid-State Electron.* **37**(8), 1477–1483 (1994).

[196] R. Šaltis, Soviet Patent 248063, "Gate modulated meter of weak pulse signals with the alternating coefficient of filling" (in Russian), *Byulleten Izobretenij* No. 23, 1969.

[197] V. Aninkevičius, V. Bareikis, J. Liberis, A. Matulionis, and P. S. Kop'ev, "Real-space-transfer noise and diffusion in GaAs/AlGaAs heterostructure," in: *Proceedings of the International Conference on Noise in Physical Systems and 1/f Fluctuations—ICNF'91, Kyoto, 1991*, T. Musha, S. Sato, and M. Yamamoto, eds., Ohmsha, Tokyo, 1991, pp. 183–186.

[198] M. de Murcia, E. Richard, J. M. Perraudin, A. Boyer, A. Benvenuti, and J. Zimmermann, "Detection and evaluation of self-heating effects in n^+nn^+ $Al_xGa_{1-x}As$ devices by noise temperature measurements," *Semicond. Sci. Technol.* **10**(4), 515–522 (1995).

[199] J. Vindevoghel, Y. Leroy, C. Bruneel, and J. Zimmermann, "An original measurement of high-field effect on microwave conductivity of semiconductors," *Rev. Sci. Instrum.* **45**(7), 920–921 (1974).

[200] D. Gasquet, J. C. Vaissière, and J. P. Nougier, "New method for wide band measurement of noise temperature of one-port networks at high pulsed bias," in: *Sixth International Conference on Noise in Physical Systems, Gaithersburg, 1981, Proceedings*, P. H. E. Meijer, R. D. Mountain, and R. J. Soulen, Jr., eds., National Bureau of Standards, Washington, D.C., 1981, pp. 305–308.

[201] V. Aninkevičius, V. Bareikis, R. Katilius, J. Liberis, I. Matulionienė, A. Matulionis, P. Sakalas, and R. Šaltis, "Γ–X Intervalley scattering time constant for GaAs estimated from hot-electron noise spectroscopy data," *Phys. Rev.*, **B53**(11), 6893–6895 (1996).

[202] S. Ašmontas, G. Valušis, J. Liberis, and L. Subačius, "Noise temperature in compensated n-type InSb:Cr," *Fiz. Tekhn. Poluprovodn.* **24**(12), 2214–2216 (1990) [*Sov. Phys.—Semicond.* **24**(12), 1373 (1990)].

[203] V. Bareikis, R. Barkauskas, A. Galdikas, and R. Katilius, "Ambipolar diffusion of hot charge carriers and noise in semiconductors," *Fiz. Tekhn. Poluprovodn.* **14**(9), 1760–1767 (1980) [*Sov. Phys.—Semicond.* **14**(9), 1046 (1980)].

[204] A. Dargys and T. Banys, "Dependence of the phenomenological energy relaxation time on electric field in n-Si and n-Ge at 77 K," *Phys. Stat. Sol. (b)* **52**(2), 699–706 (1972).

[205] A. Dargys and J. Kundrotas, *Handbook on Physical Properties of Ge, Si, GaAs and InP*, Science and Encyclopedia Publishers, Vilnius, 1994.

[206] V. Bareikis, A. Galdikas, R. Miliušytė, and A. Matulionis, "Calculation of noise in p-type Ge in a high electric field by the Monte-Carlo method," *Fiz. Tekhn. Poluprovodn.* **13**(6), 1123–1126 (1979) [*Sov. Phys.—Semicond.* **13**(6), 658 (1979)].

[207] D. Gasquet, B. Azais, J. C. Vaissiere, and J. P. Nougier, "Generation–recombination noise in p-Si at 77 K," in: *Noise in Physical Systems and 1/f Noise–1985, Proceedings of the 8th International Conference on Noise in Physical Systems and the 4th International Conference on 1/f Noise*, Rome, 1985, A. D'Amico and P. Mazzetti, eds., North-Holland, Amsterdam, 1986, pp. 231–233.

[208] L. Reggiani, P. Lugli, and V. Mitin, "Generalization of Nyquist–Einstein relationship to conditions far from equilibrium in nondegenerate semiconductors," *Phys. Rev. Lett.* **60**(8), 736–739 (1988).

[209] L. Reggiani, T. Kuhn, and L. Varani, "Noise and correlation functions of hot carriers in semiconductors," *Appl. Phys.* **A54**(1), 411–427 (1992).

[210] L. Varani and L. Reggiani, "Microscopic theory of electronic noise in semiconductor unipolar structures," *Rivista del Nuovo Cimento*, **17**, ser. 3, no. 7, 1–110 (1994).

[211] J. W. Holm-Kennedy and K. S. Champlin, "Warm-carrier microwave transport in n Si," *J. Appl. Phys.* **43**(4), 1889–1903 (1972).

[212] C. Canali, F. Nava, and L. Reggiani, "Drift velocity and diffusion coefficients from time-of-flight measurements," in: *Hot-Electron Transport in Semiconductors*, L. Reggiani, ed., *Topics in Applied Physics*, Vol. 58, Springer, Berlin, 1985, pp. 87–112.

[213] C. Jacoboni, F. Nava, C. Canali, and G. Ottaviani, "Electron drift velocity and diffusivity in germanium," *Phys. Rev.*, **B24**(2), 1014–1026 (1981).

[214] C. Jacoboni and L. Reggiani, "Bulk hot-electron properties of cubic semiconductors," *Adv. Phys.* **28**(4), 493–553 (1979).

[215] D. N. Mirlin, V. F. Sapega, I. Ya. Karlik, and R. Katilius, "Hot photoluminescence spectroscopy investigations of L-valley splitting and intervalley scattering in uniaxially stressed gallium arsenide," *Solid State Commun.* **61**(12), 799–802 (1987).

[216] B. K. Ridley and T. B. Watkins, "The possibility of negative resistance effects in semiconductors," *Proc. Phys. Soc.* **78**(2), 293–304 (1961).

[217] C. Hilsum, "Transferred electron amplifiers and oscillators," *Proc. IRE* **50**(2), 185–189 (1962).

[218] J. B. Gunn and B. J. Elliott, "Measurement of the negative differential mobility of electrons in GaAs," *Phys. Lett.* **22**(4), 369–371 (1966).

[219] J. B. Gunn, "Microwave oscillation of current in III–V semiconductors," *Solid State Commun.* **1**(4), 88–91 (1963).

[220] M. Shur, *GaAs Devices and Circuits*, Plenum Press, New York, 1985.

[221] M. P. Show, H. L. Grubin, and P. R. Solomon, *The Gunn–Hilsum Effect*, Academic Press, New York, 1979.

[222] B. G. Bosh and R. W. H. Eagelman, *Gunn Effect Electronics*, Pritman Publishing House, London, 1975.

[223] J. Požela and A. Reklaitis, "Diffusion coefficient of hot electrons in GaAs," *Solid State Commun.* **27**(11), 1073–1077 (1978).

[224] J. Graffeuil, J. F. Sautereau, G. Blasquez, and P. Rossel, "Noise temperature in GaAs epi-layer for FET's," *IEEE Trans. Electron Devices*, **ED–25**(6), 596–599 (1978).

[225] G. Hill, P. N. Robson, and W. Fawcett, "Diffusion and the power spectral density of velocity fluctuations for electrons in InP by Monte-Carlo methods," *J. Appl. Phys.* **50**(1), 356–360 (1979).

[226] C. Hammar and B. Vinter, "Diffusion of hot electrons in n indium phosphide," *Electron. Lett.* **9**(1), 9–10 (1973).

[227] D. N. Mirlin and V. I. Perel', "Hot-electron photoluminescence under continuous-wave pumping," in: *Spectroscopy of Nonequilibrium Electrons and Phonons*,

C. V. Shank and B. P. Zakharchenya, eds., *Modern Problems in Condensed Matter Sciences*, Vol. 35, North-Holland, Amsterdam, 1992, pp. 269–325.

[228] V. Aninkevičius, V. Bareikis, J. Liberis, A. Matulionis, and P. Sakalas, "Comparative analysis of microwave noise in GaAs and AlGaAs/GaAs channels," *Solid-State Electron.* **36**(9), 1339–1343 (1993).

[229] A. Matulionis, J. Požela, and A. Reklaitis, "Drift velocity oscillations in n-GaAs at 77 K," *Phys. Stat. Sol. (a)* **31**(1), 83–87 (1975).

[230] V. Bareikis, J. Liberis, A. Matulionis, R. Miliušytė, and P. Sakalas, "Long-range fluctuations of hot electrons in GaAs and InP at 80 K," in: *Noise in Physical Systems*, 10th International Conference, Budapest, 1989, *Proceedings*, A. Ambrózy, ed., Akadémiai Kiadó, Budapest, 1990, pp. 53–56.

[231] K. Seeger, *Semiconductor Physics*, Springer, Vienna, 1973, p. 170.

[232] J. G. Ruch and W. Fawcett, "Temperature dependence of of the transport properties of gallium arsenide determined by a Monte Carlo method," *J. Appl. Phys.* **41**(9), 3843–3849 (1970).

[233] A. Matulionis, J. Požela, and A. Reklaitis, "Monte Carlo treatment of electron–electron collisions," *Solid State Commun.* **16**(10/11), 1133–1137 (1975).

[234] J. Požela and A. Reklaitis, "Electron transport properties in GaAs at high electric fields," *Solid-State Electron.* **23**(9), 927–933 (1980).

[235] P. Lugli and D. K. Ferry, "Effect of electron–electron scattering on Monte Carlo studies of transport in submicron semiconductor devices," *Physica* **117–118 B&C**(1), 251–253 (1983).

[236] P. Lugli and D. K. Ferry, "Degeneracy in the Ensemle Monte Carlo Method for high-field transport in semiconductors," *IEEE Trans. Electron Devices* **ED–32**(11), 2431–2437 (1985).

[237] R. Brunetti, C. Jacoboni, A. Matulionis, and V. Dienys, "Effect of interparticle collisions on energy relaxation of carriers in semiconductors," *Physica* **134B&C**(1–3), 369–373 (1985).

[238] A. Hasegawa, K. Miyatsuji, K. Taniguchi, and C. Hamaguchi, "A new Monte Carlo simulation of hot electron transport with electron–electron scattering," *Solid-State Electron.* **31**(3/4), 547–550 (1988).

[239] L. Reggiani, P. Lugli, S. Gantsevich, V. Gurevich, and R. Katilius, "Diffusion and fluctuations in a nonequilibrium electron gas with electron-electron collisions," *Phys. Rev.* **B40**(18), 12209–12214 (1989).

[240] M. Moško and A. Mošková, "Ensemble Monte Carlo simulation of electron–electron scattering: Improvements of conventional methods," *Phys. Rev.* **B44**(19), 10794–10803 (1991).

[241] V. Gružinskis, E. Starikov, and P. Shiktorov, "Conservation equations for hot carriers — I. Transport models," Solid-State Electron. **36**(7)1055–1066 (1993).

[242] P. Bordone, L. Reggiani, L. Varani, and T. Kuhn, "Hot-phonon effect on noise and diffusion in GaAs," *Semicond. Sci. Technol.* **9**(5S), 623–626 (1994).

[243] L. Varani, "Contribution of interparticles correlations to electronic noise in semiconductors," in: *Noise in Physical Systems and 1/f Fluctuations, Proceedings of the 13th International Conference*, Palanga, Lithuania, 1995, V. Bareikis and R. Katilius, eds., World Scientific, Singapore, 1995, pp. 203–212.

[244] P. Shiktorov, V. Gružinskis, E. Starikov, L. Reggiani, and L. Varani, "Noise temperature of n^+nn^+ GaAs structures," *Phys. Rev.* **B54**(12), 8821–8832 (1996).

[245] L. Reggiani, E. Starikov, P. Shiktorov, V. Gružinskis, and L. Varani, "Modelling of small-signal response and electronic noise in semiconductor high-field transport," *Semicond. Sci. Technol.* **12**(2), 141–156 (1997).

[246] A. Matulionis, R. Raguotis, and R. Katilius, "Interparticle collisions and hot-electron velocity fluctuations in GaAs at 80 K," *Phys. Rev.* **B56**(4), 2052–2057 (1997).

[247] R. Katilius, J. Liberis, A. Matulionis, R. Raguotis, P. Sakalas, J.-P. Nougier, J. C. Vaissière, L. Varani, and L. Rota, "Noise and electron diffusion in doped n-type GaAs at heating electric fields," *Phys. Rev.* **B60**(16), 11487–11493 (1999).

[248] T. Zubkutė, private communication (April, 2000).

[249] A. Matulionis, J. Liberis, I. Matulionienė, P. Gottwald, J. Karanyi, B. Szentpáli, H. L. Hartnagel, K. Mutamba, and A. Sigurdardöttir, "Hot-electron diffusion coefficient in standard doped InP channels," in: *21st Workshop on Compound Semiconductor Devices and Integrated Circuits—WOCSDICE 97, Scheveningen, The Netherlands, 1997*, Proceedings, Eindhoven University of Technology, 1997, pp. 61–62.

[250] V. Bareikis, J. Liberis, A. Matulionis, R. Miliušytė, J. Požela, and P. Sakalas, "Length dependent hot electron noise in doped GaAs," *Solid-State Electron.* **32**(12), 1647–1650 (1989).

[251] V. Bareikis, J. Liberis, A. Matulionis, R. Miliušytė, J. Požela, and P. Sakalas, "Impurity resonant scattering of hot electrons in GaAs," in: *20th International Conference on the Physics of Semiconductors, Thessaloniki, 1990, Proceedings*, E. M. Anastassakis and J. D. Joannopoulos, eds., World Scientific, Singapore, 1990, Vol. 3, pp. 2479–2482.

[252] R. T. Bate, "Evidence for selenium donor level above the principal conduction band edge in GaSb," *J. Appl. Phys.* **33**(1), 26–28 (1962).

[253] W. Paul, "Impurity levels associated with multi-conduction bands," in: *IX International Conference on the Physics of Semiconductors, Moscow, 1968, Proceedings*, S. M. Ryvkin, ed., Nauka, Leningrad, 1968, Vol. 1, pp. 16–26.

[254] M. di Marco, Q. H. Wang, and J. G. Swanson, "Optoelectronic modulation spectroscopy (OEMS)—a new tool for device investigations," *Mater. Sci. Eng.*, **B20**, 225–231 (1993).

[255] D. J. Wolford, "Electronic states in semiconductors under high pressures," in: *18th International Conference on the Physics of Semiconductors, Stockholm, 1986, Proceedings*, O. Engström, ed., World Scientific, Singapore, 1987, Vol. 2, pp. 1115–1123.

[256] V. Bareikis, R. Katilius, and A. Matulionis, "Noise in confined structures of doped semiconductors," in: *Noise in Physical Systems and 1/f Fluctuations, Proceedings of the 12th International Conference*, St. Louis, 1993, P. H. Handel and A. L. Chung, eds., *AIP Conference Proceedings*, Vol. 285, AIP Press, New York, 1993, pp. 181–186.

[257] A. Matulionis and T. Smertina, "Resonant impurity scattering and fluctuations of hot electrons in GaAs: A Monte Carlo study," *Lithuanian J. Phys.* **35**(1), 64–71 (1995).

[258] A. Matulionis, J. Liberis, I. Matulionienė, T. Zubkutė, M. de Murcia, and F. Pascal, "Microwave- and low-frequency fluctuations caused by DX-centres in GaAs and AlGaAs," in: *Noise in Physical Systems and 1/f Fluctuations, Proceedings of the 14th International Conference*, Leuven, 1997, C. Claeys and E. Simoen, eds., World Scientific, Singapore, 1997, pp. 453–456.

[259] J. A. Kash and J. C. Tsang, "Secondary emission studies of hot carrier relaxation in polar semiconductors," *Solid-State Electron.* **31**(3/4), 419–424 (1988).

[260] J. A. Kash and J. C. Tsang, "Nonequilibrium phonons in semiconductors," in: *Spectroscopy of Nonequilibrium Electrons and Phonons*, C. V. Shank and B. P. Zakharchenya, eds., *Modern Problems in Condensed Matter Sciences*, Vol. 35, North-Holland, Amsterdam, 1992, pp. 113–167.

[261] V. Bareikis, K. Kibickas, J. Liberis, A. Matulionis, R. Miliušytė, J. Paršeliūnas, J. Požela, and P. Sakalas, "Velocity overshoot and suppression of diffusivity and microwave noise in short n^+-n-n^+ structures of GaAs," in: *High-Speed Electronics, Proceedings of the International Conference*, Stockholm, 1986, B. Källbäck and H. Beneking, eds., Springer, Berlin, 1986, pp. 28–31.

[262] K. J. Schmidt-Tiedemann, "Tensor theory of the conductivity of warm electrons in cubic semiconductors," *Philos. Res. Repts.* **18**(4), 338–360 (1963).

[263] V. Bareikis and R. Katilius, "Fluctuation spectroscopy of hot electrons in semiconductors," in: *Noise in Physical Systems, 10th International Conference, Budapest, 1989, Proceedings*, A. Ambrózy, ed., Akadémiai Kiadó, Budapest, 1990, pp. 189–199.

[264] V. Bareikis, J. Liberis, A. Matulionis, R. Miliušytė, J. Požela, and P. Sakalas, "Size dependent diffusion and microwave noise in GaAs and InP," in: *19th International Conference on the Physics of Semiconductors, Warsaw 1988, Proceedings*, W. Zawadzki, ed., Institute of Physics, Polish Academy of Science, Warsaw, 1988, Vol. 2, pp. 1427–1430.

[265] H. Fröhlich and B. V. Paranjape, "Dielectric breakdown in solids," *Proc. Phys. Soc. (London)* **B69**(433), 21–32 (1956).

[266] V. Bareikis, K. Kibickas, J. Liberis, R. Miliushyte, and P. Sakalas, "Noise in short n^+-n-n^+ GaAs diodes," in: *Noise in Physical Systems and 1/f Noise–1985, Proceedings of the 8th International Conference on Noise in Physical Systems and the 4th International Conference on 1/f Noise*, Rome, 1985, A. D'Amico and P. Mazzetti, eds., North-Holland, Amsterdam, 1986, pp. 203–206.

[267] V. Bareikis, J. Liberis, A. Matulionis, R. Miliušytė, and P. Sakalas, "Noise in micron-length GaAs n^+-n-n^+ structures," *Fiz. Tekhn. Poluprovodn.* **21**(5), 916–919 (1987) [*Sov. Phys.—Semicond.* **21**(5), 558 (1987)].

[268] J. G. Ruch, "Electron dynamics in short channel field-effect transistors," *IEEE Trans. Electron Devices* **ED-19**(5), 652–654 (1972).

[269] A. Matulionis, J. Požela, and A. Reklaitis, "Monte Carlo calculations of hot-electron transient behaviour in CdTe and GaAs," *Phys. Stat. Sol. (a)* **35**(1), 43–48 (1976).

[270] D. Junevičius and A. Reklaitis, "Monte Carlo particle investigation of noise in short n^+-n-n^+ GaAs diodes," *Electron. Lett.* **24**(21), 1307–1308 (1988).

[271] A. Matulionis and R. Miliušytė, "The length of the passive near-cathode space in GaAs and InP," *Lietuvos fizikos rinkinys* **26**(3), 293–297 (1986) [*Sov. Phys.—Collection*, **26**(3), 28–31 (1986)].

[272] V. Bareikis, J. Liberis, A. Matulionis, P. Sakalas, and M. Capizzi, "Effect of hydrogen on hot electron noise in short samples of GaAs," *Semicond. Sci. Technol.* **8**(10), 1829–1833 (1993).

[273] J. Shah, "Ultrafast luminescence spectroscopy of semiconductors: Carrier relaxation, transport and tunneling," in: *Spectroscopy of Nonequilibrium Electrons and Phonons*, C. V. Shank and B. P. Zakharchenya, eds., *Modern Problems in Condensed Matter Sciences*, Vol. 35, North-Holland, Amsterdam, 1992, pp. 57–112.

[274] C. V. Shank and P. Becker, "Femtosecond processes in semiconductors," in: *Spectroscopy of Nonequilibrium Electrons and Phonons*, C. V. Shank and B. P. Zakharchenya, eds., *Modern Problems in Condensed Matter Sciences*, Vol. 35, North-Holland, Amsterdam, 1992, pp. 215–243.

[275] S. Zollner, S. Gopalan, and M. Cardona, "Short-range deformation-potential interaction and its application to ultrafast processes in semiconductors," *Semicond. Sci. Technol.* **7**(3), B137–B143 (1992).

[276] V. Aninkevičius, V. Bareikis, R. Katilius, J. Liberis, I. Matulionienė, A. Matulionis, P. Sakalas, and R. Šaltis, "Hot electron noise in GaAs at extremely high electric fields," in: *Noise in Physical Systems and 1/f Fluctuations, Proceedings of the 13th International Conference*, Palanga, Lithuania, V. Bareikis and R. Katilius, eds., World Scientific, Singapore, 1995, pp. 173–176.

[277] J. R. Chelikowsky and M. L. Cohen, "Nonlocal pseudopotential calculationd for the electronic structure of eleven diamond and zinc-blende semiconductors," *Phys. Rev.* **B14**(2), 556–582 (1976).

[278] Yu. V. Gulyaev, "Criterion for the drift electric instability in semiconductors," *Fiz. Tekhn. Poluprovodn.* **3**(2), 246–252 (1969) [*Sov. Phys.—Semicond.* **3**(2), 203 (1969)].

[279] K. T. Compton, "On the motions of electrons in gases," *Phys. Rev.* **22**(4), 333–346 (1923).

[280] R. Stratton, "The influence of interelectronic collisions on conduction and breakdown in covalent semiconductors," *Proc. Roy. Soc. (London)* **A 242**, 355–373 (1957).

[281] R. Stratton, "The influence of interelectronic collisions on conduction and breakdown in polar crystals," *Proc. Roy. Soc. (London)*, **A246**, 406–422 (1958).

[282] Sh. M. Kogan, "On the theory of photoconductivity based on the variation of mobility of current carriers," *Fiz. Tverd. Tela (Leningrad)* **4**(7), 1891–1896 (1962) [*Sov. Phys.—Solid State* **4**(7), 1386 (1963)].

[283] T. M. Lifshits, A. Ya. Oleinikov, and A. Ya. Shulman, "On the electron gas energy relaxation mechanisms in n-type InSb at helium temperatures," *Phys. Stat. Sol.* **14**(2), 511–521 (1966).

[284] Yu. I. Ravich, B. A. Efimova, and V. I. Tamarchenko, "Scattering of current carriers and transport phenomena in lead chalcogenides. I. Theory," *Phys. Stat. Sol. (b)* **43**(1), 11–33 (1971).

[285] Yu. I. Ravich, B. A. Efimova, and V. I. Tamarchenko, "Scattering of current carriers and transport phenomena in lead chalcogenides. II. Experiment", *Phys. Stat. Sol. (b)* **43**(2), 453–469 (1971).

[286] R. Katilius, A. Matulionis, and R. Raguotis, "Cross correlation of hot electron velocities in doped GaAs: Monte Carlo simulation," *Lithuanian J. Phys.* **36**(6), 494–501 (1996).

[287] E. Kuphal, A. Schlachetzki, and A. Pöcker, "Incorporation of Sn into epitaxial GaAs grown from the liquid phase," *Appl.Phys.* **17**(1), 63–72 (1978).

[288] J. S. Blakemore, "Semiconducting and other major properties of gallium arsenide," *J. Appl. Phys.* **53**(10), R123–R181 (1982).

[289] D. A. Anderson and N. Apsley, "The Hall effect in III–V semiconductor assessment," *Semicond. Sci. Technol.* **1**(3), 187–202 (1986).

[290] I. B. Levinson and G. E. Mazhuolyte, "Effect of interelectron collisions on the electron distribution function in a strong electric field," *Zh. Eksp. Teor. Fiz.* **50**(4), 1048–1054 (1966) [*Sov. Phys.—JETP* **23**(4), 697 (1966)].

[291] R. Katilius, J. Liberis, A. Matulionis, R. Raguotis, and P. Sakalas, "Nonlinear transport and fluctuation characteristics of doped semiconductors," *Nonlinear Analysis: Modelling and Control*, No. 2, Institute of Mathematics and Informatics, Vilnius, 1998, pp. 35–42.

[292] R. Katilius, A. Matulionis, R. Raguotis, and I. Matulionienė, "Transport and fluctuations in nonlinear dissipative systems: Role of interparticle collisions," *Nonlinear Analysis: Modelling and Control*, No. 4, Institute of Mathematics and Informatics, Vilnius, 1999, pp. 31–86.

[293] J. L. Thobel, A. Sleiman, and R. Fauquembergue, "Determination of diffusion coefficients in degenerate electron gas using Monte Carlo simulation," *J. Appl. Phys.* **82**(3), 1220–1226 (1997).

[294] T. Ando, A. B. Fowler, and F. Stern, "Electronic properties of two-dimensional systems," *Rev. Mod. Phys.* **54**(2), 437–672 (1982).

[295] H. L. Störmer, "Novel physics in two dimension with modulation-doped heterostructures," *Surface Sci.* **142**, 130–146 (1984).

[296] C. Weisbuch and B. Vinter, *Quantum Semiconductor Structures*, Academic Press, San Diego, 1991.

[297] B. K. Ridley, "Hot electrons in low-dimensional structures," *Rep. Prog. Phys.* **54**(2), 169–256 (1991).

[298] Z. S. Gribnikov, K. Hess, and G. A. Kosinovsky, "Nonlocal and nonlinear transport in semiconductors: Real-space transfer effects," *J. Appl. Phys.* **77**(4), 1337–1373 (1995).

[299] P. M. Koenraad, "Electron mobility in delta-doped layers," in: *Delta-Doping of Semiconductors*, E. F. Schubert, ed., Cambridge University Press, Cambridge, 1996, pp. 407–443.

[300] J. H. Davies, *The Physics of Low-Dimensional Semiconductors*, Cambridge University Press, Cambridge, 1997.

[301] T. J. Drummond, W. T. Masselink, and H. Morkoç, "Modulation-doped GaAs/(Al, Ga)As heterojunction field-effect transistors: MODFETs," *Proc. IEEE* **74**(6), 773–822 (1986).

[302] M. Jaros, *Physics and Applications of Semiconductor Microstructures*, Clarendon Press, Oxford, 1989.

[303] S. Luryi, "Device building blocks," in: *High-Speed Semiconductor Devices*, S. M. Sze, ed., John Wiley & Sons, New York, 1990, p. 57–136.

[304] S. J. Pearton and N. J. Shah, "Heterostructure field-effect transistors," in: *High-Speed Semiconductor Devices*, S. M. Sze, ed., John Wiley & Sons, New York, 1990, p. 283–334.

[305] M. J. Kelly, *Low-Dimensional Semiconductors: Materials, Physics, Technology, Devices*, Oxford University Press, Oxford, 1995.

[306] D. E. Aspnes, C. G. Olson, and D. W. Lynch, "Ordering and absolute energies of the L_6^c and X_6^c conduction band minima in GaAs," *Phys. Rev. Lett.* **37**(12), 766–769 (1976).

[307] D. E. Aspnes, "GaAs lower conduction-band minima: Ordering and properties," *Phys. Rev.* **B14**(12), 5331–5343 (1976).

[308] G. Bastard and J. A. Brum, "Electronic states in semiconductor heterostructures," *IEEE J. Quantum Electron.* **QE-22**(9), 1625–1644 (1986).

[309] A. Zrenner, F. Koch, and K. Ploog, "Subband physics for a 'realistic' δ-doping layer," *Surface Sci.* **196**, 671–676 (1988).

[310] L. Ioriatti, "Thomas-Fermi theory of δ-doped semiconductor structures: Exact analytical results in the high-density limit," *Phys. Rev.* **B41**(12), 8340–8344 (1990).

[311] M. H. Degani, "Electron energy levels in a δ-doped layer in GaAs," *Phys. Rev.* **B44**(11), 5580–5584 (1991).

[312] A. Zrenner, H. Reisinger, F. Koch, and K. Ploog, "Electron subband structure of a $\delta(z)$-doping layer in n-GaAs," in *Proceedings of the 17th International Conference on the Physics of Semiconductors, San Francisco, 1984*, J. P. Chadi and W. A. Harrison, eds., Springer, New York, 1984, pp. 325–328.

[313] V. Aninkevičius, *Hot Electron Fluctuations in Quantum Well Structures*, Doctoral dissertation, Semiconductor Physics Institute, Vilnius, 1997.

[314] Y. Kwon, D. Pavlidis, T. L. Brock, and D. C. Streit, "Experimental and theoretical characteristics of high performance pseudomorphic double heterojunction InAlAs/In$_{0.7}$Ga$_{0.3}$As/InAlAs HEMT's," *IEEE Trans. Electron Devices* **ED-42**(6), 1017–1025 (1995).

[315] T. Enoki, K. Arai, A. Kohzen, and Y. Ishii, "Design and characteristics of InGaAs/InP composite-channel HFET's," *IEEE Trans. Electron Devices* **ED-42**(8), 1413–1418 (1995).

[316] H. L. Störmer, R. Dingle, A. C. Gossard, W. Wiegmann, and M. D. Sturge, "Two-dimensional electron gas at a semiconductor–semiconductor interface," *Solid State Commun.* **29**(10), 705–709 (1979).

[317] D. K. Ferry, *Semiconductors*, Macmillan, New York, 1991, p. 165.

[318] F. Stern and S. Das Sarma, "Electron energy levels in GaAs–Ga$_{1-x}$Al$_x$As heterojunctions," *Phys. Rev.* **B30**(2), 840–848 (1984).

[319] P. M. Koenraad, F. A. P. Blom, C. J. G. M. Langerak, M. R. Leys, J. A. A. J. Perenboom, J. Singleton, S. J. R. M. Spermon, W. C. van der Vleuten, A. P. J. Voncken, and J. H. Wolter, "Observation of high mobility and cyclotron resonance in 20 Å silicon δ-doped GaAs grown by MBE at 480 C," *Semicond. Sci. Technol.* **5**(8), 861–866 (1990).

[320] S. Adachi, "GaAs, AlAs, and Al$_x$Ga$_{1-x}$As: Material parameters for use in research and device applications," *J. Appl. Phys.* **58**(3), R1–R29 (1985).

[321] A. Matulionis, V. Aninkevičius, and J. Liberis, "Interwell-transfer noise in InP-based InGaAs channels caused by electron tunneling at equilibrium and low electric fields," in: *23rd Workshop on Compound Semiconductor Devices and Integrated Circuits—WOCSDICE'99, Chantilly, 1999*, pp. 55–56.

[322] K. Hess and G. J. Iafrate, "Hot electrons in semiconductor heterostructures and superlattices," in: *Hot-Electron Transport in Semiconductors*, L. Reggiani, ed., *Topics in Applied Physics*, Vol. 58, Springer, Berlin, 1985, pp. 201–226.

[323] V. Bareikis, R. Katilius, and A. Matulionis, "High-frequency noise in heterostructures," in: *Noise in Physical Systems and 1/f Fluctuations, Proceedings of the 13th International Conference*, Palanga, Lithuania, V. Bareikis and R. Katilius, eds., World Scientific, Singapore, 1995, pp. 14–21.

[324] A. Matulionis, "Hot-electron noise in HEMT channels and other 2-DEG structures," in: *5th European Gallium Arsenide and Related III-V Compounds Applications Symposium, Bologna, 1997, Conference Proceedings*, University of Bologna, 1997, pp. 165–174.

[325] A. Matulionis, V. Aninkevičius, and J. Liberis, "Hot-electron velocity fluctuations in two-dimensional electron gas channels," *Microelectronics Reliability* (October 2000); see also *15th International Conference on Noise in Physical Systems and 1/f Fluctuations, Hong Kong 1999*, C. Surya, ed., Bentham Press, London, 1999, pp. 231–236.

[326] C. F. Whiteside, G. Bosman, and H. Morkoç, "Velocity fluctuation noise measurements on AlGaAs–GaAs interfaces," *IEEE Trans. Electron Devices* **ED-34**(12), 2530–2534 (1987).

[327] J. Gest, H. Fawaz, H. Kabbaj, and J. Zimmermann, "Microwave hot electron noise power and two-dimensional electron diffusion coefficient in AlGaAs-GaAs MODFETs," in: *Proceedings of the International Conference on Noise in Physical Systems and 1/f Fluctuations-ICNF'91, Kyoto, 1991*, T. Musha, S. Sato, and M. Yamamoto, eds., Ohmsha, Tokyo, 1991, pp. 291–295.

[328] J. Zimmermann and Y. Wu, "Diffusion coefficients of two-dimensional electron gas in heterojunctions," *Solid-State Electron.* **31**(3/4), 367–370 (1988).

[329] V. Bareikis, V. Aninkevičius, J. Liberis, T. Lideikis, A. Matulionis, P. Sakalas, G. Treideris, and R. Katilius, "Hot electron noise in GaAs epilayer and heterojunction structures," in: *Proceedings of the 1991 International Semiconductor Device Research Symposium, Charlottesville*, University of Virginia, 1991, pp. 469–472.

[330] V. Aninkevičius, V. Bareikis, R. Katilius, P. S. Kop'ev, M. R. Leys, J. Liberis, and A. Matulionis, "Hot-electron noise and diffusion in AlGaAs/GaAs," *Semicond. Sci. Technol.* **9**(5S), 576–579 (1994).

[331] V. Aninkevičius, V. Bareikis, R. Katilius, J. Liberis, I. Matulionienė, A. Matulionis, P. S. Kop'ev, and V. M. Ustinov, "Transverse tunnelling time constant estimated from hot-electron noise in GaAs-based heterostructure," *Solid State Commun.* **98**(11), 991–995 (1996).

[332] A. P. Heberle, X. Q. Zhou, A. Tackeuchi, W. W. Rühle, and K. Köhler, "Dependence of resonant electron and hole tunnelling times between quantum wells on barrier thickness," *Semicond. Sci. Technol.* **9**(5S), 519–522 (1994).

[333] A. Matulionis, V. Aninkevičius, J. Berntgen, D. Gasquet, J. Liberis, and I. Matulionienė, "QW-shape-dependent hot-electron velocity fluctuations in InGaAs-based heterostructures," *Phys. Stat. Sol. (b)* **204**(1), 453–455 (1997).

[334] A. Matulionis, V. Aninkevičius, J. Liberis, I. Matulionienė, J. Berntgen, K. Heime, and H. L. Hartnagel, "Hot-electron energy relaxation, noise, and lattice strain in InGaAs quantum well channels," *Appl. Phys. Lett.* **74**(13), 1895–1897 (1999).

[335] J. Liberis, V. Aninkevičius, I. Matulionienė, A. Matulionis, J. Berntgen, B. Henle, and E. Kohn, "Doping-dependent microwave noise in InP lattice-matched InGaAs channels," *Lithuanian J. Phys.* **38**(4), 401–408 (1998).

[336] V. Aninkevičius, J. Liberis, I. Matulionienė, A. Matulionis, P. Sakalas, B. Henle, E. Kohn, and J. Berntgen, "Hot-electron noise in InAlAs/InGaAs/InAlAs quantum wells," in: *Noise in Physical Systems and 1/f Fluctuations, Proceedings of the 14th International Conference*, Leuven, 1997, C. Claeys and E. Simoen, eds., World Scientific, Singapore, 1997, pp. 71–74.

[337] L. Ardaravičius and J. Liberis, in: *Proceedings of 33th Lithuanian National Conference on Physics*, Vilnius, 1999, pp. 74–75.

[338] L. Ardaravičius and J. Liberis "Anisotropy of hot-electron noise in InAlAs/InGaAs/InAlAs quantum wells," *Lithuanian Journal of Physics*, **40**(5), (2000).

[339] A. Matulionis, J. Liberis, V. Aninkevičius, L. Ardaravičius, and I. Matulionienė, "Fluctuations and ultrafast processes of dissipation in 2DEG channels," in: *24th Workshop on Compound Semiconductor Devices and Integrated Circuits held in Europe — WOCSDICE 2000, Aegean Sea, Greece, 2000*, pp. V-3–V-4.

[340] A. Matulionis, V. Aninkevičius, J. Berntgen, H. L. Hartnagel, K. Heime, and J. Liberis, "A new experimental technique for phonon engineering: Characterization of InGaAs-based 2DEG channels," in: *22nd Workshop on Compound Semiconductor Devices and Integrated Circuits — WOCSDICE'98, Zeuthen, Germany, 1998*, pp. 147–148.

[341] J. X. Yang, J. Li, C. F. Musante, and K. S. Yngvesson, "Microwave mixing and noise in the two-dimensional electron gas medium at low temperatures," *Appl. Phys. Lett.* **66**(15), 1983–1985 (1995).

[342] H. Fukui, "Optimal noise figure of microwave GaAs MESFET's," *IEEE Trans. Electron Devices* **ED-26**(7), 1032–1037 (1979).

[343] T. Tsukishima and C. K. McLane, "Correlation of density fluctuations and diffusion in a plasma," *Phys. Rev. Lett.* **17**(17), 900–901 (1966).

[344] C. Van den Broeck and L. Brenig, "Fluctuating kinetic equation for a two-component Boltzmann gas," *Phys. Lett.* **73A**(4), 298–302 (1979).

[345] T. R. Kirkpatrick and E. G. D. Cohen, "Light scattering in a fluid far from equilibrium," *Phys. Lett.* **78A**(4), 350–353 (1980).

[346] T. R. Kirkpatrick and E. G. D. Cohen, "Pair correlation function near a convective instability," *Phys. Lett.* **88A**(1), 44–47 (1982).

[347] L. Brenig and C. Van den Broeck, "Stochastic hydrodynamic theory for one-component systems," *Phys. Rev.* **A21**(3), 1039–1048 (1980).

[348] H. Ueyama, "The stochastic Boltzmann equation and hydrodynamic fluctuations," *Journal of Statistical Physics* **22**(1), 1–26 (1980).

[349] H. Ueyama, "Light scattering from a fluid in the stationary state caused by a temperature gradient," *J. Phys. Soc. Japan* **51**(11), 3443–3448 (1982).

[350] A.-M. S. Tremblay, M. Arai, and E. D. Siggia, "Fluctuations about simple nonequilibrium steady states," *Phys. Rev.* **A23**(3), 1451–1480 (1981).

[351] E. E. Salpeter, "Electron density fluctuations in a plasma," *Phys. Rev.* **120**(5), 1528–1535 (1960).

[352] M. S. Grewal, "Effects of collisions on electron density fluctuations in plasmas," *Phys. Rev.* **134**(1A), A86–A93 (1964).

[353] K. Jain, S. Lai, and M. V. Klein, "Electronic Raman scattering and the metal-insulator transition in doped silicon," *Phys. Rev.* **B13**(12), 5448–5464 (1976).

[354] M. Chandrasekhar, M. Cardona, and E. O. Kane, "Intraband Raman scattering by free carriers in heavily doped n-Si," *Phys. Rev.* **B16**(8), 3579–3595 (1977).

[355] V. A. Voitenko and I. P. Ipatova, "Theory of quasi-elastic scattering of light from electrons in semiconductors with non-parabolic dispertion law," *Zh. Eksp. Teor. Fiz* **97**(1), 224–233 (1990) [Sov. Phys.—JETP **70**(1), 125 (1990)].

[356] P. A. Lebwohl and P. J. Price, "Direct microscopic simulation of Gunn-domain phenomena," *Appl. Phys. Lett.* **19**(12), 530–532 (1971).

[357] R. J. Malik, T. R. AuCoin, R. L. Ross, K. Board, C. E. C. Wood, and L. F. Eastman, "Planar-doped barriers in GaAs by molecular beam epitaxy," *Electron. Lett.* **16**(22), 836–838 (1980).

[358] N. R. Couch and M. J. Kearney, "Hot-electron properties of GaAs planar-doped barrier diodes," *J. Appl. Phys.* **66**(10), 5083–5085 (1989).

[359] W. N. Jiang and U. K. Mishra, "Current flow mechanisms in GaAs planar-doped-barrier diodes with high built-in fields," *J. Appl. Phys.* **74**(9), 5569–5574 (1993).

[360] J. Liberis, A. Matulionis, P. Sakalas, R. Šaltis, L. Dózsa, B. Szentpáli, V. Van Tuyen, H. L. Hartnagel, K. Mutamba, A. Sigurdardöttir, and A. Vogt, "Microwave noise in unipolar diodes with nanometric barriers," in: *Noise in Physical Systems and 1/f Fluctuations, Proceedings of the 14th International Conference*, Leuven, 1997, C. Claeys and E. Simoen, eds., World Scientific, Singapore, 1997, pp. 67–70.

[361] V. Gružinskis, J. Liberis, A. Matulionis, P. Sakalas, E. Starikov, P. Shiktorov, B. Szentpáli, V. Van Tuyen, and H. L. Hartnagel, "Competition of shot noise and hot-electron noise in GaAs planar-doped barrier diode," *Appl. Phys. Lett.* **73**(17), 2488–2490 (1998).

[362] R. W. Hokney and J. W. Eastwood, *Computer Simulation Using Particles*, McGraw-Hill, New York, 1981.

[363] J. Liberis, A. Matulionis, P. Sakalas, R. Šaltis, L. Dózsa, B. Szentpáli, and V. Van Tuyen, "Microwave noise in GaAs planar-doped barrier diodes," *Lithuanian J. Phys.* **38**(2), 203–206 (1998).

[364] J. Mateos, T. González, D. Pardo, P. Tadyszak, F. Danneville, and A. Cappy, "Noise analysis of 0.1 μm gate MESFETs and HEMTs," *Solid-State Electron.* **42**(1), 79–85 (1998).

INDEX

Absolute temperature, 4, 8, 9
Acceptors, 165, 247
　plane of, 248
　residual, 189, 190
Acoustic phonon, 64, 67–70, 97–100, 104
　runaway terminated by, 122, 132
Active region (of electron energies), 70, 72, 104
　penetration into, 72–74, 104
Additional (interelectron-collision-caused)
　　correlation, 37–41, 78, 155–160, 177–179
　tensor, 52, 53, 151, 152, 239
Admittance, 9, 26
　real (dissipative) part, 9, 85
Alberigi–Quaranta, 82
AlAs, 183, 184
Al-free heterostructure, 210, 214–216, 222
AlGaAs, 183, 185
AlGaAs/δ-GaAs/AlAs/GaAs heterostructure, 204–206
　hot-electron noise temperature, 204–206
　real-space transfer fluctuations, 205
　　suppression of, 206
　spectral intensity of velocity fluctuations, 205, 206
　subband structure, 205
　transverse tunneling, 205
　time constant, 206, 207
AlGaAs/GaAs heterostructures, 191–194, 197–204
　2DEG channels, 197

　hot electron noise temperature, 199, 200, 204
　Monte Carlo simulation of noise, 202
　spectral intensity of velocity fluctuations, 201–203
　band structure, 192, 200
　real-space transfer
　　noise, 198–204
　　relaxation time, 203, 225
AlGaAs/InGaAs HBT, 10
Anisotropy
　of current fluctuations, 68, 71, 100–112
　of electron diffusivity, 82, 113–115, 151
　of hot-electron velocity fluctuations, 82
　of noise, 79, 97, 217
　of small-signal conductivity, 68, 151
Averaging, 1
　over ensemble of systems, 18, 27, 28, 162
　over initial deviations, 28
　over time, 1, 27, 28, 162

Balance equations, 155
Bareikis, 82, 84
Barrier, 4, 7, 14, 194, 247
　confining, 196
　electrostatic, 18
　height, 181, 194–196, 198–200, 210, 248
　heterojunction, 6, 181
　layer, 194
　Schottky, 4, 6, 7, 247
Bate, 127

282 INDEX

Bias current, 69
Black body, 8, 95, 250
 radiation, 26, 95
Boltzmann, 17, 43
Boltzmann constant, 8, 26
Boltzmann (classical, nondegenerate) statistics, 18, 29, 43, 50, 75
Boltzmann (kinetic) equation, 12–14, 16–24, 55
 applicability of, 28, 36, 40, 42, 49
 for response, 23, 232
 linear, 22, 50, 51
 linearized, 21, 43, 44, 55, 232
 nonlinear, 23, 52, 53
 space- and time-dependent, 238
 stationary (time-independent), 20, 21, 57
 time-dependent, 18, 43
 with fluctuations, 44
Boltzmann–Langevin equation, 44, 45, 47, 233, 238
 random-force term in, 44
Brillouin zone, 182
 for GaAs, 182
Brooks–Herring model, 123
Brownian motion, 26

Callen–Welton relations, *see* Fluctuation-dissipation theorem
Canali, 82
CdS, scattering of light by free electrons in, 243
Channel, 14
 2DEG, 84
 conductive, 6
 high-mobility, 199, 216
 lattice-matched, 10
 low-mobility, 199, 216
Chaotic motion, 1
 of mobile electrons, 81, 98
 velocity, 64
Chapman–Enskog procedure, 55, 60, 61, 237, 238, 243
Charge carriers, 1, 2. *See also* Electrons *and* Holes
Circuit, *see also* Equivalent circuit
 low-noise, 2, 3
 noise measuring, 86
 transistor-containing 7, 9
 two-port 9, 10
Classical mechanics, 42
Coaxial-type gated setup, 92–95
Collision(s), 17
 duration, 42, 128
 electron-lattice, 20, 22, 32, 147, 154
 hole-hole, 159
 integral, 18–20

 one-particle, 18–20, 32
 two-particle, 17–19, 21, 22, 39
 interelectron (electron-electron), 17–22, 29, 37–41, 45–48, 52–54, 147–180
 conservation of energy and quasi-momentum in, 19, 40, 47
 operator, 22, 43, 159, 232
 linearized, 38, 42, 147
 with impurities, 17, 70, 71, 147
 with phonons, 17, 64, 147
 quasi-elastic, 64, 70, 77
 with thermal bath, 18, 48
Common Anion Rule, 192
Compensation degree, 165, 178
Compton, 148
Conductance, 4, 89
 dissipative part of, 4
Conduction, 7
 two-channel, 195
Conduction band, 5, 12, 81, 190
 minima, 182, 183
 secondary, 183
 offset, 184, 192–195, 210
 of Ge, 75
 of Si, 75
 valleys, 106, 116, 182
Conductivity, 24, 54, 64, 170–172, 176, 242
 chord, 69, 149, 153
 differential, 24, 69, 149–155, 173
 anisotropy of, 68, 151, 177
 longitudinal, 72, 150, 153, 243
 negative, 72
 of many-valley semiconductor, 75
 tensor of, 155, 173, 241
 ohmic, 36, 92
 real (dissipative) part, 33, 76
 small-signal, 24, 67, 86, 91, 149–155
 longitudinal, 74, 150
 transverse, 71, 150
 tensor, 171
Confinement potential, 191
Contact, 4
 nonohmic, 4
 ohmic, 6, 91, 114, 191, 247
Continuity equation, 13, 59, 239, 240, 241
Convective noise, 64–68, 99–102, 105, 112, 130, 212, 221
Conwell–Weisskopf model, 123
Correlated occupancy of two states, 29, 37–40
Correlation function, 44, 47
 equal-time, 29, 31, 35, 37–40, 236
 equation for, 39
 two-point, 231, 235, 236
 of fluctuations

INDEX   **283**

in nonequilibrium gas, 77
time-displaced, 27, 28, 30, 38, 133, 162, 254
   Fourier transform of, 30, 33
   two-point, 231
Correlation
   in nonequilibrium gas, 77, 78
   of noise sources, 14, 229, 254
Correlation tensor, *see* Additional correlation tensor
Coulomb interaction, 154, 166, 237, 241
   of confined electrons, 188
      among themselves, 188
      with other charges, 188, 189
   potential, 154
      screened, 154, 166
Coupling constant
   electron-phonon, 70, 80
   intervalley,
      in GaAs, 120, 124, 135
      in InP, 120
Crystal lattice, 17
   dielectric constant tensor, 232, 241
Cross-correlation, 29
   equal-time, 29, 157
Cross-section of electromagnetic wave scattering, 243, 244
   differential, 241
      in collisionless limit, 242
   integral, 242
Current density, 5, 7, 21, 149
   steady 5, 17, 20, 21, 27, 65, 81, 85
Current fluctuations, 1, 27. *See also* Spectral intensity of current fluctuations
Current-carrying state, 4, 5, 17, 64–66, 75
Current-voltage characteristic, 21, 67–72, 165, 169
   linear, 69
   nonlinear, 155, 179
   saturation, 72
   sublinear, 67, 100
   superlinear, 67, 71
Cutoff frequency, 83, 220
   for electron temperature fluctuations, 221
   in 2DEG channels, 223
   in bulk semiconductors, 223, 224
   of convective noise, 220, 221
   of fast and ultrafast kinetic processes, 220–227
   of hot-electron "thermal" noise, 221

Das Sarma, 193
Davis, 82
Davydov, 148
   distribution, 66, 78

Davydov-type operator, 158
De Broglie wavelength, 12, 231
Debye screening length, 235–237
Deconfinement of 2DEG electrons, 198. *See also* Real-space transfer
Degani, 194
Degeneracy effects in electron gas, 168
Degrees of freedom, 26
   macroscopic, 26
   quasi-macroscopic, 244
Density
   of electrons, 21, 31, 56, 148, 154, 157, 165
      fluctuations, 240, 243
      in jth valley, 75, 105, 118
      in jth subband, 189
      in jth well, 196, 202, 217, 224
      local, 56, 233
      sheet, 190
      spatial gradient, 51, 61, 86, 113, 175, 240
   of acceptors, 188
   of donors, 168, 188
   of holes, 159, 160
   of ionized impurities, 165, 168
   of states, 186, 190
   of two-dimensional electron gas, 6, 195, 200, 207, 210, 217
Device
   active, 5
   low-noise, 10
   models, 12, 228, 248, 253
Diagnostics of solid-state plasma, 26, 80, 243
Diagram technique, 77, 78
Dielectric constant, 188, 232
   longitudinal, 234
   static, 232
   tensor, 232, 241
Diffusion, 50–61, 76
   near equilibrium, 50
   in system of intercolliding carriers, 52
   of electrons in gases, 52, 82
Diffusion coefficient(s), 61, 151, 157–160, 238–243
   of hot carriers (electrons), 14, 51–54, 79, 82, 86, 113
      anisotropy, 82, 113, 151, 176
      in doped GaAs, 175–180
      longitudinal, 80, 82, 113, 151, 157, 177
      transverse, 80, 82, 113, 151, 157, 177
   tensor of, 51–56, 61, 113, 150, 155, 238–241
Diode
   planar-doped-barrier, 247
      barrier height, 247
      GaAs Monte Carlo model, 248

Diode (*Continued*)
  noise properties, 248–252
  Schottky, 247
Dissipative system, 25
Distribution of electrons, 5, 17, 24, 28, 76, 147
  actual, 27
    coordinate-dependent, 230, 231
    averaged, 27, 47
    initial, 56
      smooth (nearly uniform), 56
    in quasi-momentum space, 17, 20, 52, 61, 147, 154, 156
Distribution function, 14, 16–23, 70, 148, 151, 154, 157
  density-dependent, 157, 158, 179
  equilibrium, 17, 19, 20, 36, 40, 41, 47
  fluctuations of, 43
  in jth valley, 75
  Maxwellian, 148, 151, 154, 168, 170, 171, 174
  needle-shaped, 20, 78
  nonequilibrium, 5, 16, 17, 20, 41, 53, 76, 77, 242
  non-stationary (time-dependent), 18, 56
  of electron energies, 66, 148, 151, 154, 156, 170
  space-and time-dependent, 55
  stationary, 20–23, 51, 55–57, 231
  uniform (homogeneous), 17, 52, 55, 56, 231
Donor, 165, 188, 195
  plane, 195
Doping
  critical, 194
  effect on hot-electron noise, 123–127, 137, 165, 207
  planar, 188, 204, 207, 209, 247
  selective, 191, 198
  uniform, 123, 165
Drift-diffusion, 52
  equation, 13, 55, 59–61
    with fluctuations, 14, 79, 237–241
Drifted Maxwellian distribution, *see* Maxwellian distribution, displaced
Drift of density fluctuation, 240
  Doppler-shift due to, 242
Drift velocity of electrons, 19, 21, 51, 155, 162
  differential, 52, 60, 238
  fluctuations of, 85, 91, 162
  in quantum well, 213, 221
    jth, 202, 224
  in jth valley, 75, 105
  steady-state value, 21, 32, 51, 98
Drude-type formula, 63, 64
DX-levels, 139

Effective electron temperature, *see* Electron temperature
Effective mass, 63, 186, 221
Eigenfunction of relaxation operator, 22, 57
Eigenvalues of energy, 186
Einstein, 26
  relation, 50, 54, 76, 77
  summation convention, 24
Electric field
  DC, 20
  distribution of, in diode, 252
  extremely high, in GaAs, 142–145
  local fluctuating, 136, 240
  self-consistent, 232, 239
  small AC, 23
  weakly heating, 150
Electromagnetic wave scattering, *see* Scattering of electromagnetic wave
Electron affinity, 192
Electron energy, 19, 63, 67, 152
  mean, 20, 65, 83, 98, 233, 243
  relaxation, 90, 98
Electron-electron collisions, *see* Collisions, interelectron
Electron confinement in quantum well channels, 6, 185–196
Electron cooling, 98
Electron density gradient, 51, 61, 86, 113, 175, 240
Electron energy distribution, *see* Distribution function of electron energies
Electron gas in semiconductor, 2, 17, 40, 175, 238
  hot, 5, 17, 40, 175, 239
  nondegenerate, 50
  weakly heated, 150, 154, 156–158
Electron mobility, 3, 6, 11, 64, 67, 165, 166, 178
  dependence on electron energy, 65
  differential, 75
    in individual valley, 75
  in 2DEG, 200, 210, 214
  low-field (ohmic), 50, 92, 165
  small-signal, 97
Electrons
  hot, *see* Hot electrons
  warm, 150, 152
Electron temperature, 14, 77, 79, 147–157, 170–174, 176–180, 190
  approximation, 152, 157, 159, 167, 170–172, 179
  fluctuations of, 147, 151–153, 161, 170
  space-dependent, 243, 244
  relaxation, 148

relaxation time, 149, 152, 155
steady-state value, 149
Electron velocity, 21
Electron wave-function, 182, 185. *See also* Envelope wave function
Electronic subband in quantum well, *see* Subband
Emitted noise power, 7, 89
Energy band
  gap, 181–185
  offset (edge discontinuity), 184, 192
  parabolic, 100, 103
  spherical, 100, 101, 103
  structure, 25, 106, 182
    of compound semiconductors, 182
    of heterostructures, 184, 192, 195, 200, 205
Energy loss by electrons, 17, 149
  on acoustic phonons, 98
  on optical phonons, 98, 167
  in GaAs channels, 174, 179
  interelectron-collision-dependent, 167
Ensemble-average, 18, 26, 27
Envelope wave function, 186–191
Equal-time correlation function, *see* Correlation function
Equipartition law, 8, 26
Equivalent circuit
  approach, 11
    empirical, 11
    physics-based, 11, 229
  lumped element, 228
  noise sources for, 227–229
Equivalent noise temperature, *see* Noise temperature
Erlbach, 82
Evolution
  of equal-time correlation, 39
  of small deviations from steady state, 21
Exchange-correlation potential energy, 193
Extra correlation, *see* Additional correlation
Extraneous local random currents, 238, 240
  Fourier transforms of correlation functions of, 238, 241

Fermi
  energy, 190, 195
  particles, 29
Fick's diffusion coefficient, *see* Diffusion coefficient
Fick's law, 175, 176
Fischetti, 12
Fluctuation-diffusion relation, *see* Price relation

Fluctuation-dissipation theorem, 13, 16, 26, 33–36, 76. *See also* Nyquist theorem
  second, 14, 239, 244
Fluctuation(s), 1, 2, 13–15, 24–28, 37–40, 50–54, 230–244
  $1/f$ (flicker), 4, 81
  classical, 14, 26, 27, 41, 43
  due to transverse tunneling, 204
  hydrodynamic, 14, 237
  in collision rate, 47, 48
  in drifted Maxwellian approximation, 154
  in electric circuits, 1
  in electron distribution, 26, 232, 240
  in equilibrium state, 33–36, 41, 46, 76
  in nonequilibrium systems, 2, 14, 25, 26, 41–43, 62, 76–80
  in number of mobile electrons, 5, 108, 237
  in occupation numbers, 26, 27, 34, 44, 47
  intersubband, 207, 208
  intervalley, 75, 105, 246
  in streaming motion regime, 72
  in weakly heated electron gas, 156
  of carrier density, 79, 233, 239, 244
    collisionless limit, 234
    in jth valley, 106
  of charge density, 135, 230, 243, 246
  of current, 1, 27. *See also* Spectral intensity of current fluctuations
  of distribution function, 27, 30–33, 41, 44, 47, 77, 164, 230
    in quasi-momentum space, 30, 65, 77, 230
  of drift velocity, 63, 64
  of electron temperature, 79, 161, 212, 244
    space-dependent, 79, 243
  of energy, 64–66, 68, 100, 101
  of hot electrons, 7, 15, 62, 77, 81–83, 91, 197
    at extremely high fields, 142
    effect of doping on, 127
  of observables, 2, 26, 27, 32
  of quasi-momentum, 65
  spatially-inhomogeneous, 23, 230–244
    experimental detection, 241
    long-wavelength low-frequency, 14, 237–244
  spectroscopy, *see* Noise spectroscopy
Fluctuational drift-diffusion equation, *see* Drift-diffusion equation with fluctuations
Fourier analysis of random process, 30
Fourier transform, 30, 31, 33, 231, 232, 235, 239
  of equal-time two-point correlation function, 235, 236
  of time-displaced correlation function, 30

Fourier transform (*Continued*)
  of time-displaced two-point correlation function, 231
Fröhlich, 148
Frequency(ies)
  bandwidth, 7
  classical, 69
  cutoff, 83, 102, 105, 220–227
Fukui, 228

GaAs, n-type
  band structure, 182
    Γ–L–X model, 116, 120, 124, 136, 248, 253
  Brillouin zone, 182
  doped, 124, 161, 176
  electron temperature, 117, 171
  field-effect transistor (MESFET), 6, 253
  hot-electron noise, 116–127
    length-dependent, 128–146
    Monte Carlo simulation, 117, 125, 135
  intervalley coupling constant, 120
  intervalley noise, 116–122, 124–126, 133–136, 142–146
    Monte Carlo simulation, 117, 124, 135
    thermally quenched, 120, 121
  intervalley transfer time
    Γ–L, 141
    Γ–X, 141
  negative differential conductivity, 117
  noise-speed tradeoff, 134
  noise temperature, longitudinal, 117–122
    doping-dependent, 125, 126, 171
  planar-doped, 188, 207
    noise temperature, 207
  planar-doped-barrier diode, 247
    spectral intensity of current fluctuations, 248–250, 252
  resonant impurity scattering, 127, 141
  spectral intensity of velocity fluctuations, 117, 121, 166
    doping-dependent, 123–125
Gantsevich, 14, 37, 77
Gas of neutral molecules, 17, 40, 50, 55, 61
Gate, 6
  current fluctuations, 254
  T-shaped, 255, 256
Gated radiometer, 86, 87, 94
Generation–recombination noise, 5, 66, 79, 108, 119
Ge
  hot-electron diffusion, 113
    Monte Carlo simulation, 114, 115
    time-of-flight measurement, 114, 115

  microwave noise measurements, 96–107
n-type, 82
  convective noise, 111
  energy relaxation time, 111
  hot-electron "thermal" noise, 111
  intervalley noise, 111
  intervalley relaxation time, 110
  noise temperature, 82, 107
  spectral intensity of current fluctuations, 111
p-type, 82
  convective noise, 99, 100
  energy relaxation time, 99, 101
  hot-hole "thermal" noise, 97
  noise temperature, 96, 97
  spectral intensity of current fluctuations, 100–105
Gibbs distribution, 16
Grand canonical ensemble, 29
Green, 37
Gunn, 82
Gunn effect, 117
Gurevich, 14, 77

Half-width of line of scattered light, 242
Harmonic perturbation, 23
Hartree approximation, 188, 191
Hashitsume, 27, 77
Holes, 1, 13, 100, 159
  heavy, 100, 159
  light, 100
Heterojunction, 4, 183–185
  barrier, 184, 198
  lattice-matched, 183
Heterostructure, 3, 4
  field-effect transistor (HFET), 3
  subcritically doped, 194
  supercritically doped, 194
High electron mobility transistor (HEMT), 6
High-speed electronics, 1–3
Hot electron(s), 5, 17, 20, 50, 62, 86, 243
  confined, 198
  fluctuation(s), *see* Fluctuations of hot electrons
  intervalley repopulation, 106
  luminescence, 21, 183
  noise, *see* Noise of hot electrons
Hot hole(s), 82, 97, 99, 109, 114
Hurst, 82
Hydrodynamic equations for gas, 55, 61
Hydrodynamic fluctuations, 14
  in gases and fluids, 14, 244

Image potential energy, 193, 194
Impact ionization, 5, 144, 251, 253

energy, 145
noise, 144, 252
Impedance, 7–9, 13, 87
　matched, 7
　real (dissipative) part 9, 89
Impurities, ionized, 70, 97, 104, 123, 127, 154, 165
InAlAs, 185
InAlAs/InGaAs/InAlAs HEMT, 6
　cutoff frequency, 255
　noise figure, 10, 255, 256
　　extrinsic, 255
　　intrinsic, 255
　noise simulation, 256
InAlAs/InGaAs/InAlAs/InP, 191, 209–219
　band diagram, 195
　convective noise, 212, 213
　energy relaxation time, 213–216, 218, 222
　hot electron "thermal" noise, 218
　interwell relaxation time, 211, 218, 225
　lattice-matched, 209, 210
　noise temperature, 210, 211, 216, 217
　　longitudinal, 210–212, 216–218
　　transverse, 218
　real-space transfer, 209, 211, 217
　　relaxation time, 211, 218, 225
　real-space transfer noise, 211
　　length-dependent, 211, 212
　strained, 209, 210
InP/InGaAs/InP channels, 209, 210
　energy relaxation time, 214–216, 222
InGaAs, 185
InGaP, 185
InP, n-type, 3, 183
　band structure, 116–118
　hot-electron noise, 119, 120, 122, 125
　　length-dependent, 131, 132, 136
　intervalley coupling constant, 119, 120
　intervalley noise, 118, 122, 125
　negative differential conductivity, 117
　noise temperature, longitudinal, 118, 122, 125
　effect of doping, 125
　runaway of lucky electrons, 122, 123
　　acoustic phonon terminated, 123
　spectral intensity of current fluctuations, 120
InP lattice-matched HEMT, 10
InP/InGaAs heterojunction, 185
InSb, n-type, 82, 97
　　spectral intensity of current fluctuations, 102
Interelectron scattering, see Collisions, interelectron
Interface states, 191

Intersubband transfer, 13, 206, 207
Intervalley scattering, 75
Intervalley transfer, 5, 75, 105, 116, 224
Interwell transfer, see Real-space transfer
Ioriatti, 191
Isotropic medium, 65, 97, 150–153

Jacoboni, 12
Johnson noise, see Thermal noise
Joule effect, 86, 90, 108, 204

Kadomtsev, 14, 27, 46, 77
Katilius, 14, 77
Kinetic (Boltzmann) theory, 14, 16, 17
Kinetic coefficients, 24, 26, 33, 50, 239, 241
　frequency dependence, 24, 67–69, 149
Kinetic correlation, see Additional correlation
Kinetic equation, 14, 17, 55. See also Boltzmann equation
　with fluctuations, 14, 77
Kinetic processes, 5, 16, 25, 166, 219, 244
　fast and ultrafast, 146, 197, 220
　　cutoff frequencies, 5, 220–227
Kinetic theory
　of fluctuations, 14, 25–43, 77–79
　　in non-equilibrium state, 25–42, 230–237
　　limits of applicability, 42, 43
　of non-equilibrium processes, 16–24
Kino, 82
Kogan, 14, 37, 47, 77, 78, 244
Kramers–Kronig relations, 35

Langevin method, 14, 44–49, 237–244
Lattice constant data, 183
Lattice polarization, 154
Lax, 14, 77
Lead chalcogenides, 154
Length, critical, 134, 137
Leontovich, 37, 77
Light scattering in solids, 21, 241–243
　from electronic excitations, 241–243
　　in collision-controlled regime, 242
Local density approximation, 188, 191
Lorentzian, 63–65, 105, 108, 152, 220, 242
Low-noise high-speed operation, 11
Low-dimensional structure, 5
Ludwig, 37
Luminescence, 21, 120, 183
　spectroscopy, high-pressure, 127

Many-particle system, 16, 28
Matching, 87
Matched load, 8, 85
Matulionienė, 82

288   INDEX

Maxwellian distribution, 19, 148, 151, 154, 170
　displaced (drifted), 19, 154–157
Mean free path, 61, 232, 237, 243
Mean free time, 42, 232, 237
Measurements
　of noise power, 90–95
　of noise temperature, 87, 93
　of small-signal response, 91
　of thermal walkout, 90
　pulsed radiometric, 87
Microwave noise spectroscopy, *see* Noise spectroscopy, microwave
Microwave radiometer, 86, 94
Microwave technique for noise measurements, 81, 86, 98
　waveguide-type pulsed, 82, 197
　nanosecond pulsed, 95, 142
　coaxial, 93
Mismatch of sample and load, 87
Mixed crystals, 184, 185
Mobile electrons, 15, 17, 20
Mobility of carriers, *see* Electron mobility
Modulation radiometer, 86, 94
Moglestue, 12
Monte Carlo method,
　procedure, 12, 80, 161–164, 167, 245, 246
　　coupled to Poisson equation, 136, 246, 248, 254
　　simulation of noise
　　　in AlGaAs/GaAs, 202
　　　in AlGaAs/GaAs HEMT, 254, 255
　　　in devices, 245–256
　　　in GaAs, 117, 118, 124, 134, 135
　　　in GaAs, doped, 161, 166–174
　　　in GaAs MESFET, 254, 255
　　　in Ge, n-type, 115
　　　in Ge, p-type, 97, 102–104
　　　in InGaAs-channel HEMT, 255, 256
　　　in InP, 133
　　　in InSb, n-type, 97
　　　in planar-doped-barrier diode, 248
　　　in Si, n-type, 109, 111

Nava, 82
Nyquist, 4, 26
Nyquist theorem (relation), 4, 8, 26, 32–36, 54
Noise, 1–4, 13, 26
　$1/f$ (flicker), 4, 81
　"colored" at microwave frequencies, 7, 69
　excess, 2
　generation-recombination, 4, 5, 7, 66, 81, 86, 108, 166
　in current-carrying state, 4, 5, 17, 64, 75
　　longitudinal, 65

　　transverse, 65
　　transverse in 2DEG plane, 217, 218
　microwave, 5, 10, 69, 86
　modeling, 13–15. *See also* Monte Carlo method
　shot, 4, 5, 7, 13, 91, 250
　spectra at microwave frequencies, 13, 62–80, 83, 103, 108, 119, 211
　thermal, 4, 5, 8, 26
Noise characterization, 7–9, 84, 241–244
　two-terminal device, 7–9, 84
　two-port device, 9
Noise figure
　of two-port circuit, 9
　of transistor
　　calculated, 255, 256
　　measured, 256
　　minimum 10, 228, 255, 256
Noise of hot electrons, 4, 26
　convective, 64–68, 99–102, 105, 112, 130, 212, 221
　　suppression, 130
　doping-dependent, 123–126
　due to real-space transfer, 198, 199, 202, 209, 211, 216
　　at low electric field, 216
　　suppression, 205, 206, 209–212
　experimental results for
　　AlGaAs/GaAs 2DEG channels, 200–204, 206
　　GaAs, 117–119, 121–124, 126, 129–131, 134, 135, 137
　　GaAs, doped, 140–143, 166, 169
　　GaAs, δ-doped, 207
　　GaAs planar-doped-barrier diode, 248–252
　　Ge, n-type, 107, 111, 113, 115
　　Ge, p-type, 97–100, 102, 105
　　InAlAs/InGaAs/InAlAs/InP channels, 209–219
　　InP, 119–120, 125, 131, 132, 136
　　InSb, 102
　　Si, n-type, 107–109, 111, 112
　　Si, p-type, 101, 114
　intersubband, 198, 207
　intervalley, 66, 75, 76, 83, 105–108, 201
　　doping-stimulated, 139–141
　　suppression, 133, 134, 138, 201
　　suppression by real-space transfer, 201
　　transverse components, 106
　intravalley, 75, 76, 105, 106
　measurement technique, 81–95
　resonant impurity influence, 127
　short-channel effect on, 128–141
　stimulated with doping, 139

"thermal", 65, 97, 105, 212, 218, 220
  thermal quenching of, 121, 204, 207
  transverse-tunneling-related, 205, 206
  suppression, 205
Noise suppression,
  in planar-doped-barrier diode, 250
  in 2DEG channels, 201, 205, 212
  in short channels, 128–139, 205, 206, 212
Noise power, 4, 7, 26, 85–87, 92–95
  at equilibrium, 8
  available, 4, 7, 8, 26, 85, 89, 92, 93
  emitted into waveguide, 87, 89
  measurements, 86, 93
Noise spectroscopy, microwave, 25, 62, 78, 80, 81
  E-spectroscopy, 83, 99–105, 110–112, 116–124, 197–219
  L-E-spectroscopy, 84, 128–146
  ω-spectroscopy, 62–80, 83, 103, 108, 119, 211
Noise temperature, 7–9, 85–87, 89, 90, 95, 113
  of hot electrons, 8, 68–72, 74. *See also*
    Noise of hot electrons
    anisotropy, 68, 71, 73, 97, 153–156
    longitudinal, 68, 71, 72, 82, 91, 94, 153–156, 168–173
    transverse, 68, 71, 82, 96–98, 153–156, 168–173
    transverse in 2DEG plane, 217, 218
    length-dependent, 133–137
    nearly isotropic, 68, 167–173
  of hot holes, 82
Noise-speed tradeoff, 10, 128
  in short channels, 128, 134, 146
Noise-to-signal ratio, 9
Noisy element, 9
Nonequilibrium interparticle correlation, *see*
  Additional correlation
Nonequilibrium state, 2, 5, 17, 55
  steady, 20, 50, 55, 231
Nonlinear element, 11
Nonuniform processes, 50
Nougier, 82, 84

Observables, 2, 20, 26, 161
  actual values, 26
  average values, 2, 26, 27
  fluctuations of, 27
  space independent, 27
Occupation number of one-electron state, 26–29, 37–39, 78
  average, 27, 29, 37
  fluctuations, 26, 27, 30, 34
Ohmic behavior, 70, 99
  deviations from, 21, 167, 168, 178
Ohmic contact, *see* Contact
Onsager regression hypothesis, 28
Onsager principle, 33, 34
Operator, 18, 42, 43
  Davydov type, 158
  electron-lattice collision, 22, 32
  interelectron collision, 22, 38, 40, 158
  linear, 18, 32
  of Boltzmann equation, linearized, 21, 28, 31, 37, 42, 56, 148
  of relaxation *see* Relaxation operator
  of spatially homogeneous response, 23, 232
  of spatially inhomogeneous response, 232, 233, 236
Optical phonon, 70, 98, 103–105, 122, 131, 144, 170
  absorption, 70
  confined, 216
  emission
    by hot carriers, 70–72, 103, 104, 167, 172, 213, 215
    time in GaAs, 127, 144
Optoelectronic modulation spectroscopy, 127

Paranjape, 148
Passive region (of electron energies), 70–74
  boundary, 72
  weak scattering inside, 74
Phonons, *see* Acoustic phonons *and* Optical phonons
Physical kinetics, 26, 77
Planar doping (δ-doping), 188, 204, 207, 209, 247
Poisson
  equation, 55, 188, 232, 240, 241
  probability distribution, 29
  random process, 48, 49
  statistics of collisions, 47
Polarization vectors of incident and scattered waves, 241
Pospieszalski, 228
  model, 14, 228
Power
  absorbed, 67
  consumed, 86, 98, 108
Požela (Pozhela), 82, 84
Price, 14, 51, 77
Price (fluctuation-diffusion) relation, 50–52, 77, 82, 86, 113, 175–180
  violation by interelectron collisions, 52, 77, 151–160, 175–180
Pseudomorphic layer, 185

290 INDEX

Quantization of carrier energy, 181
Quantized structure, 12
Quantum confinement of mobile electrons, 5, 181, 185, 188, 197
    plane of, 181
Quantum mechanics, 42, 77
Quantum statistics, 43
Quantum well (QW), 3, 5, 181, 185–196
    depth, 190
    empty, 185
    high-mobility, 195
    in planar-doped GaAs, 191
    occupied, 188
    of finite depth, 187
    rectangular, 184, 186, 187
    triangular, 187, 188
Quasi-elastic scattering, *see* Scattering of electrons, nearly elastic
Quasi-hydrodynamic Langevin random forces, 79
Quasi-hydrodynamic description of fluctuations, 14, 79, 244
Quasi-momentum of carrier, 13, 17, 27, 35, 42, 63, 64, 148
    relaxation time, 64, 67–69, 147, 154, 158, 163
Quasi-momentum space, 17, 21, 38, 47, 57, 61, 73
Quasi-momentum-coordinate space, 17

Rabinovich model, 70, 72
Radiometer for noise measurement, 86
    coaxial, 92–94
    nanosecond-time-domain, 95
    waveguide-type pulsed, 86–91, 94
Rayleigh, 26
Ramo–Shockley theorem, 85, 105
Random fluxes in quasi-momentum space, 44–48
    spectral intensity, 45–48, 233
Random variable, 30
Real-space transfer, 13, 198, 211, 216, 219, 225
    noise, 198, 216
        in AlGaAs/GaAs channels, 199–207
        in InAlAs/InGaAs/InAlAs/InP channels, 209–212, 217
Redistribution of electrons in coordinate space, 55, 61, 232, 237, 240
Reflection coefficient, 93
Reggiani, 82
Relaxation, 13, 21, 40, 63, 81, 148, 212
    dielectric, 175
    in quasi-momentum space, 13, 56
    intervalley, 13
    mechanisms of, 68, 158
    of carrier energy, 5, 64, 99, 147, 155, 212, 220
        supported by interelectron collisions, 172
    of current, 64
    of distribution function, 64
    of quasi-momentum, 63, 64
    of small deviations from a steady distribution, 22
Relaxation operator, 22, 28, 38, 44, 52–58
Relaxation time, 13, 15, 56, 61, 81
    energy-independent, 63, 69
    for intervalley transfer, 75, 80
    intravalley, 75
    of electron temperature, 149, 152, 155
    of energy, 64–71, 99, 147, 154
        dependence on energy, 67, 68
    of quasi-momentum, 64, 67–69, 147, 154, 158, 163
        dependence on energy, 67, 68
Relaxation time, effective, 97, 98, 105
    for intervalley transfer, 105, 110, 111, 225
        in GaAs, 145, 146, 225
        in Ge, 110, 225
        in Si, 110, 111, 225
    for interwell transfer, 203, 211, 225
        due to tunneling, 205, 225
    interelectron, 147–151, 154, 157–159, 163
    of energy, 98–101, 118, 213–216, 222
    of quasi-momentum, 97
Resonance due to streaming motion, 74, 103, 104
    frequency, 104
Resonant impurity levels, 127
Response, 9, 11–13, 21–24, 238
    current, 24
    linear, 21–24, 50
    modeling, 11–13
    operator, 23, 232
    small-signal, 11, 91, 227
    to spatially inhomogeneous perturbation, 52, 232
Rolland, 82
Ruch, 82
Runaway noise suppression, 131

Scattering, 11, 25, 37, 38
    interelectron, *see* Collisions, interelectron
    intervalley, 75
    intravalley, 75
    mechanisms, 15, 156, 161
    of electromagnetic waves, 23, 79, 230, 241–244

by non-equilibrium charge carriers, 79,
    241–244
  collision-controlled, 23, 242–244
  from solid-state plasma, 230, 241–244
  of electrons (of carriers), 25
    by acoustic phonons, 64, 67–71, 97–104,
      122
    by ionized impurities, 71, 97, 104,
      123–125, 166
    by optical phonons, 78, 98, 103, 159, 166
    by resonant impurities, 127, 140
    by thermal bath, 38, 66, 149, 156, 161
    in passive region, 72–74
    nearly elastic (quasi-elastic), 64–72,
      98–102, 151, 155–159
  of electron quasi-momentum, 64, 71, 147,
    154
Schottky
  barrier, 4, 227
  diode, 247
    spectral intensity of current fluctuations,
      248
  formula, 4, 250
Schrödinger equation, 12, 186, 188
Schrödinger-Poisson equations, 12, 188, 195,
  199
Screening
  length in nonequilibrium system, 241
Self-consistent electric field, 55, 249
  of fluctuations, 232, 239, 249
Semiclassical kinetic methods, 12
Semiconductor, 17, 25, 30, 62
  anisotropic, 106
  compound, 116, 165, 183, 184
  device, 5
  doped, 71, 123, 136–146, 147, 161, 175
    lightly, 30–32, 62, 96, 117, 137, 175
  elementary, 96
  isotropic, 65, 106, 171
  many-valley, 106, 116
  one-valley (single-valley), 64, 82, 103, 135
  pure, 62, 70, 71
Semiconductor structures, 5, 11–15
  short, 128
  containing two-dimensional electron gas, 181
Sensitivity of conductance to electron heating,
  150, 167, 173, 178
  coefficient of, 150, 154, 173
  weak, 154, 174
Short channel
  effect on noise, 128, 129
    doping-dependent, 136
Sheet density
  of acceptors, 248

  of donors, 189, 195, 196
  of mobile electrons, 181, 189, 193–196, 200,
    210
Shot noise, 4, 91, 228
Shubnikov–de Haas oscillations, 189, 191
Shulman, 14, 37, 47, 77, 78, 244
Si, 75, 183, 185
  n-type, 82, 83, 108
    E-spectroscopy, 112
    convective noise, 109, 112
    generation-recombination fluctuations, 109
    hot-electron "thermal" noise, 109, 112
    intervalley noise, 109–112
    intervalley relaxation time, 109–112
    Monte Carlo simulation of fluctuations,
      109
    noise temperature, 107
    relaxation times, 108–110
    spectral intensity of current fluctuations,
      108, 112
  p-type
    diffusion coefficient of hot holes, 114
    energy relaxation time, 99, 101
    spectral intensity of current fluctuations,
      101
Small-signal
  processing, 2
  response, 21
    measurement, 91
Smoluchowski, von, 26
Source and drain, 6
Space charge density fluctuations, 135, 246, 249
Spectral density, see Spectral intensity
Spectral intensity, 13, 30, 84
  of current fluctuations, 4, 31–35, 41, 42,
    51–54, 70–74, 85
    at thermal equilibrium, 8, 34–36, 50, 54,
      63
    convective contribution, 64–68
    in AlGaAs/GaAs 2DEG channels,
      201–203, 206
    in GaAs, 117, 121, 124, 126, 129–131,
      135
    in GaAs, doped, 124, 141, 166
    in GaAs planar-doped-barrier diode, 248,
      249, 252
    in Ge, n-type, 111
    in Ge, p-type, 100–105
    in InP, 120, 131, 136
    in InSb, 102
    in many-valley semiconductor, 75
    in Si, n-type, 108, 109, 112
    in Si, p-type, 101
    length-dependent, 129, 135

Spectral intensity (*Continued*)
    longitudinal, 65, 69, 74, 152
    tensor of, 32, 36, 42, 152
    transverse, 69, 152
    of distribution function fluctuations, 30–35, 41, 45
        spatially inhomogeneous, 231–233
    of drift-velocity fluctuations, 86
    of electron density fluctuations, 233–237, 240
        collision-controlled, 240
        in collisionless limit, 234
    of extraneous local random currents, 238
    of fluctuations, 30
    of voltage fluctuations 9
Störmer et al, 191
Standing wave ratio, 87, 92
    meter (SWR-meter), 88, 95
Statistical physics, 26, 77
Stern, 193
Stochastic damped waves
    of electron density, 242, 244
    of electron temperature, 244
Stochastic excitations, 242
    long-wavelength low-frequency, 242
Stochastic equations, *see* Langevin method
Stratton, 148
Streaming motion of carriers, 72–74, 82, 103
Stueckelberg's property, 19, 39, 46
Subband, 181, 185–189
    bottom of, 190
    energies, 187–191, 200
    occupation, 190, 191
Suppression of noise, *see* Noise suppression
Surface charge, 190
Surface states, 189, 194, 196, 247

Temperature, *see* Absolute temperature, Electron temperature, *and* Noise temperature
Tensor
    contracted, 232, 240
    symmetric, 36
Theory of fluctuations, 14, 15, 26, 42, 62, 76, 147
    in nonequilibrium state, 14, 25–32, 37–43, 62–80
    macroscopic, 237–241
    spatially inhomogeneous, 230
Theory of linear operators (general), 59
Thermal bath, 13, 16–19, 22, 38, 148
    interaction of electrons with, 13, 16–18, 63, 149, 163
Thermal equilibrium, 2, 4, 8, 19, 26, 33–36

    distribution, 17, 36, 40, 47
Thermal motion, 26
Thermal noise, 4, 5, 13, 26
Thermal walkout, 3, 86, 90, 204
Thermodynamic potentials, 16
Thermodynamics, 17, 26
Thomas-Fermi approximation, 188, 191
Thomson radius of electron, 241
Time-of-flight technique, 82, 113–115
Time-space scale, 232, 238
Time-reversal symmetry in equilibrium, 33
Transfer
    band-to-band, 5
    intersubband, 206, 207
    intervalley, 5, 75, 105, 116, 224, 225
    interwell, 218, 219, 224, 225
    of electron energy to lattice, 159
    real-space, 5, 218, 219, 224, 225
Transistor, 2, 5, 10
    bipolar junction (BJT), 3, 10
    cutoff frequency of, 6, 255
    field-effect (MESFET), 2–7, 10, 14, 191, 253
        current correlation functions, 254
        equivalent circuit, 227, 228
        Monte Carlo simulation of noise properties, 253, 254
        noise figure, 10, 14, 228, 254, 255
    high-electron-mobility (HEMT), 6, 10–12, 191, 253–256
        noise figure, 10, 14, 254–256
        pseudomorphic PHEMT, 191
    heterobipolar (HBT), 3, 10
    intrinsic, 227, 228
        equivalent noise temperatures, 228, 229
        noise figure, 254–256
Transit time, 84, 130, 206
Transition probability, 18–20, 43, 45, 47
    for collision with thermal bath, 18, 19
    for electron-electron collision, 18–20
    for intervalley transfer, 75, 105
    from active to passive region, 70, 71
    of optical-phonon absorption, 70
    of optical-phonon emission, 70
Transmission
    coefficient, 89
    line, 91
Tunneling
    transverse, 198, 204
        at equilibrium, 216
Two-dimensional electron gas (2DEG) 6, 8, 12, 181, 185, 188
    channel, 6, 191
        high mobility, 195
        relaxation times in, 222, 225

sheet density, 181, 189, 193–196, 200, 210
    critical, 195, 196
Two-terminal sample, 84, 91, 199

Vaitkevičiūtė, 82
Valleys, 75
    anisotropic, 106
    ellipsoidal, 106
    equivalent, 75, 76, 83, 106
    nonequivalent, 83, 106, 116
    spherical, 106
Valence band offset, 184, 192
Velocity of mass center of mobile electrons, 85, 162
    fluctuations, 83–86, 105, 162

Velocity overshoot, 134
Violation of Price relation, *see* Price relation

Wagner, 82
Wannier, 13, 51
Warm electrons, 154–159, 169, 173, 177–180
Waveguide technique, 85, 87, 91, 95
Wave-vector space (**k**-space), 182. *See also*
    Quasi-momentum space
Wiener–Khintchine theorem, 30

X-band microwave radiometer, 86

Zrenner et al., 189